Adaptive Control of Parabolic PDEs

Adaptive Control of Parabolic PDEs

Andrey Smyshlyaev
and Miroslav Krstic

PRINCETON UNIVERSITY PRESS
PRINCETON AND OXFORD

Published by Princeton University Press, 41 William Street,
Princeton, New Jersey 08540

In the United Kingdom: Princeton University Press,
6 Oxford Street, Woodstock, Oxfordshire OX20 1TW

press.princeton.edu

Library of Congress Cataloging-in-Publication Data
Smyshlyaev, Andrey.
Adaptive control of parabolic PDEs / Andrey Smyshlyaev and Miroslav Krstic.
p. cm.
Includes bibliographical references and index.
ISBN 978-0-691-14286-9 (hardcover : alk. paper)
1. Differential equations, Parabolic.
2. Distributed parameter systems. 3. Adaptive control systems.
I. Krstic, Miroslav. II. Title.
QA374.S59 2010
515′.3534—dc22
2009048242

British Library Cataloging-in-Publication Data is available

This book has been composed in Times

Printed on acid-free paper. ∞

Typeset by S R Nova Pvt Ltd, Bangalore, India

Printed in the United States of America

10 9 8 7 6 5 4 3 2 1

Contents

Preface

Both *control of PDEs* (partial differential equations) and *adaptive control* are challenging fields. In control of PDEs, the challenge comes from the system dynamics being infinite dimensional. In adaptive control, the challenge comes from the need to design feedback for a plant whose parameters may be highly uncertain (much more than a few percent or a few dozen percent relative to a nominal value) and which may be open-loop unstable, requiring simultaneous control and learning. This requirement, in turn, results in adaptive controllers being nonlinear even when the plants are linear. Because of this, the detailed behavior of adaptive feedback systems (e.g., transient performance) is very hard to predict.

With challenges come controversies. The 1980s were a decade of controversies about adaptive control in which a community that was used to the predictability of linear feedback systems and to the availability of linear performance limitation results like Bode's was faced, perhaps for the first time, with a feedback problem that was fundamentally nonlinear. The controversies were fueled not only by the lack of understanding of particular adaptive feedback schemes but also by the lack of appreciation for the nature of the problem itself—stabilization in the presence of large parametric uncertainty. The controversies somewhat subsided in the early 1990s, when new adaptive schemes, based on the method of integrator backstepping, emerged with which one can systematically reduce the performance bounds (in L^∞- and L^2-norms). However, the unpredictability of adaptive controllers (in a detailed sense, rather than in a "bulk," performance-bound sense) in the absence of "persistency of excitation," namely, the dependence of the transient and asymptotic behaviors of adaptive feedback schemes on initial conditions, is fundamental to the problem. One could say that this difficulty is the point where the engineering appeal of adaptive control starts coming into question and its mathematical beauty begins. Despite its stormy adolescence, adaptive control continues to be strong in its middle years, with plenty of intellectual mileage left, and with a growing number of industrial applications, including in fields where it had traditionally been considered to be too risky, such as aerospace systems and flight control.

In the case of control of PDEs, no controversy of the same magnitude has existed; however, this field has chronically suffered from the inability to enthuse engineers. The first reason is, again, the mathematical difficulty associated with PDEs. The second is the belief that finite-dimensional control design tools should suffice, since many (or most, though not all) PDEs are dominated by some finite-dimensional dynamic behavior. Hence, some form of model reduction, Galerkin approximation, or spatial discretization should suffice. If only it were this simple! If it were, the control engineers would be self-sufficient in the area of control of PDEs, starting with the quick step of reducing the problem to an ODE and then applying a finite-dimensional design. Unfortunately, neither is this step quick—in some cases it involves techniques that researchers in other fields, such as fluid dynamics and solid mechanics, spend entire careers honing—nor is the rest easy (and mathematically straightforward), even if the designer has successfully completed this first step of PDE→ODE reduction.

The question of approximating a PDE by an ODE for the purpose of control design is the same chicken-or-egg question as the question of model (dimension) reduction for control design for ODEs. One can never be sure that model reduction for the plant (based on open-loop considerations) will lead to a control design (performed on the reduced model) that will be effective for the original (unreduced) plant. For PDEs, this question translates into how to perform spatial discretization in such a way that, upon finite-dimensional control design, and upon refinement of the discretization grid, the control gains converge as the spatial discretization step goes to zero. This is a highly nontrivial question. Very intuitive, "obvious," and standard discretization techniques used for PDE *simulation* (without control) lead to divergent control gain kernels upon grid refinement. It often takes completely non-obvious discretization ideas to achieve convergence of gains (though convergence of signals may be achieved even when the gains are not convergent). The moral of this story is that a priori approximation of a PDE by an ODE for the purpose of feedback design is not a problem that is solvable in general. One can easily approximate the PDE after the controller has been designed, but it is not safe to do so for the purpose of designing a controller.

Hence, it is crucial to develop control tools in the continuum domain, for the PDEs, and invariant of model reduction choices.

* * *

What does the book cover? This book undertakes a major task, that of solving adaptive control problems for PDEs. The difficulty of this task is even greater than the sum of the difficulties of the two components, namely, (1) control of PDEs and (2) adaptive control (for ODEs).

Indeed, until a few years ago, most researchers would have considered adaptive control of infinite-dimensional systems (and PDEs in particular) almost a misnomer because adaptive control requires an explicit parametrization of the control gains (either directly or indirectly, through the plant parameters). Within the framework of classical control methods for PDEs, such as linear quadratic optimal control, this approach would literally require that an operator Riccati equation (infinite-dimensional and nonlinear, though algebraic, independent of time) be solved at each time step.

With the method of backstepping, which we recently introduced for boundary control of PDEs, the need for solving a new operator Riccati equation at each time step for each new parameter estimate is eliminated. With the backstepping approach the control gains are explicitly parametrized in terms of both the spatial variables and the plant parameters. One need only run online estimation of the plant parameters. Computation of the gains is a matter of plugging the plant parameters into an explicit formula, or at most of solving a linear problem of a very easy kind, such as an integral equation in the spatial variable (derived from a linear hyperbolic PDE that characterizes the control gains) at each time step.

This book focuses on adaptive control of *parabolic* PDEs. This is the most forgiving class of PDEs, at least in terms of modeling errors, because only a finite number of eigenvalues of the plant can be in the right half plane or on the imaginary axis. Hyperbolic PDEs require additional tools as compared to those presented in this book and contain peculiarities related primarily to these equations being second order in time, rather than being infinite-dimensional. A successful and thorough development for hyperbolic PDEs will likely require years of additional research. For this reason we restrict our attention to parabolic PDEs.

The efforts in adaptive control of PDEs began in the late 1980s and continued throughout the 1990s. The leading authors in these efforts were Balas, Bentsman, Demetriou, Duncan, Kobayashi, Orlov, Pasik-Duncan, and a few others.

The backstepping approach for PDEs, which started crystallizing for nonadaptive problems around the year 2002, allows us to make a major leap forward methodologically and to design adaptive controllers for previously intractable classes of problems. The emphasis in this book is fivefold:

- unstable PDEs,
- boundary control (and boundary sensing),
- infinite relative degree problems,
- simultaneous lack of state measurement and parameter knowledge (adaptive output feedback),
- functional unknown parameters (infinite-dimensional parameter vector).

We of course proceed pedagogically, tackling these problems one at a time and putting them all together in one of the final chapters (Chapter 13).

On the methodological level, the book contains an ambitious and exhaustive catalog of approaches to adaptive control. Each of these approaches arises from different parametrizations of the system and different methods for parameter identifier design. Our book mirrors the categorization of methods introduced in the second author's 1995 classic monograph [68]. Three methods for identifier design are employed:

- Lyapunov based,
- passivity or observer based,
- swapping based.

In addition, two parametrizations are used:

- u-model (the original plant model),
- w-model (the error system model, after the application of the backstepping transformation).

This large collection of methods and parametric models creates a large number of combinations, each of which is a possible adaptive control scheme. We don't pursue all of them, but we do pursue a representative subset of combinations.

The book is structured as follows. It has two major parts: Part I introduces the underlying nonadaptive controller and observer designs (highlighting the problems in which the controller and observer gains can be found explicitly) and Part II presents the adaptive designs. Roughly, the first third of the book is nonadaptive, whereas the remaining two thirds are adaptive.

* * *

Whom is the book for? The book should be of interest to any and every researcher who has worked on adaptive control of ODEs. It presents surprisingly accessible solutions to adaptive control problems for PDEs, which one might expect to be abstract and dry. On the contrary: a researcher, student, or engineer in adaptive control will find a wealth of examples that are solved and presented in full analytical and numerical detail, graphically illustrated, and with intuition provided that helps understand the results.

Researchers in PDEs and, more broadly, mathematics (including nonlinear functional analysis) will find the book filled with problems of a new kind. The problems abound in mathematical challenges that do not arise in any other field except control theory, and more specifically, parameter-adaptive control. Adaptive control of *linear* PDEs is a unique stepping stone toward the challenging world of nonlinear PDEs. In adaptive control of linear PDEs the overall system is nonlinear, owing to the presence of the parameter identifier dynamics and its state feeding into the controller. However, this nonlinearity is sufficiently benign and globally tractable, unlike the nonlinearity one encounters in problems like 3D Navier-Stokes equations or nonlinear reaction-diffusion systems that exhibit finite-time blowup.

In addition to engineers who are control specialists, some chemical engineers and process dynamics researchers should find the book of interest, as it covers a broad class of parabolic PDEs, including the reaction-advection-diffusion systems with spatially varying coefficients that arise in process dynamics.

Likewise, mechanical and aerospace engineers dealing with various mechanics problems of a parabolic kind (thermal, fluid, combustion, and various other diffusion-dominated problems), even if control is not their focus, should find the book of interest, since the methods employed are not from the generic control catalog (LQR, pole placement) but are very structure-specific and PDE-specific. Physicists should find the book of interest for the same reason.

Even a researcher with no interest whatsoever in control systems in general can enjoy the book on two levels:

1. The methods for control developed in this book "morph" the system dynamics in an uncommonly intuitive way. For example, these methods are capable of removing *spatially distributed*, domain-wide reaction effects using inputs only from the boundary of the PDE domain. This is only an example of a much more general capability of the method to radically alter the nature of the PDE dynamics within the domain, acting only through boundary conditions.
2. A component of adaptive control design that is deemphasized in the book (for the sake of focus on control—but is still significant) is parameter estimation, namely, *system identification for PDEs*. A researcher who cares about this topic alone would find the book useful. We tackle problems with both constant and

spatially varying parameters, and problems where the full state of the PDE is measured and where only a boundary value is measured (adaptive observers).

Regarding the background required to read this book, relatively little is needed beyond calculus. The book is self-contained in terms of most of the nonlinear stability analysis tools, special functions background, or functional analysis techniques and inequalities employed.

Acknowledgments. We would like to thank Ole Morten Aamo, Saverio Bolognani, and Yuri Orlov for their contributions to Chapters 6, 12, and 13. We sincerely appreciate Andrea Serrani's, Gang Tao's, and Kartik Ariyur's generous commentary on the book. We would like to thank Joseph Bentsman, Michael Demetriou, Tyrone Duncan, Hartmut Logemann, Bozenna Pasik-Duncan, Stewart Townley, and John Wen for their inspiring early work on system identification and adaptive control of infinite-dimensional systems. We appreciate Brian Anderson's and Steve Morse's encouraging feedback when we were taking our first steps in adaptive control of PDEs in late 2004. We thank Laurent Praly and Irena Lasiecka for their interest and technical input into our work. We are grateful to Petar Kokotovic for his encouragement and support. We thank Cymer Inc., General Atomics, Bosch, the National Science Foundation, and the Office of Naval Research for their support. We dedicate this book to our daughters, Katerina, Victoria, and Alexandra, and our wives, Olga and Angela.

La Jolla, California
June 2009

Andrey Smyshlyaev
Miroslav Krstic

Chapter One

Introduction

1.1 PARABOLIC AND HYPERBOLIC PDE SYSTEMS

This book investigates problems in control of partial differential equations (PDEs) with unknown parameters. Control of PDEs alone (with known parameters) is a complex subject, but also a physically relevant subject. Numerous systems in aerospace engineering, bioengineering, chemical engineering, civil engineering, electrical engineering, mechanical engineering, and physics are modeled by PDEs because they involve fluid flows, thermal convection, spatially distributed chemical reactions, flexible beams or plates, electromagnetic or acoustic oscillations, and other distributed phenomena. Model reduction to ordinary differential equations (ODEs) is often possible, but model reduction approaches suitable for simulation (in the absence of control) may lead to control designs that are divergent upon grid refinement.

While ODEs represent a class of dynamic systems for which general, unified control design can be developed, at least in the linear case, this is not so for PDEs. Even in dimension one (1D), different classes of PDEs require fundamentally different approaches in analysis and in control design. Three basic classes of PDEs are often considered to be the fundamental, distinct classes:

- parabolic PDEs (including reaction-diffusion equations),
- first-order hyperbolic PDEs (including transport equations),
- second-order hyperbolic PDEs (including wave equations).

In addition to these basic classes, many other classes and individual PDEs of interest exist, some formalized in the Schrödinger, Ginzburg-Landau, Korteweg–de Vries, Kuramoto-Sivashinsky, Burgers, and Navier-Stokes equations.

Designing adaptive controllers for PDEs requires not only mastering the nonadaptive designs for PDEs (with known parameters) but also developing a design approach that is parametrized in such a way that the controllers can be made parameter-adaptive. This requirement excludes most design approaches. For example, optimal control approaches require solving operator Riccati equations, which cannot be done continuously in real time, whereas pole placement–based approaches are parametrized in terms of the open-loop system's eigenvalues and not in terms of the plant parameters (as required for *indirect* adaptive control) or in terms of controller parameters (as required for *direct* adaptive control). In this book we build on the *backstepping* approach to control of PDEs [70], as it leads to feedback laws that are explicitly or nearly explicitly parametrized in terms of the plant parameters.

At the present time the backstepping approach [70] is well developed for a rather broad set of classes of PDEs, including the three basic classes of PDEs—parabolic, first-order hyperbolic, and second-order hyperbolic. The backstepping approach allows adaptive control development for PDEs in all of these classes. However, owing to the vast number of design possibilities, the development of adaptive controllers for PDEs using the backstepping approach has so far been mostly limited to parabolic PDEs. The reason for this is that this class is complex enough to be representative of many (but not all) of the mathematical challenges one faces when dealing with PDEs but does not include the idiosyncratic challenges such as those arising from second-order-in-time derivatives in second-order hyperbolic PDEs, where the adaptive issue may be secondary to the analysis issues specific to this PDE class.

Though our focus in this book is on parabolic PDEs, which we have chosen as the benchmark class for the development of adaptive controllers for PDE systems, adaptive control designs for parametrically uncertain hyperbolic PDE systems with boundary actuation are starting to emerge. Examples of such designs are the design in [63] for an unstable wave equation, as a representative of second-order hyperbolic PDEs, and in [18, 19] for ODE systems with an actuator delay of unknown length, as a representative of first-order hyperbolic PDEs.

1.2 THE ROLES OF PDE PLANT INSTABILITY, ACTUATOR LOCATION, UNCERTAINTY STRUCTURE, RELATIVE DEGREE, AND FUNCTIONAL PARAMETERS

The field of adaptive control of infinite-dimensional systems is a complex landscape in which problems of vastly different complexity can be considered, depending on not only the class of PDE systems but also the following three properties of the system:

- *Stability of the open-loop system.* Open-loop unstable systems present greater challenges than open-loop stable systems. In this book we focus on PDEs that are open-loop unstable.
- *Actuator location.* An immense difference exists between problems with distributed control, where independent actuation access is available at each point in the PDE domain, and problems with localized actuation, such as *boundary control*, where the dimensionality of the state is higher than the dimensionality of the actuation. A similar difference exists between problems with distributed sensing and boundary sensing. In this book we focus on PDEs with boundary control. In cases where we extend our designs from the full-state feedback case to the output-feedback case, we consider boundary sensing.
- *Uncertainty structure.* One of the key challenges in adaptive control is compensating parametric uncertainties that are *not matched* by the control, namely, uncertainties that cannot be directly canceled by the control but require some state transformation, or the solution of some Bezout equation, or some other nontrivial construction through which the control gains access to uncertainties

that are not collocated with the control input. In this book we focus exclusively on problems with unmatched parametric uncertainties.

- *Relative degree.* Historically, the greatest challenge in adaptive control has been associated with extending adaptive control from systems of relative degree one to systems of relative degree two, and then to systems of relative degree three and higher. In PDE control problems the relative degree can be infinite. We aim for problems of this type in this book, as they introduce issues that are beyond what is encountered in adaptive control of ODE systems. Our benchmark system in this regard is the unstable reaction-diffusion PDE of dimension one, where control is applied at one boundary and sensing is conducted at the opposite boundary.
- *Spatially constant or functional parameters?* Even when a PDE has a finite number of spatially constant parameters, the designer may face a challenging problem if the PDE is unstable and the actuation is applied from the boundary. However, an entirely different level of challenge arises when the parametric uncertainties are functional, namely, when the unknown parameters depend on the spatial variables. This book contains solutions to such problems for unstable reaction-diffusion plants with boundary control and sensing.

1.3 CLASS OF PARABOLIC PDE SYSTEMS

As we have indicated earlier, this book focuses on parabolic PDEs as a good benchmark class for adaptive control design for PDEs with boundary actuation, open-loop instability, possibly functional parametric uncertainties, and the possibility of occurrence of infinite relative degree when the input and output are non-collocated.

A good representative example of such a parabolic PDE is the reaction-diffusion equation. However, in this book we develop designs that are applicable to the following class of linear parabolic partial integro-differential equation (P(I)DE):

$$u_t(x,t) = \varepsilon(x)u_{xx}(x,t) + b(x)u_x(x,t) + \lambda(x)u(x,t) + g(x)u(0,t) + \int_0^x f(x,y)u(y,t)\,dy, \quad 0 < x < L. \tag{1.1}$$

This system is supplied with two boundary conditions. The $x = 0$ boundary of the domain is uncontrolled and is of either Dirichlet or Robin type:

$$u(0,t) = 0 \quad \text{or} \quad u_x(0,t) = -qu(0,t). \tag{1.2}$$

The other end, $x = 1$, is controlled, either with Dirichlet or with Neumann actuation:

$$u(L,t) = U(t) \quad \text{or} \quad u_x(L,t) = U(t), \tag{1.3}$$

where $U(t)$ is the control input.

The PDE (2.1)–(2.3) serves as a model for a variety of systems with thermal, fluid, and chemically reacting dynamics. Applications include, among others, chemical tubular reactors [15, 36], solid propellant rockets [16], thermal-fluid convection problems [14], flows past bluff bodies [103], steel casting [56], and cardiac pacing devices [37]. In general, spatially varying coefficients come from applications with non-homogeneous materials or unusually shaped domains, and can also arise from linearization.

The g- and f-terms are not as common as reaction, advection, and diffusion terms but are important nonetheless (see Chapter 13) and have the "spatially causal" structure, which makes them tractable by our method.

Most of the book covers 1D, real-valued plants. However, in Chapters 6 and 13 we extend the approach to complex-valued plants, and in Chapter 9 we deal with an example of a 3D problem.

Two main characteristic features of the plant (1.1)–(1.3) are:

- The plant is *unstable* for large ratios $\lambda(x)/\varepsilon(x)$, $g(x)/\varepsilon(x)$, $f(x, y)/\varepsilon(x)$ or large positive q.
- The plant is actuated from the *boundary*. In many applications this is the only physically reasonable way of placing actuators.

In this book we consider two types of measurement scenarios, either full-state feedback or *boundary* sensing, where the measured output can be one of the following four:

$$u(0,t), \quad u_x(0,t), \quad u(L,t), \quad u_x(L,t).$$

1.4 BACKSTEPPING

The backstepping approach in control of nonlinear ODEs "matches" nonlinearities unmatched by the control input by using a combination of a diffeomorphic change of coordinates and feedback. In this book we pursue a continuum equivalent of this approach and build a change of variables, which involves a Volterra integral operator (which has a lower-triangular structure, just as backstepping transformations for nonlinear ODEs do) that "absorbs" the destabilizing terms acting in the domain and allows the control to completely eliminate their effect acting only from the boundary.

While the backstepping method is certainly not the first solution to the problems of boundary control of PDEs, it has several distinguishing features. First of all, it takes advantage of the structure of the system, resulting in a problem of solving a linear hyperbolic PDE for the gain kernel, an object much easier, both conceptually and computationally, than operator Riccati equations arising in linear quadratic regulator (LQR) approaches to boundary control. Second, the well-posedness difficulties are circumvented by transforming the plant into a simple heat equation. The control problem is solved essentially by calculus, making a design procedure clear and constructive and the analysis easy, in contrast to standard abstract approaches (semigroups, etc.). Third, for a number of physically relevant problems, the observer/controller kernels can be found in closed form, that is, as explicit functions of the spatial variable. Finally, unlike other methods, the backstepping approach naturally extends to adaptive control problems.

1.5 EXPLICITLY PARAMETRIZED CONTROLLERS

One of the most significant features of the design approach presented in this book is that for a certain class of physically relevant parabolic 1D PDEs, the control laws can be obtained in *closed form*, which is not the case with other methods (LQR, pole placement). *One* of the subclasses for which explicit controllers can be obtained is

$$u_t(x,t) = \varepsilon u_{xx}(x,t) + bu_x(x,t) + \lambda u(x,t) + ge^{\alpha x}u(0,t), \tag{1.4}$$

$$u_x(0,t) = -qu(0,t), \tag{1.5}$$

$$u(L,t) = U(t) \quad \text{or} \quad u_x(L,t) = U(t), \tag{1.6}$$

for arbitrary values of six parameters $\varepsilon, b, \lambda, g, \alpha$, and q. The explicit controllers, in turn, allow us to find explicit closed-loop solutions. Some of the PDEs within the class (1.4)–(1.6) are not even explicitly solvable in open loop; however, we solve them in closed form in the presence of the backstepping control law. While we prove closed-loop well-posedness for the general class separately from calculating the explicit solutions for the subclasses, these solutions serve to illustrate how well-posedness issues can be pretty much bypassed by a structure-specific control approach, allowing the designer to concentrate on control-relevant issues such as the feedback algorithm and its performance.

Since the backstepping observers are dual to the backstepping controllers, they are also available in closed form for the same class of plants (1.4)–(1.6).

1.6 ADAPTIVE CONTROL

In many distributed parameter systems the physical parameters, such as the Reynolds, Rayleigh, Prandtl, or Péclet numbers, are unknown because they vary with operating conditions. While adaptive control of finite-dimensional systems is an advanced field that has produced adaptive control methods for a very general class of linear time-invaria systems, adaptive control techniques have been developed for only a few classes of PDEs restricted by relative degree, stability, or domain-wide actuation assumptions. There are two sources of difficulties in dealing with PDEs with parametric uncertainties. The first difficulty, which also exists in ODEs, is that even for linear plants, adaptive schemes are nonlinear. The second difficulty, unique to PDEs, is the absence of parametrized families of controllers. The backstepping approach removes this second difficulty, opening the door for the use of a wealth of certainty equivalence and Lyapunov techniques developed for finite-dimensional systems.

In this book we develop the first constructive adaptive designs for unstable parabolic PDEs controlled from a boundary. We differentiate between two major classes of schemes:

- Lyapunov schemes,
- certainty equivalence schemes.

Within the certainty equivalence class, two types of identifier designs are pursued:

- passivity-based identifiers,
- swapping-based identifiers.

Each of those designs is applicable to two types of parametrizations:

- the plant model in its original form (which we refer to as the u-model),
- a transformed model to which a backstepping transformation has been applied (which we refer to as the w-model).

Hence, a large number of control algorithms result from combining different design tools—Lyapunov schemes, w-passive schemes, u-swapping schemes, and so on.

Our presentation culminates in Chapter 13 with the adaptive output tracking design for the system with scalar input and output, infinite-dimensional state, and infinitely many unknown parameters.

1.7 OVERVIEW OF THE LITERATURE ON ADAPTIVE CONTROL FOR PARABOLIC PDES

Early efforts to design adaptive controllers for distributed parameter systems used tuning of a scalar gain to a high level to stabilize some classes of (relative degree one) infinite-dimensional plants [58, 83, 85] (see also the survey by Logemann and Townley [84] for additional references).

Early identifiability results include [24, 57, 61, 90, 98]. A monograph by Banks and Kunisch [8] focuses on approximation methods for least squares estimation of elliptic and parabolic equations. Online parameter estimation schemes were developed by Demetriou and Rosen [32, 33], Baumeister, Scondo, Demetriou, and Rosen [9], and Orlov and Bentsman [94].

Model reference adaptive control (MRAC)–type schemes were designed by Hong and Bentsman [48, 49], Bohm, Demetriou, Reich, and Rosen [12], and Bentsman and Orlov [11]. These schemes successfully identify functional, spatially varying parametric uncertainties under the assumption that control is distributed in the PDE domain. A sliding mode observer of a spatial derivative has been designed in [93] to get a state-derivative free MRAC scheme. Demetriou and Rosen [34] developed robust parameter estimation schemes using parameter projection and sigma modification. Variable structure MRAC is considered in Demetriou and Rosen [35].

Adaptive observers for structurally perturbed and slowly time-varying infinite-dimensional systems have been developed in [27] and [28].

Positive real tools have been employed by Wen and Balas [120] and by Demetriou and co-authors [25, 30, 31].

Adaptive linear quadratic control with least squares estimation was pursued by Duncan, Maslowski, and Pasik-Duncan [38] for stochastic evolution equations with unbounded input operators and stable uncontrolled dynamics.

Nonlinear PDEs have also received some attention. Liu and Krstic [81] and Kobayashi [59] considered a Burgers equation with various parametric uncertainties; Kobayashi [60] also considered the Kuramoto-Sivashinsky equation.

Jovanovic and Bamieh [52] designed adaptive controllers for nonlinear systems on lattices, which include applications like infinite vehicular platoons or infinite arrays of microcantilevers. Experimentally validated adaptive controllers for distributed parameter systems were presented by Dochain [36] and de Queiroz, Dawson, Agarwal, and Zhang [102].

1.8 INVERSE OPTIMALITY

In this book, for the first time for PDEs, we develop *inverse optimal* controllers. These controllers minimize meaningful cost functionals that penalize both the state and the control. Unlike LQR controllers for PDEs, our controllers avoid the problem of having to solve operator Riccati equations.

For control of PDEs with parametric uncertainties, solving Riccati equations on a real-time basis at each time step for each new plant parameter estimate is out of the question. Even if one were to indulge in solving a new Riccati equation at each time step, one would not be rewarded for the effort, as the certainty equivalence combination of the standard parameter estimation schemes with the linear quadratic optimal control formula does not possess any optimality property (in fact, even its transient performance can be unpredictably poor, with its stability proof being among the most complicated of any adaptive control scheme [51]). The backstepping approach leads to *adaptive inverse optimal* controllers that minimize cost functionals simultaneously penalizing parameter estimation error, state, and control transients.

1.9 ORGANIZATION OF THE BOOK

The book is divided into two parts. In Part I we present nominal, nonadaptive controllers and observers for the class of plants (1.1)–(1.3). In Part II we design three different adaptive schemes for state and output feedback stabilization and present a concept of inverse optimality for control of PDEs.

In Chapter 2, the core chapter of Part I, we introduce the backstepping method for boundary control of parabolic PDEs. We derive the equations for the control gain and show that they have a unique solution, which is found by the method of successive approximations.

In Chapter 3 we develop closed-form, ready-to-implement boundary controllers for a six-parameter class of PDEs (1.4)–(1.6). The solutions of the closed-loop systems are also given explicitly.

In Chapter 4 we develop backstepping observers for problems in which only boundary measurements are available. These observers are dual to the backstepping state feedback controllers developed in Chapters 2 and 3.

In Chapter 5 we combine the backstepping observers with the backstepping controllers to obtain the solution to the output feedback problem with boundary measurement and actuation, with actuator and sensor located on the same or the opposite boundaries.

In Chapter 6 we extend the designs developed in Chapters 2–5 to the case of "parabolic-like" plants with a complex-valued state (such plants can also be viewed as two coupled PDEs). We consider two cases: the linearized Ginzburg-Landau equation, which models vortex shedding in bluff body flows, and its special case, the Schrödinger equation. In this chapter we also show how to apply the backstepping method to problems in a semi-infinite domain (typical for bluff body flows) in such a way that the gain kernels have a compact support, resulting in a partial state feedback.

Part II of the book opens with Chapter 7, in which we give a bird's-eye view of different approaches to adaptive control of PDEs. This chapter is meant to serve as an entry point into adaptive control for a nonexpert reader. We proceed through a series of benchmark examples, skipping the detailed proofs of closed-loop stability (which are presented later in Chapters 8–13) and focusing on a broader context for the individual approaches and the trade-offs between them. This chapter also contains several original adaptive schemes not pursued in full detail in the rest of the book.

In Chapters 8–10 we develop three different adaptive schemes for plants with constant coefficients. These schemes are based on the nominal controllers designed in Chapter 3. In Chapter 8 we present the *Lyapunov* approach. It is the most complicated method of the three; however, it ensures the best transient performance properties and has the lowest dynamic order. The uncommon Lyapunov function, which contains a logarithmic weight on the plant state, results in a normalization of the update law by a norm on the plant state, preventing overly fast adaptation. In Chapter 9 we introduce the *passivity-based* (or "observer-based") approach. This approach is appealing for its simplicity: it employs an observer in the form of a copy of the plant, plus a stabilizing error term. This chapter also provides a design example for a 3D problem. In Chapter 10 we present the *swapping* approach (often called simply the gradient method), the most commonly used identification method in finite-dimensional adaptive control. Filters of the "regressor" and of the measured part of the plant are implemented to convert a dynamic parametrization of the problem into a static one where standard gradient and least squares estimation techniques can be used. This method has a higher dynamic order than the passivity-based method because it uses one filter per unknown parameter instead of just one filter. However, it allows for least squares estimation and can be used in output feedback problems.

In Chapter 11 we design the adaptive controllers for the reaction-advection-diffusion plants with *spatially varying* parameters. The control gain is computed through an approximate solution of a linear PDE or through a limited number of recursive integrations. We show robustness of the proposed scheme with respect to an error in the online gain computation.

Chapter 12 is a bridge between Chapters 8 and 10 and Chapter 13. We develop output-feedback adaptive controllers for two benchmark parabolic PDEs motivated by a model of thermal instability in solid propellant rockets. Both benchmark plants are unstable, are of infinite relative degree, and are controlled from the boundary. This chapter is the only chapter in the book in which we prove the convergence of the parameter estimates to the true parameter values.

In Chapter 13 we solve a problem of *output-feedback* stabilization of plants with unknown reaction, advection, and diffusion parameters. Both sensing and actuation are performed at the boundary, and the unknown parameters are allowed to be spatially varying. We construct a special transformation of the original system into the PDE analog of *observer canonical form*, with unknown parameters multiplying the measured output. We implement input and output filters to convert dynamic parametrization of the problem into a static one and then use the gradient estimation algorithm. We also solve the problem of the adaptive output tracking of a desired reference trajectory prescribed at the boundary.

In Chapter 14 we modify the designs of Chapter 2 to incorporate optimality in addition to stabilization for plants with known parameters. We also propose an adaptive control law to reduce the control effort and derive explicit bounds on the state and control. In the second part of the chapter we deal with an unknown parameter case. We present an *inverse optimal adaptive* controller, in other words, an adaptive controller made inverse optimal. No attempt is made to enforce the parameter convergence, but the estimation transients are penalized simultaneously with the state and control transients.

1.10 NOTATION

For the function of two variables $v(x,t)$, $x \in [0,1]$, $t \in [0,\infty)$:

$\|v(t)\|$	$\left(\int_0^1 v(x,t)^2\,dx\right)^{1/2}$
$\|v(t)\|_{H^1}$	$\left(\int_0^1 v(x,t)^2\,dx + \int_0^1 v_x(x,t)^2\,dx\right)^{1/2}$
$v(x)$, $\|v\|$	$v(x,t)$, $\|v(t)\|$ (dependence on time is often dropped in the proofs)

For the function of time $f(t)$, $t \in [0,\infty)$:

$f \in \mathcal{L}_1$	$\int_0^\infty \|f(t)\|\,dt < \infty$
$f \in \mathcal{L}_2$	$\int_0^\infty \|f(t)\|^2\,dt < \infty$
$f \in \mathcal{L}_\infty$	$\sup_{t\in[0,\infty)} \|f(t)\| < \infty$

Other notation:

$(x,y) \in \mathcal{T}$	$0 < y < x \le 1$
l_1, l_2	generic functions of time that belong to $\mathcal{L}_1$, $\mathcal{L}_2$, respectively
J_0, J_1, J_2	Bessel functions of the first kind of order 0, 1, 2, respectively
I_0, I_1, I_2	modified Bessel functions of the first kind of order 0, 1, 2, respectively

PART I
Nonadaptive Controllers

Chapter Two

State Feedback

In this chapter we introduce the method of *backstepping* and present nominal designs that will serve as the basis for all the designs in the rest of the book.

2.1 PROBLEM FORMULATION

We consider the following class of linear parabolic partial integro-differential equations (P(I)DEs):

$$u_t(x,t) = \varepsilon(x)u_{xx}(x,t) + b(x)u_x(x,t) + \lambda(x)u(x,t) + g(x)u(0,t) + \int_0^x f(x,y)u(y,t)\,dy \tag{2.1}$$

for $x \in (0, L)$, $t > 0$, with boundary conditions

$$u_x(0,t) = -qu(0,t), \tag{2.2}$$

$$u(L,t) = U(t), \tag{2.3}$$

where $U(t)$ is the control input. We make the following assumptions:

$$\varepsilon(x) > 0 \text{ for all } x \in [0, L], \quad \varepsilon \in C^3[0, L], \quad q \in \mathbb{R}, \quad b \in C^2[0, L], \quad \lambda,\ g \in C^1[0, L], \quad f \in C^1([0, L] \times [0, L]). \tag{2.4}$$

For large positive q, $\lambda(x)$, $g(x)$, or $f(x, y)$, the plant (2.1) with $U(t) = 0$ is unstable; therefore, the objective is to stabilize the equilibrium $u(x,t) \equiv 0$.

Before we proceed with the control design, let us introduce the change of variables (the so-called gauge transformation):

$$\bar{u}(\bar{x},t) = u(x,t)\left(\frac{\bar{\varepsilon}}{\varepsilon(x)}\right)^{1/4} \exp\left\{\int_0^x \frac{b(s)}{2\varepsilon(s)}ds\right\}, \tag{2.5}$$

$$\bar{x} = \sqrt{\bar{\varepsilon}}\int_0^x \frac{ds}{\sqrt{\varepsilon(s)}}, \tag{2.6}$$

where

$$\bar{\varepsilon} = \left(\int_0^L \frac{ds}{\sqrt{\varepsilon(s)}}\right)^{-2} \tag{2.7}$$

is the new diffusivity coefficient. Note that there is an implicit coordinate change within the transformation (2.5) as the term on the left, $\bar{u}(\bar{x},t)$, is a function of $\bar{x}$, while the term on the right is a function of x. One can show that the modified plant has the following parameters:

$$\bar{b}(\bar{x}) = 0,$$

$$\bar{\lambda}(\bar{x}) = \lambda(x) + \frac{\varepsilon''(x)}{4} - \frac{b'(x)}{2} - \frac{3}{16}\frac{(\varepsilon'(x))^2}{\varepsilon(x)} + \frac{1}{2}\frac{(b(x)\varepsilon'(x))}{\varepsilon(x)} - \frac{1}{4}\frac{b^2(x)}{\varepsilon(x)},$$

$$\bar{q} = q\sqrt{\frac{\varepsilon(0)}{\bar{\varepsilon}}} - \frac{b(0)}{2\sqrt{\bar{\varepsilon}\varepsilon(0)}} - \frac{\varepsilon'(0)}{4\sqrt{\bar{\varepsilon}\varepsilon(0)}},$$

$$\bar{g}(\bar{x}) = g(x)\left(\frac{\varepsilon(0)}{\varepsilon(x)}\right)^{1/4} \exp\left\{\int_0^x \frac{b(s)}{2\varepsilon(s)} ds\right\},$$

$$\bar{f}(\bar{x}, \bar{y}) = f(x, y)\left(\frac{\varepsilon^3(y)}{\varepsilon(x)\bar{\varepsilon}^2}\right)^{1/4} \exp\left\{\int_y^x \frac{b(s)}{2\varepsilon(s)} ds\right\},$$

$$\bar{U}(t) = U(t)\left(\frac{\bar{\varepsilon}}{\varepsilon(L)}\right)^{1/4} \exp\left\{\int_0^L \frac{b(s)}{2\varepsilon(s)} ds\right\}. \tag{2.8}$$

We now drop the bars for notational convenience and write the plant as

$$u_t(x, t) = \varepsilon u_{xx}(x, t) + \lambda(x)u(x, t) + g(x)u(0, t) + \int_0^x f(x, y)u(y, t)\, dy, \tag{2.9}$$

$$u_x(0, t) = -qu(0, t), \tag{2.10}$$

$$u(1, t) = U(t). \tag{2.11}$$

Thus, the effect of the transformation (2.5)–(2.6) is to eliminate $b(x)$ from the original problem (2.1), to make $\varepsilon(x)$ constant, and to scale the domain to [0, 1]. Note that this transformation does not affect the dynamics of the system.

Equation (2.9) is in fact a P(I)DE, but for convenience we abuse the terminology and call it a PDE for the rest of the book.

The cases of Neumann actuation and Dirichlet boundary condition at the uncontrolled end will be considered in Sections 2.6 and 2.7.

2.2 BACKSTEPPING TRANSFORMATION AND PDE FOR ITS KERNEL

The natural objective for a feedback is to eliminate all unwanted terms from the equation. In other words, we want the closed-loop system to have the dynamics of the following "target" system

$$w_t(x, t) = \varepsilon w_{xx}(x, t) - cw(x, t), \quad x \in (0, 1), \tag{2.12}$$

$$w_x(0, t) = 0, \tag{2.13}$$

$$w(1, t) = 0, \tag{2.14}$$

which is exponentially stable for $c > -\varepsilon\pi^2/4$ (as will be shown in Section 2.5). The design parameter c sets the desired decay rate of the closed-loop system.

Remark 2.1. Note that the above choice for the target system is good for several reasons. First, it is a simple, well-studied heat equation, which allows us to avoid any issues with well-posedness of the closed-loop system. Second, this equation is explicitly solvable, so that the exact closed-loop eigenvalues are known and explicit closed-loop solutions can be obtained. Finally, as will be shown in the adaptive part of the book, to solve adaptive output tracking problems one needs to generate a reference trajectory only for the simple target system, not for the complicated original system.

The main idea of the backstepping method is to use coordinate transformation

$$w(x,t) = u(x,t) - \int_0^x k(x,y)u(y,t)\,dy \tag{2.15}$$

along with boundary feedback

$$U(t) = \int_0^1 k(1,y)u(y,t)\,dy \tag{2.16}$$

to transform the system (2.9)–(2.11) into the system (2.12)–(2.14).

The most characteristic feature of the Volterra integral transformation (2.15) is that it is "spatially causal," that is, for a given x, the right-hand side of (2.15) depends only on the values of u in the interval $[0,x]$. Another important property of the Volterra transformation is that it is invertible (as will be shown later), so that stability of the target system translates into stability of the closed-loop system consisting of the plant plus boundary feedback.

Our goal now is to find the function $k(x,y)$, which we call the "gain kernel." At this point it is not clear what class of functions k belongs to. We show that given assumptions (2.4), it is twice continuously differentiable with respect to its arguments.

We start deriving a condition on k by differentiating (2.15) with respect to time:[1]

$$\begin{aligned}
w_t(x,t) =& \, u_t(x,t) - \int_0^x k(x,y)\left\{\varepsilon u_{yy}(y,t) + \lambda(y)u(y,t)\right. \\
& \left. + g(y)u(0,t) + \int_0^y f(y,\xi)u(\xi,t)\,d\xi\right\} dy \\
=& \, u_t(x,t) - \varepsilon k(x,x)u_x(x,t) + \varepsilon k(x,0)u_x(0,t) + \varepsilon k_y(x,x)u(x,t) \\
& -\varepsilon k_y(x,0)u(0,t) - \int_0^x \left(\varepsilon k_{yy}(x,y) + \lambda(y)k(x,y)\right) u(y,t)\,dy \\
& - u(0,t)\int_0^x k(x,y)g(y)\,dy \\
& - \int_0^x u(y,t)\left(\int_y^x k(x,\xi)f(\xi,y)\,d\xi\right) dy,
\end{aligned} \tag{2.17}$$

---

[1] We use the following notation: $k_x(x,x) = k_x(x,y)|_{y=x}$, $k_y(x,x) = k_y(x,y)|_{y=x}$, $\frac{d}{dx}k(x,x) = k_x(x,x) + k_y(x,x)$.

and space:

$$w_{xx}(x,t) = u_{xx}(x,t) - u(x,t)\frac{d}{dx}k(x,x) - k(x,x)u_x(x,t)$$
$$- k_x(x,x)u(x,t) - \int_0^x k_{xx}(x,y)u(y,t)\,dy. \tag{2.18}$$

Substituting (2.17) and (2.18) into equations (2.12)–(2.13) and using (2.9)–(2.10), we obtain the following equation:

$$\begin{aligned} 0 = &\int_0^x \{\varepsilon k_{xx}(x,y) - \varepsilon k_{yy}(x,y) - (\lambda(y)+c)k(x,y) + f(x,y)\}\, u(y,t)\,dy \\ &- \int_0^x u(y,t)\int_y^x k(x,\xi) f(\xi,y)\,d\xi\,dy \\ &+ \left\{\lambda(x) + c + 2\varepsilon\frac{d}{dx}k(x,x)\right\} u(x,t) \\ &+ \left\{g(x) - \int_0^x k(x,y)g(y)\,dy - \varepsilon k_y(x,0) - \varepsilon q k(x,0)\right\} u(0,t). \end{aligned} \tag{2.19}$$

For this equation to be verified for all $u(x,t)$, the following PDE for $k(x,y)$ must be satisfied:

$$\begin{aligned} \varepsilon k_{xx}(x,y) - \varepsilon k_{yy}(x,y) = &(\lambda(y)+c)k(x,y) - f(x,y) \\ &+ \int_y^x k(x,\xi) f(\xi,y)\,d\xi, \qquad (x,y)\in\mathcal{T} \end{aligned} \tag{2.20}$$

with boundary conditions

$$\varepsilon k_y(x,0) = -\varepsilon q k(x,0) + g(x) - \int_0^x k(x,y)g(y)\,dy, \tag{2.21}$$

$$\frac{d}{dx}k(x,x) = -\frac{1}{2\varepsilon}(\lambda(x)+c). \tag{2.22}$$

Here we denote $\mathcal{T} = \{x, y : 0 < y < x < 1\}$. The one remaining condition on k is obtained using the boundary condition (2.13):

$$w_x(0,t) = u_x(0,t) - k(0,0)u(0,t) = (-q - k(0,0))u(0,t) = 0, \tag{2.23}$$

which is verified for all $u(0,t)$ when $k(0,0) = -q$. Using this condition to integrate (2.22) from 0 to x, we obtain

$$k(x,x) = -q - \frac{1}{2\varepsilon}\int_0^x (\lambda(y)+c)\,dy. \tag{2.24}$$

The conditions (2.20), (2.21), and (2.24) form a P(I)DE with rather peculiar boundary conditions: one is on the characteristic (Goursat type) and the other is nonlocal, that is, it contains an integral term. Note that (2.20) is a second-order integro-differential equation, just like the plant itself. However, it is in a different class of systems—hyperbolic rather than parabolic.

In the next two sections we prove well-posedness of this PDE and derive a bound on its solution.

2.3 CONVERTING THE PDE INTO AN INTEGRAL EQUATION

The first step in proving well-posedness of (2.20), (2.21), (2.24) is to convert this PDE into an integral equation. We introduce the change of variables

$$\xi = x + y, \qquad \eta = x - y, \tag{2.25}$$

and denote

$$G(\xi, \eta) = k(x, y) = k\left(\frac{\xi + \eta}{2}, \frac{\xi - \eta}{2}\right), \tag{2.26}$$

transforming problem (2.20)–(2.24) into the following PDE:

$$\begin{aligned} 4\varepsilon G_{\xi\eta}(\xi, \eta) = {} & a\left(\frac{\xi - \eta}{2}\right) G(\xi, \eta) - f\left(\frac{\xi + \eta}{2}, \frac{\xi - \eta}{2}\right) \\ & + \int_{(\xi-\eta)/2}^{(\xi+\eta)/2} G\left(\frac{\xi + \eta}{2} + \tau, \frac{\xi + \eta}{2} - \tau\right) f\left(\tau, \frac{\xi - \eta}{2}\right) d\tau \end{aligned} \tag{2.27}$$

for $(\xi, \eta) \in \mathcal{T}_1$ with boundary conditions

$$\varepsilon G_{\xi}(\xi, \xi) = \varepsilon G_{\eta}(\xi, \xi) - \varepsilon q G(\xi, \xi) + g(\xi) - \int_0^{\xi} G(\xi + \tau, \xi - \tau) g(\tau)\, d\tau, \tag{2.28}$$

$$G(\xi, 0) = -q - \frac{1}{4\varepsilon} \int_0^{\xi} a\left(\frac{\tau}{2}\right) d\tau. \tag{2.29}$$

Here we introduced $\mathcal{T}_1 = \{\xi, \eta : 0 < \xi < 2,\ 0 < \eta < \min(\xi, 2 - \xi)\}$ and $a(\tau) = \lambda(\tau) + c$.

Integrating (2.27) with respect to η from 0 to η and using (2.29), we obtain

$$\begin{aligned} G_{\xi}(\xi, \eta) = {} & -\frac{1}{4\varepsilon} a\left(\frac{\xi}{2}\right) + \frac{1}{4\varepsilon} \int_0^{\eta} a\left(\frac{\xi - s}{2}\right) G(\xi, s)\, ds \\ & + \frac{1}{4\varepsilon} \int_0^{\eta} \int_{\xi}^{\xi+\eta-s} G(\tau, s) f\left(\frac{\tau - s}{2}, \xi - \frac{\tau + s}{2}\right) d\tau\, ds \\ & - \frac{1}{4\varepsilon} \int_0^{\eta} f\left(\frac{\xi + \tau}{2}, \frac{\xi - \tau}{2}\right) d\tau. \end{aligned} \tag{2.30}$$

Integrating (2.30) with respect to ξ from η to ξ gives

$$G(\xi,\eta)=G(\eta,\eta)-\frac{1}{4\varepsilon}\int_{\eta}^{\xi}a\left(\frac{\tau}{2}\right)d\tau+\frac{1}{4\varepsilon}\int_{\eta}^{\xi}\int_{0}^{\eta}a\left(\frac{\tau-s}{2}\right)G(\tau,s)\,ds\,d\tau$$
$$+\frac{1}{4\varepsilon}\int_{\eta}^{\xi}\int_{0}^{\eta}\int_{\mu}^{\mu+\eta-s}G(\tau,s)f\left(\frac{\tau-s}{2},\mu-\frac{\tau+s}{2}\right)d\tau ds\,d\mu$$
$$-\frac{1}{4\varepsilon}\int_{\eta}^{\xi}\int_{0}^{\eta}f\left(\frac{s+\tau}{2},\frac{s-\tau}{2}\right)d\tau\,ds. \tag{2.31}$$

To find $G(\eta,\eta)$, we use (2.28):

$$\frac{d}{d\xi}G(\xi,\xi)=G_{\xi}(\xi,\xi)+G_{\eta}(\xi,\xi)$$
$$=2G_{\xi}(\xi,\xi)+qG(\xi,\xi)-\frac{1}{\varepsilon}g(\xi)+\frac{1}{\varepsilon}\int_{0}^{\xi}G(\xi+s,\xi-s)g(s)\,ds. \tag{2.32}$$

Using (2.30) with $\eta=\xi$, we can write (2.32) in the form of a differential equation for $G(\xi,\xi)$:

$$\frac{d}{d\xi}G(\xi,\xi)=qG(\xi,\xi)-\frac{1}{2\varepsilon}\int_{0}^{\xi}f\left(\frac{\xi+\tau}{2},\frac{\xi-\tau}{2}\right)d\tau$$
$$-\frac{1}{2\varepsilon}a\left(\frac{\xi}{2}\right)+\frac{1}{2\varepsilon}\int_{0}^{\xi}a\left(\frac{\xi-s}{2}\right)G(\xi,s)\,ds$$
$$+\frac{1}{2\varepsilon}\int_{0}^{\xi}\int_{\xi}^{2\xi-s}G(\tau,s)f\left(\frac{\tau-s}{2},\xi-\frac{\tau+s}{2}\right)d\tau\,ds$$
$$-\frac{1}{\varepsilon}g(\xi)+\frac{1}{\varepsilon}\int_{0}^{\xi}G(\xi+s,\xi-s)g(s)\,ds. \tag{2.33}$$

Integrating (2.33) using the variation of constants formula and substituting the result into (2.31), we obtain an integral equation for G,

$$G(\xi,\eta)=G^{1}(\xi,\eta)+F[G](\xi,\eta), \tag{2.34}$$

where G^1 and $F[G]$ are given by

$$G^{1}(\xi,\eta)=-qe^{q\xi}-\frac{1}{4\varepsilon}\int_{\eta}^{\xi}a\left(\frac{\tau}{2}\right)d\tau-\frac{1}{2\varepsilon}\int_{0}^{\eta}e^{q(\eta-\tau)}\left[a\left(\frac{\tau}{2}\right)+2g(\tau)\right]d\tau$$
$$-\frac{1}{4\varepsilon}\int_{\eta}^{\xi}\int_{0}^{\eta}f\left(\frac{s+\tau}{2},\frac{s-\tau}{2}\right)d\tau\,ds$$
$$-\frac{1}{2\varepsilon}\int_{0}^{\eta}e^{q(\eta-\tau)}\int_{0}^{\tau}f\left(\frac{\tau+s}{2},\frac{\tau-s}{2}\right)ds\,d\tau \tag{2.35}$$

and

$$
\begin{aligned}
F[G](\xi,\eta) = & \frac{1}{2\varepsilon}\int_0^\eta e^{q(\eta-\tau)}\int_0^\tau a\left(\frac{\tau-s}{2}\right)G(\tau,s)\,ds\,d\tau \\
& +\frac{1}{4\varepsilon}\int_\eta^\xi\int_0^\eta a\left(\frac{\tau-s}{2}\right)G(\tau,s)\,ds\,d\tau \\
& +\frac{1}{2\varepsilon}\int_0^\eta\int_s^{2\eta-s} e^{q\left(\eta-\frac{\tau+s}{2}\right)} g\left(\frac{\tau-s}{2}\right)G(\tau,s)\,d\tau\,ds \\
& +\frac{1}{4\varepsilon}\int_\eta^\xi\int_0^\eta\int_\mu^{\mu+\eta-s} f\left(\frac{\tau-s}{2},\mu-\frac{\tau+s}{2}\right)G(\tau,s)\,d\tau\,ds\,d\mu \\
& +\frac{1}{2\varepsilon}\int_0^\eta e^{q(\eta-\mu)}\int_0^\mu\int_\mu^{2\mu-s} f\left(\frac{\tau-s}{2},\mu-\frac{\tau+s}{2}\right) \\
& \times G(\tau,s)\,d\tau\,ds\,d\mu. \qquad (2.36)
\end{aligned}
$$

We have thus proved the following statement.

Lemma 2.1. *Any $G(\xi,\eta)$ satisfying* (2.27)–(2.29) *also satisfies integral equation* (2.34).

2.4 ANALYSIS OF THE INTEGRAL EQUATION BY SUCCESSIVE APPROXIMATION SERIES

Using the result obtained in the previous section, we can now compute a uniform bound on the solutions by the method of successive approximations.[2]

Starting with $G^0 = 0$, let us set up a recursion,

$$G^{n+1} = G^1 + F[G^n], \quad n = 0, 1, 2, \ldots, \qquad (2.37)$$

and denote

$$\bar{\lambda} = \sup_{x\in[0,1]} |\lambda(x)|, \quad \bar{g} = \sup_{x\in[0,1]} |g(x)|, \quad \bar{f} = \sup_{(x,y)\in[0,1]\times[0,1]} |f(x,y)|. \qquad (2.38)$$

Let us estimate the differences between consecutive terms $\Delta G^n = G^{n+1} - G^n$, which are obtained through recursion

$$\Delta G^{n+1} = F[\Delta G^n], \quad n = 0, 1, 2, \ldots. \qquad (2.39)$$

[2]The results of this section can also be obtained using the argument that the operator F is compact and does not have (-1) as an eigenvalue, which means that the operator on the right-hand side of (2.34) is bounded invertible. However, we use the successive approximations approach because it can be used for finding an approximate solution to the kernel by symbolic calculation, because the expressions for successive terms are used to derive explicit controllers in Chapter 3, and because this approach yields a quantitative bound on the kernel (such a bound on the kernel of the inverse transformation is needed in the inverse optimal design in Chapter 14).

We have

$$
\begin{aligned}
|\Delta G^0(\xi,\eta)| &\le |q|e^{|q|} + \frac{1}{4\varepsilon}(\bar{\lambda}+c)(\xi-\eta) + \frac{1}{2\varepsilon}(\bar{\lambda}+c+2\bar{g})\eta e^{|q|} \\
&\quad + \frac{1}{4\varepsilon}\bar{f}\eta^2 e^{|q|} + \frac{1}{4\varepsilon}\bar{f}(\xi-\eta)\eta \\
&\le \frac{1}{\varepsilon}(\bar{\lambda}+c+\bar{f}+\bar{g})(1+e^{|q|}) + |q|e^{|q|} \equiv M.
\end{aligned}
\tag{2.40}
$$

Suppose that

$$
|\Delta G^n(\xi,\eta)| \le M^{n+1}\frac{(\xi+\eta)^n}{n!}. \tag{2.41}
$$

Then,

$$
\begin{aligned}
|\Delta G^{n+1}(\xi,\eta)| &\le \frac{1}{4\varepsilon}M^{n+1}\frac{1}{n!}\Bigg\{(\bar{\lambda}+c)\int_\eta^\xi \int_0^\eta (\tau+s)^n\, ds\, d\tau \\
&\quad +2(\bar{\lambda}+c)\int_0^\eta e^{q(\eta-\tau)}\int_0^\tau (\tau+s)^n\, ds\, d\tau \\
&\quad +2\bar{g}\int_0^\eta \int_s^{2\eta-s} e^{q(\eta-\frac{\tau+s}{2})}(\tau+s)^n\, d\tau\, ds \\
&\quad +\bar{f}\int_\eta^\xi \int_0^\eta \int_\mu^{\mu+\eta-s} (\tau+s)^n\, d\tau\, ds\, d\mu \\
&\quad + 2\bar{f}\int_0^\eta e^{q(\eta-\mu)}\int_0^\mu \int_\mu^{2\mu-s} (\tau+s)^n\, d\tau\, ds\, d\mu\Bigg\} \\
&\le \frac{1}{4\varepsilon}M^{n+1}\frac{1}{n!}\Bigg\{2(\bar{\lambda}+c)\frac{(\xi+\eta)^{n+1}}{n+1} + 2(1+e^{|q|})(\bar{\lambda}+c) \\
&\quad \times\frac{(\xi+\eta)^{n+1}}{n+1} + 4(1+e^{|q|})\bar{g}\frac{(\xi+\eta)^{n+1}}{n+1} + 2\bar{f}\frac{(\xi+\eta)^{n+1}}{n+1} \\
&\quad + 2(1+e^{|q|})\bar{f}\frac{(\xi+\eta)^{n+1}}{n+1}\Bigg\} \\
&\le M^{n+2}\frac{(\xi+\eta)^{n+1}}{(n+1)!}.
\end{aligned}
\tag{2.42}
$$

So, by induction, (2.41) is proved. Note also that $G^n(\xi,\eta)$ is $C^2(\overline{\mathcal{T}_1})$, which follows from (2.35)–(2.36) and the assumption (2.4). Therefore, the series

$$
G(\xi,\eta) = \lim_{n\to\infty} G^n(\xi,\eta) = \sum_{n=0}^{\infty} \Delta G^n(\xi,\eta) \tag{2.43}
$$

converges absolutely and uniformly in $\mathcal{T}_1$, and its sum G is a twice continuously differentiable solution of equation (2.34) with a bound $|G(\xi,\eta)| \le M\exp(M(\xi+\eta))$.

The uniqueness of this solution can be proved by the following argument. Suppose $G'(\xi, \eta)$ and $G''(\xi, \eta)$ are two different solutions of (2.34). Then $\delta G(\xi, \eta) = G'(\xi, \eta) - G''(\xi, \eta)$ satisfies the homogeneous integral equation (2.39) in which ΔG^n and ΔG^{n+1} are changed to δG. Using the above result of boundedness, we have $|\delta G(\xi, \eta)| \leq 2Me^{2M}$. Using this inequality in the homogeneous integral equation and following the same estimates as in (2.42), we get the result that $\delta G(\xi, \eta)$ satisfies for all n

$$|\delta G(\xi, \eta)| \leq 2M^{n+1}e^{2M}\frac{(\xi + \eta)^n}{n!} \to 0 \text{ as } n \to \infty. \tag{2.44}$$

Thus, $\delta G \equiv 0$, which means that (2.43) is a unique solution to (2.34). By direct substitution we can check that it is also a unique (by Lemma 2.1) solution to PDE (2.27)–(2.29). Thus we have proved the following result.

THEOREM 2.1. *The equation* (2.20) *with boundary conditions* (2.21)–(2.24) *has a unique $C^2(\overline{\mathcal{T}})$ solution. The bound on the solution is*

$$|k(x, y)| \leq Me^{2Mx}, \tag{2.45}$$

where M is given by (2.40).

To prove stability of the closed-loop system, we need to prove that the transformation (2.15) is invertible. The proof that for (2.15) an inverse transformation with bounded kernel exists can be found in [6] and [80], and can also be inferred from [92, p. 254]. The other way to prove it is to directly find and analyze the PDE for the kernel of the inverse transformation. We take this route because we need the inverse kernel for explicit solutions of the closed-loop system and for estimates in Chapter 14. Let us denote the kernel of the inverse transformation by $l(x, y)$. The transformation itself has the form

$$u(x, t) = w(x, t) + \int_0^x l(x, y)w(y, t)\, dy. \tag{2.46}$$

Substituting (2.46) into equations (2.12)–(2.14) and using (2.9)–(2.11), we obtain the following PDE governing $l(x, y)$:

$$\begin{aligned} \varepsilon l_{xx}(x, y) - \varepsilon l_{yy}(x, y) = &- (\lambda(x) + c)l(x, y) - f(x, y) \\ &- \int_y^x l(\tau, y) f(x, \tau)\, d\tau \end{aligned} \tag{2.47}$$

for $(x, y) \in \mathcal{T}$ with boundary conditions

$$\varepsilon l_y(x, 0) = -\varepsilon q l(x, 0) + g(x), \tag{2.48}$$

$$l(x, x) = -q - \frac{1}{2\varepsilon}\int_0^x (\lambda(y) + c)\, dy. \tag{2.49}$$

This hyperbolic P(I)DE is a little bit simpler than the one for k (the boundary condition (2.48) does not contain an integral term) but has a very similar structure. So we can apply the same approach of converting the PDE to an integral equation and use a method of successive approximations to show that the inverse kernel exists and has the same properties as we proved for the direct kernel.

Theorem 2.2. *The equation* (2.47) *with boundary conditions* (2.48)–(2.49) *has a unique $C^2(\overline{\mathcal{T}})$ solution. The bound on the solution is*

$$|l(x, y)| \leq M e^{2Mx}, \tag{2.50}$$

where M is given by (2.40).

2.5 STABILITY OF THE CLOSED-LOOP SYSTEM

To establish stability of the closed-loop system we first give a stability result for the target system (2.12)–(2.14).

Lemma 2.2. *For any initial data $w_0 \in H^1(0, 1)$ compatible with boundary conditions, the system* (2.12)–(2.14) *has a unique classical solution for $t > 0$ and is exponentially stable at the origin in H^1:*

$$\|w(t)\|_{H^1} \leq e^{-\left(c+\varepsilon\frac{\pi^2}{4}\right)t}\|w_0\|_{H^1}. \tag{2.51}$$

Proof. It is well known that the system (2.12)–(2.14) is well posed; see, for example, [71, Chap. 4]. To show stability we differentiate the Lyapunov function

$$V(t) = \frac{1}{2}\|w(t)\|^2 + \frac{1}{2}\|w_x(t)\|^2 \tag{2.52}$$

along the solutions of the target system to get

$$\begin{aligned}\dot V(t) &= \int_0^1 w(x,t) w_t(x,t)\,dx + w_x(x,t) w_t(x,t)|_0^1 - \int_0^1 w_{xx}(x,t) w_t(x,t)\,dx \\ &= -\varepsilon\|w_x(t)\|^2 - \varepsilon\|w_{xx}(t)\|^2 - c(\|w(t)\|^2 + \|w_x(t)\|^2) \\ &\leq -\left(2c + \varepsilon\frac{\pi^2}{2}\right) V(t),\end{aligned} \tag{2.53}$$

where the last inequality follows from the Poincaré-type inequalities (B.3) and (B.4) (see Appendix B). From (2.52) and (2.53) we get (2.51). □

Combining Lemma 2.2 with Theorems 2.1 and 2.2 gives us the following result.

Theorem 2.3. *For any initial data $u_0(x) \in H^1(0, 1)$ compatible with boundary conditions* (2.10)–(2.16)*, the system* (2.9)–(2.10) *with boundary control* (2.16)*, where $k(x, y)$ is the solution of* (2.20), (2.21), (2.24)*, has a unique classical solution $u \in C^{2,1}([0, 1] \times (0, \infty))$ and is exponentially stable at the origin:*

$$\|u(t)\|_{H^1} \leq C e^{-\left(c+\frac{\pi^2}{4}\right)t}\|u_0\|_{H^1}, \tag{2.54}$$

where C is a positive constant independent of u_0.

Proof. First, we note that the closed-loop system is well posed, as follows from Lemma 2.2 and (2.46). Using the transformations (2.1) and (2.2), Theorems (2.15) and (2.46), and Lemma 2.2, we have

$$\begin{aligned}\|u\| &\leq (1+Me^{2M})\|w\| \leq (1+Me^{2M})e^{-\left(c+\varepsilon\frac{\pi^2}{4}\right)t}\|w_0\| \\ &\leq (1+Me^{2M})^2 e^{-\left(c+\varepsilon\frac{\pi^2}{4}\right)t}\|u_0\|\end{aligned} \tag{2.55}$$

and

$$\begin{aligned}\|u_x\| &\leq \|w_x\| + (|q|+\bar{\lambda})\|w\| + \max_{(x,y)\in\mathcal{T}} |l_x(x,y)|\|w\| \\ &\leq e^{-\left(c+\varepsilon\frac{\pi^2}{4}\right)t}\left[\|w_{0x}\| + \left(|q|+\bar{\lambda}+\max_{(x,y)\in\mathcal{T}}|l_x(x,y)|\right)\|w_0\|\right] \\ &\leq e^{-\left(c+\varepsilon\frac{\pi^2}{4}\right)t}\left[\|u_{0x}\| + \left(|q|+\bar{\lambda}+\max_{(x,y)\in\mathcal{T}}|k_x(x,y)|\right)\|u_0\|\right] \\ &\quad + e^{-\left(c+\varepsilon\frac{\pi^2}{4}\right)t}\left(|q|+\bar{\lambda}+\max_{(x,y)\in\mathcal{T}}|l_x(x,y)|\right)(1+Me^{2M})\|u_0\|.\end{aligned} \tag{2.56}$$

From (2.55) and (2.56) we get (2.54). □

With the backstepping method employing target systems in the simple heat equation form, it is possible to write the solution of the closed-loop system (2.9)–(2.11), (2.16) explicitly, in terms of the initial condition $u_0(x)$ and the kernels $k(x,y)$ and $l(x,y)$. We start with the closed-form solution to the target system (2.67)–(2.69),

$$w(x,t) = 2\sum_{n=0}^{\infty} e^{-(c+\varepsilon\mu_n^2)t}\cos(\mu_n x)\int_0^1 w_0(\xi)\cos(\mu_n\xi)\,d\xi, \tag{2.57}$$

where $\mu_n = \pi(n+1/2)$. The initial condition $w_0(x)$ can be calculated explicitly from $u_0(x)$ using the transformation (2.15):

$$w_0(x) = u_0(x) - \int_0^x k(x,y)u_0(y)\,dy. \tag{2.58}$$

Substituting (2.57) and (2.58) into the inverse transformation (2.46) and changing the order of integration, we obtain the following result.

Theorem 2.4. *The solution of the closed-loop system* (2.9)–(2.11), (2.16) *is*

$$u(x,t) = \sum_{n=0}^{\infty} e^{-(c+\varepsilon\mu_n^2)t}\phi_n(x)\int_0^1 \psi_n(\xi)u_0(\xi)\,d\xi, \tag{2.59}$$

where

$$\phi_n(x) = 2\left(\cos(\mu_n x) + \int_0^x l(x,y)\cos(\mu_n y)\,dy\right), \tag{2.60}$$

$$\psi_n(x) = \cos(\mu_n x) - \int_x^1 k(y,x)\cos(\mu_n y)\,dy, \tag{2.61}$$

and $\mu_n = \pi(n+1/2)$.

Using the relationship between Volterra kernels,

$$l(x,y)-k(x,y)=\int_y^x k(x,\xi)l(\xi,y)\,dy, \tag{2.62}$$

it is easy to show that $\{\phi_n\}_{n=0}^{\infty}$ and $\{\psi_n\}_{n=0}^{\infty}$ are orthogonal bases. It is interesting to note that although infinitely many eigenvalues cannot be arbitrarily assigned, our controller is able to assign all of them to the particular location $-(c+\varepsilon\mu_n^2)$. The eigenfunctions of the closed-loop system are assigned to $\phi_n(x)$.

2.6 DIRICHLET UNCONTROLLED END

The case of the Dirichlet boundary condition at the uncontrolled end has to be considered separately.

Consider the plant

$$u_t(x,t)=\varepsilon u_{xx}(x,t)+\lambda(x)u(x,t)+g(x)u_x(0,t)+\int_0^x f(x,y)u(y,t)\,dy, \tag{2.63}$$

$$u(0,t)=0, \tag{2.64}$$

$$u(1,t)=U(t). \tag{2.65}$$

Since $u(0,t)=0$, it does not make sense to consider the term $g(x)u(0,t)$ in (2.63), and we replace it with $g(x)u_x(0,t)$. Using the transformation

$$w(x,t)=u(x,t)-\int_0^x k(x,y)u(x,t)\,dy, \tag{2.66}$$

we map (2.63)–(2.65) into the target system

$$w_t(x,t)=\varepsilon w_{xx}(x,t)-cw(x,t), \tag{2.67}$$

$$w(0,t)=0, \tag{2.68}$$

$$w(1,t)=0. \tag{2.69}$$

Substituting (2.66) into (2.67) and using (2.63), we obtain the following equation:

$$\begin{aligned}0=&\int_0^x \{\varepsilon k_{xx}(x,y)-\varepsilon k_{yy}(x,y)-(\lambda(y)+c)k(x,y)+f(x,y)\}\,u(y,t)\,dy\\&-\int_0^x u(y,t)\int_y^x k(x,\xi)f(\xi,y)\,d\xi\,dy+\left\{\lambda(x)+c+2\varepsilon\frac{d}{dx}k(x,x)\right\}u(x,t)\\&+\left\{g(x)-\int_0^x k(x,y)g(y)\,dy+\varepsilon k(x,0)\right\}u(0,t).\end{aligned} \tag{2.70}$$

For this equation to be verified for all $u(x,t)$, the following PDE for $k(x,y)$ must be satisfied:

$$\varepsilon k_{xx}(x,y) - \varepsilon k_{yy}(x,y) = (\lambda(y)+c)k(x,y) - f(x,y) + \int_y^x k(x,\xi) f(\xi,y)\,d\xi, \qquad (x,y)\in\mathcal{T}, \tag{2.71}$$

$$\varepsilon k(x,0) = -g(x) + \int_0^x k(x,y)g(y)\,dy, \tag{2.72}$$

$$k(x,x) = -\frac{1}{\varepsilon}g(0) - \frac{1}{2\varepsilon}\int_0^x (\lambda(y)+c)\,dy. \tag{2.73}$$

The kernel $l(x,y)$ of the inverse transformation

$$u(x,t) = w(x,t) + \int_0^x l(x,y)w(x,t)\,dy \tag{2.74}$$

satisfies the following PDE:

$$\varepsilon l_{xx}(x,y) - \varepsilon l_{yy}(x,y) = -(\lambda(x)+c)l(x,y) - f(x,y) - \int_y^x f(x,\xi)l(\xi,y)\,d\xi, \qquad (x,y)\in\mathcal{T} \tag{2.75}$$

$$\varepsilon l(x,0) = -g(x), \tag{2.76}$$

$$l(x,x) = -\frac{1}{\varepsilon}g(0) - \frac{1}{2\varepsilon}\int_0^x (\lambda(y)+c)\,dy. \tag{2.77}$$

Using the method of successive approximations, it is straightforward to obtain the following result.

THEOREM 2.5. *The equations* (2.71)–(2.73) *and* (2.75)–(2.77) *have a unique* $C^2(\mathcal{T})$ *solution.*

The stability result is stated in the following theorem.

THEOREM 2.6. *For any initial data $u_0 \in H^1(0,1)$ compatible with boundary conditions, the system* (2.63)–(2.65) *with boundary control* (2.16), *where $k(x,y)$ is the solution of* (2.71)–(2.73), *has a unique classical solution $u \in C^{2,1}([0,1]\times(0,\infty))$ and is exponentially stable at the origin,*

$$\|u(t)\|_{H^1} \le Ce^{-(c+\varepsilon\pi^2)t}\|u_0\|_{H^1}, \tag{2.78}$$

where C is a positive constant independent of u_0.

Proof. Very similar to the proof of Theorem 2.3. □

As in the previous section, the solutions of the closed-loop system can be obtained explicitly using the explicit solution of the target system (2.67)–(2.69) given in Appendix E and direct and inverse backstepping transformations.

THEOREM 2.7. *The solution of the closed-loop system* (2.63)–(2.65), (2.16) *is*

$$u(x,t) = \sum_{n=1}^{\infty} e^{-(c+\varepsilon\pi^2 n^2)t} \phi_n(x) \int_0^1 \psi_n(\xi) u_0(\xi)\, d\xi, \tag{2.79}$$

where

$$\phi_n(x) = 2\left(\sin(\pi n x) + \int_0^x l(x,y)\sin(\pi n y)\, dy\right), \tag{2.80}$$

$$\psi_n(x) = \sin(\pi n x) - \int_x^1 k(y,x)\sin(\pi n y)\, dy. \tag{2.81}$$

2.7 NEUMANN ACTUATION

So far we have only considered the case of Dirichlet actuation ($u(1,t)$ is controlled), which is usually the case in fluid problems where the velocity is controlled (e.g., by using microjets). In problems with thermal and chemically reacting dynamics, the natural choice is the Neumann actuation ($u_x(1,t)$, or heat flux is controlled). The Neumann controllers are obtained using the same exact transformation (2.15) as in the case of the Dirichlet actuation but with the appropriate change in the boundary condition of the target system (from Dirichlet to Neumann).

The Neumann controller is given by

$$u_x(1,t) = U(t) = k(1,1)u(1,t) + \int_0^1 k_x(1,y)u(y,t)\, dy. \tag{2.82}$$

The stability results for the plants (2.9)–(2.10) and (2.63)–(2.64) with Neumann actuation are given by the following theorems.

THEOREM 2.8. *For any initial data $u_0 \in H^1(0,1)$ compatible with boundary conditions, the system* (2.9)–(2.11) *with boundary control* (2.82), *where $k(x,y)$ is the solution of* (2.20), (2.21), (2.24), *has a unique classical solution $u \in C^{2,1}([0,1] \times (0,\infty))$ and is exponentially stable at the origin,*

$$\|u(t)\|_{H^1} \le C e^{-ct} \|u_0\|_{H^1}, \tag{2.83}$$

where C is a positive constant independent of u_0.

THEOREM 2.9. *For any initial data $u_0 \in H^1(0,1)$ compatible with boundary conditions, the system* (2.63)–(2.65) *with boundary control* (2.82), *where $k(x,y)$ is the solution of* (2.71)–(2.73), *has a unique classical solution $u \in C^{2,1}([0,1] \times (0,\infty))$ and is exponentially stable at the origin,*

$$\|u(t)\|_{H^1} \le C e^{-\left(c+\varepsilon\frac{\pi^2}{4}\right)t} \|u_0\|_{H^1}, \tag{2.84}$$

where C is a positive constant independent of u_0.

2.8 SIMULATION

In this section we present the results of numerical simulation of the closed-loop system. For the plant (2.63)–(2.65) we take $\varepsilon(x) = 1 + 0.4\sin(6\pi x)$, $\lambda = 10$, and all other coefficients are zero.

To find $k(x, y)$, we directly numerically solve the PDE (2.71)–(2.73) (an alternative would be to calculate several terms of the successive approximation series). The numerical scheme for (2.71)–(2.73) should be selected carefully, since the term with $\lambda(x)$ can cause numerical instability. We suggest the Ablowitz-Kruskal-Ladik scheme [4] (typically used for solving nonlinear Klein-Gordon equations in quantum physics), which we modified to suit the geometry and boundary conditions of the kernel PDE:

$$k_j^{i+1} = -k_j^{i-1} + k_{j+1}^i + k_{j-1}^i + a_j h^2 \frac{k_{j+1}^i + k_{j-1}^i}{2}, \tag{2.85}$$

$$k_{i+1}^{i+1} = k_i^i - \frac{h}{4}(a_i + a_{i+1}), \qquad k_1^{i+1} = 0. \tag{2.86}$$

Here $k_j^i = k((i-1)h, (j-1)h), i = 2, \ldots, N, j = 2, \ldots, i-1, a_i = (\bar{\lambda}((i-1)h) + c)/\bar{\varepsilon}$, $h = 1/N$, where N is the number of steps. The key feature of this scheme is the discretization of the $a(y)k(x, y)$-term, averaging $k(x, y)$ in the y-direction.

In Figure 2.1 (top) one can see $\varepsilon(x)$ and the corresponding control gain computed using the scheme (2.85)–(2.86) ($N = 100$).

For the closed-loop simulation, we implement the controller on a coarse grid with only six points (evenly distributed over [0,1]). As shown in Figure 2.1 (bottom), the reduced-order controller stabilizes the system.

2.9 DISCUSSION

2.9.1 Connection with Backstepping for ODEs

The backstepping technique might best be appreciated by readers familiar with the historical developments of finite-dimensional *nonlinear* control. The first systematic nonlinear control methods were the methods of optimal control, which require the "solution" of Hamilton-Jacobi-Bellman (HJB) nonlinear PDEs. The breakthrough in nonlinear control came with differential geometric theory and feedback linearization, which recognized the structure of nonlinear control systems and exploited it using coordinate transformations and feedback cancelations. In the same way that nonlinear PDEs (HJB) are more complex than what ODE control problems call for, operator Riccati equations are more complex than what boundary control problems call for, at least for the class of plants in this book. In summary, backstepping, with its linear hyperbolic PDE for the gain kernel, is unique in not exceeding the complexity of the PDE control problem that it is solving.

Let us look at the connection between ODE and PDE backstepping more closely. The backstepping method originated in the early 1990s [54, 68] and is linked to

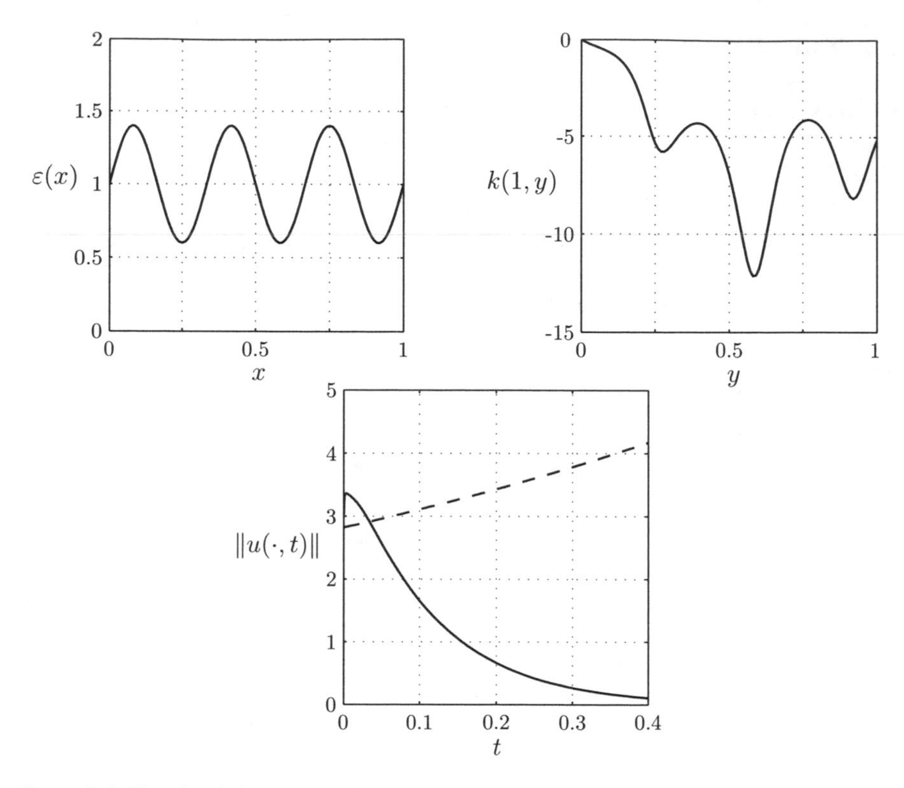

Figure 2.1 The simulation results for (2.63)–(2.65) with the controller (2.16). Top left: $\varepsilon(x)$. Top right: the control gain. Bottom: the L^2 norm of an open-loop (dashed line) and closed-loop (solid line) response.

problems of stabilization of nonlinear ODE systems. Consider the following three-state nonlinear system:

$$\dot{y}_1 = y_2 + y_1^3 \tag{2.87}$$

$$\dot{y}_2 = y_3 + y_2^3 \tag{2.88}$$

$$\dot{y}_3 = u + y_3^3. \tag{2.89}$$

Since the control input u is only in the last equation (2.89), it is helpful to view it as boundary control. The nonlinear terms y_1^3, y_2^3, y_3^3 can be viewed as nonlinear "reaction" terms, and they are clearly destabilizing, because for $u = 0$, the overall system is a "cascade" of three unstable subsystems of the form $\dot{y}_i = y_i^3$ (the open-loop system exhibits a finite-time escape instability). The control u can cancel the "matched" term y_3^3 in (2.89) but cannot cancel directly the unmatched terms y_1^3 and y_2^3 in (2.87) and (2.88). To achieve the cancelation of all three of the destabilizing

y_i^3 terms, a backstepping change of variable is constructed recursively,

$$z_1 = y_1, \tag{2.90}$$
$$z_2 = y_2 + y_1^3 + c_1 y_1, \tag{2.91}$$
$$z_3 = y_3 + y_2^3 + (3y_1^2 + 2c)y_2 + 3y_1^5 + 2cy_1^3 + (c^2 + 1)y_1, \tag{2.92}$$

along with the control law

$$u = -c_3 z_3 - z_2 - y_3^3 - (3y_2^2 + 3y_1^2 + 2c)(y_3 + y_2^3) \\ - (6y_1 y_2 + 15y_1^4 + 6cy_1^2 + c^2 + 1)(y_2 + y_1^3), \tag{2.93}$$

which converts the system (2.87)–(2.89) into

$$\dot{z}_1 = z_2 - cz_1, \tag{2.94}$$
$$\dot{z}_2 = -z_1 + z_3 - cz_2, \tag{2.95}$$
$$\dot{z}_3 = -z_2 - cz_3, \tag{2.96}$$

where the control parameter c should be chosen positive. The system (2.94)–(2.96), which can also be written as

$$\dot{z} = Az, \tag{2.97}$$

where

$$A = \begin{bmatrix} -c & 1 & 0 \\ -1 & -c & 1 \\ 0 & -1 & -c \end{bmatrix}, \tag{2.98}$$

is exponentially stable because

$$A + A^T = -cI. \tag{2.99}$$

The equality (2.99) guarantees that the Lyapunov function

$$V = \frac{1}{2} z^T z \tag{2.100}$$

has a negative definite time derivative:

$$\dot{V} = -cz^T z. \tag{2.101}$$

Hence, the target system (2.94)–(2.96) is in a desirable form, but we still have to explain the relation between the change of variable (2.90)–(2.92) and the one used for PDEs in this chapter, as well as the relation between the structures in the plant (2.87)–(2.89) and the target system (2.94)–(2.96) relative to those of the PDE plants and the target systems in this chapter.

Let us first examine the change of variables $y \mapsto z$ in (2.90)–(2.92). This change of variables is clearly of the form $z = (I - K)[y]$, where I is the identity matrix and K is a "lower-triangular" nonlinear transformation. The lower-triangular structure of K in (2.90)–(2.92) is analogous to the Volterra structure of the spatially causal integral operator $\int_0^x k(x, y)u(y)\, dy$ in our change of variable $w(x) = u(x) - \int_0^x k(x, y)u(y)\, dy$ in this chapter. Another important feature of the change of variable (2.90)–(2.92) is that it is invertible, that is, y can be expressed as a smooth function of z (to be specific, $y_1 = z_1$, $y_2 = z_2 - z_1^3 - cz_1$, and so on).

Next, let us examine the relation between the plant (2.87)–(2.89) and those studied in this chapter, such as the reaction-diffusion system $u_t = u_{xx} + \lambda u$, as well as the relation between the target systems (2.94)–(2.96) and $w_t = w_{xx}$. The analogy between the target systems is particularly transparent because both admit a simple 2-norm as a Lyapunov function, specifically,

$$\frac{d}{dt}\frac{1}{2}z^T z = -cz^T z \tag{2.102}$$

in the ODE case and

$$\frac{d}{dt}\frac{1}{2}\int_0^1 w^2(x)\,dx = -\int_0^1 w_x^2(x)\,dx \tag{2.103}$$

in the PDE case. A finer structural analogy where one might expect the z-system to be a spatial discretization of the w-system does not hold. If we discretize the PDE system $w_t = w_{xx}$, with boundary conditions $w(0,t) = w(1,t) = 0$, over a spatial grid with N points, we get the ODE system $\dot{w}_i = N^2(w_{i+1} - 2w_i + w_{i-1})$, which is different in structure from $\dot{z}_i = z_{i+1} - z_{i-1} - cz_i$, even after the N^2 factor is absorbed (into the time variable). This is where the subtle difference between ODE backstepping and PDE backstepping comes into play. The recursive procedure used for ODEs does not have a limit as the number of states goes to infinity. In contrast, the backstepping process for PDEs does have a limit, as we have proved in Section 2.4. Let us try to understand this difference by comparing the plant structure (2.87)–(2.89) with the plant structure $u_t = u_{xx} + \lambda u$. The former is dominated by a chain of integrators, while the latter is dominated by the diffusion operator. Whereas the diffusion operator is a well-defined, meaningful object, an "infinite integrator chain" is not. It is for this reason that the infinite-dimensional backstepping design succeeds only if particular care is taken to convert the unstable parabolic PDE $u_t = u_{xx} + \lambda u$ into a stable target system $w_t = w_{xx}$ that is within the same PDE class, namely, parabolic. To put it more simply, we make sure to retain the ∂_{xx} term in the target system, even though it may be tempting to go for some other target system, such as the first-order hyperbolic (transport equation-like) PDE $w_t = w_x - cw$, which is more reminiscent of the ODE target system (2.94)–(2.96). If such an attempt is made, the derivation of the PDE conditions for the kernel $k(x, y)$ would not be successful, and the matching of terms between the plant $u_t = u_{xx} + \lambda u$ and the target system $w_t = w_{xx} - cw$ would result in terms that cannot be canceled.

Finally, let us explain the meaning of the term *backstepping*. In the ODE setting this procedure is referred to as *integrator backstepping* because, as illustrated with the help of example (2.87)–(2.89), the design procedure propagates the feedback law synthesis "backward" through a chain of integrators. Upon careful inspection of the change of variables (2.90)–(2.92), the first step of the backstepping procedure is to treat the state y_2 as the control input in the subsystem $\dot{y}_1 = y_2 + y_1^3$, design the control law $y_2 = -y_1^3 - cy_1$, then "step back" through the integrator in the second subsystem $\dot{y}_2 = y_3 + y_2^3$ and design the control y_3 so that the error state $z_2 = y_2 - (-y_1^3 - cy_1)$ is forced to go to zero, thus ensuring that the state y_2 acts (approximately) as the control $y_2 = -y_1^3 - cy_1$. This backward stepping through integrators continues until one encounters the actual control u in (2.93), which in the example (2.87)–(2.89) happens after two steps of backstepping. Even though in our *continuum* version of backstepping for PDEs there are no simple integrators to step through, the analogy with the method for ODEs lies in the triangularity of the change of variable and the pursuit of a stable target system. For this reason, we retain the term *backstepping* for PDEs.

2.9.2 Comparison with Discretization-Based Backstepping

One of the possible approaches to stabilization of (2.1)–(2.3) is the following. One first discretizes the plant (e.g., with a finite-difference scheme) and then uses the finite-dimensional backstepping to transform it into the discretized version of the target system (2.12)–(2.14). The control is obtained by solving the resulting difference equation for the gain kernel and taking the inner product of the gain vector and the state. This control is then applied to the original, nondiscretized plant, and for a sufficiently fine discretization grid one achieves stability. This approach is pursued in [7].

What are the benefits of the continuum backstepping method compared with the approach described above? First, doing the design procedure without prediscretizing the plant is more elegant, and this becomes important when one goes beyond stabilization problems. For example, it would be much more difficult to prove the stability of adaptive schemes (which are nonlinear) in a discretized setting. Second, control gains obtained with discretization-based backstepping do not converge when the discretization step goes to zero, even though the controllers obtained from such gains are still stabilizing, as shown in [7]. For example, consider the simple plant

$$u_t(x,t) = u_{xx}(x,t) + \lambda u(x,t), \tag{2.104}$$

where λ is a constant and the boundary conditions are $u(0,t) = 0$, $u(1,t) = U(t)$. The simplest and most straightforward discretization of this plant is

$$\dot{u}_i = N^2(u_{i+1} - 2u_i + u_{i-1}) + \lambda u_i, \quad i = 2, \ldots, N, \tag{2.105}$$

where N is the number of steps and $u_i = u((i-1)/N, t)$, $u_1 = 0$, $u_{N+1} = U$. The backstepping control gain for this particular discretization is shown in Figure 2.2 (top). As N increases, the kernel becomes more oscillatory, though it stays in the same envelope.

It turns out that if one discretizes (2.104) in the following nonobvious way,

$$\dot{u}_i = N^2(u_{i+1} - 2u_i + u_{i-1}) + \lambda\left(\frac{1}{4}u_{i-1} + \frac{1}{2}u_i + \frac{1}{4}u_{i+1}\right), \tag{2.106}$$

then the resulting control gain is smooth, as shown in Figure 2.2 (bottom). However, it is impossible to know in advance what discretization would produce a smooth control gain for a particular problem.

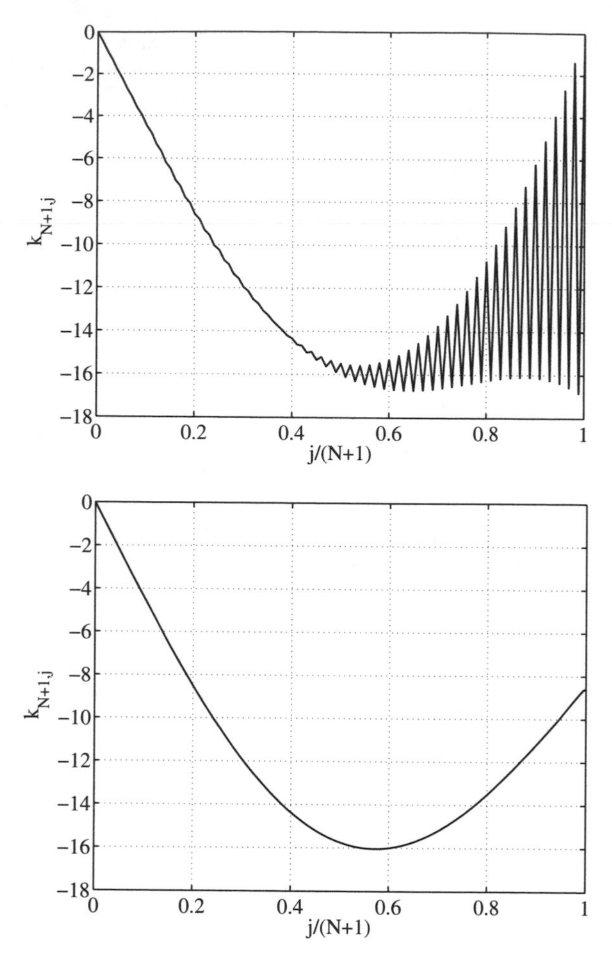

Figure 2.2 Backstepping control gain for the discretized plant (2.104) ($\lambda = 17$). Top: obtained from the discretization (2.105). Bottom: obtained from the discretization (2.106).

2.9.3 Comparison with Other Designs

The results of this chapter should be evaluated in comparison with the results of the infinite-dimensional versions of standard control approaches—pole assignment and linear quadratic regulator (LQR).

The pole assignment for parabolic PDEs originates with the paper by Triggiani [116]. On a conceptual level, the main disadvantage of the pole assignment method is that it is notoriously nonrobust. Even in the finite-dimensional case it is known that backstepping is more robust than pole placement because it converts a system into a tridiagonal Hessenberg form rather than a companion form. Furthermore, as we show in Chapter 14, backstepping controllers can be modified (without recalculating the gain kernel) to have robustness margins. From a computational point of view, backstepping design is also advantageous. The design procedure in [116] goes as follows. One first finds the unstable eigenvalues and the corresponding eigenvectors of the open-loop system. Suppose there are n of them. Then one solves the auxiliary elliptic problem and computes n integrals (which are L^2 inner products of the elliptic solution and the unstable modes). After that one solves an n-dimensional matrix equation to find the desired kernel. In contrast, the backstepping approach requires just a few successive integrations (symbolic or numerical) to obtain several terms of the series (2.43). In addition, as is shown in Chapter 3, in many cases closed-form controllers are available.

Even though the designs in this chapter are not optimal, it is possible to modify them so that the controllers become *inverse optimal*, that is, minimize a meaningful cost functional that penalizes both state and control. We explore this topic in more detail and compare the results with those of the LQR design in Chapter 14.

2.10 NOTES AND REFERENCES

Some of the basic ingredients of backstepping go back to works by Colton [23] and Seidman [106], where integral transformations are used to solve PDEs and state controllability results, but not for the design of feedback laws. It is interesting that these ideas appeared well before the development of finite-dimensional backstepping (though we became aware of them several years after the inception of our research program on backstepping control for PDEs, essentially rediscovering them, and arriving at the Volterra operator transformations from the finite-dimensional backstepping context). It is also curious that these powerful ideas were not pursued further after [23, 106].

Our very first attempt at developing continuum backstepping for PDEs was in [17], where we developed an explicit feedback law, backstepping transformation, and a Lyapunov function, but the plant's level of open-loop instability was limited. Then, in [6, 7, 13] we turned our attention to a discretization-based (in space) approach, but the approach was dependent on the discretization scheme and did not yield convergent gain kernels when the discretization step $\delta x \to 0$, though, interestingly, the control input was nevertheless convergent, as it is an inner product of the gain kernel and the measured state of the system. The continuum backstepping approach received a boost after the paper by Liu [80], where the PDE (2.20), (2.21), (2.24) was shown to be well posed for $g = f = 1/q = 0$. In [108, 110], the control design was presented for the general plant (2.1)–(2.3), and it was shown how to modify the design so that the controllers would minimize a meaningful cost functional (see Chapter 14).

It is also possible to extend the backstepping approach to certain 2D and 3D PDEs in regular geometries; see [117] for a 2D thermal convection loop, [118] for a Navier-Stokes flow in a 2D infinite channel, and [119] for a 3D magnetohydrodynamics channel flow.

The prior work on stabilization of general parabolic equations includes, among others, the results of Triggiani [116] and Lasiecka and Triggiani [73], who developed a general framework for the structural assignment of eigenvalues in parabolic problems through the use of semigroup theory. Separating the open-loop system into a finite-dimensional unstable part and an infinite-dimensional stable part, they applied feedback boundary control that stabilizes the unstable part while leaving the stable part stable. A unified treatment of both interior and boundary observations/control generalized to semilinear problems can be found in [5]. Stabilizability by boundary control in the optimal control setting is discussed by Bensoussan et al. [10] and Lasiecka and Triggiani [75]. For the Pritchard-Salamon class of state-space systems, several frequency domain stabilization results have been established (see, e.g., [26] and [82] for surveys). The placement of finitely many eigenvalues was generalized to the case of moving infinitely many eigenvalues by Russell [104]. The stabilization problem can also be approached using the abstract theory of boundary control systems developed by Fattorini [39], which results in a dynamical feedback controller (see remarks in [29, Sect. 3.5]). Extensive surveys on the controllability and stabilizability theory of linear PDEs can be found in [75, 105].

Chapter Three

Closed-Form Controllers

In this chapter we present a collection of problems for which one can obtain explicit stabilizing controllers.

One of the striking features of the backstepping control design for PDEs is that it leads to explicit feedback controllers for many physically relevant problems. Such controllers are important for several reasons. The first and most obvious benefit is that one does not have to numerically compute a solution to the gain kernel PDE. Second, the explicit gain kernels allow us to find explicit solutions to the closed-loop system, offering valuable insight into how control affects eigenvalues and eigenfunctions. Explicit solutions to gain kernel PDEs are also useful in testing numerical schemes. Last but not least, parametrized families of controllers play a crucial role in *adaptive control*. In Chapters 8–10 we design certainty equivalence–based adaptive schemes using some of the controllers presented in this chapter.

3.1 THE REACTION-DIFFUSION EQUATION

Consider the reaction-diffusion plant

$$u_t(x,t) = \varepsilon u_{xx}(x,t) + \lambda_0 u(x,t), \quad x \in (0,1), \tag{3.1}$$

$$u(0,t) = 0, \tag{3.2}$$

$$u(1,t) = U^D(t) \quad \text{or} \quad u_x(1,t) = U^N(t), \tag{3.3}$$

where $\varepsilon > 0$ and λ_0 are constants. The open-loop system (3.1)–(3.3) is unstable and for sufficiently large ratio λ_0/ε has arbitrarily many unstable eigenvalues.

The kernel PDE (2.71)–(2.73) in this case takes the following form:

$$k_{xx}(x,y) - k_{yy}(x,y) = \lambda k(x,y), \quad (x,y) \in \mathcal{T}, \tag{3.4}$$

$$k(x,0) = 0, \tag{3.5}$$

$$k(x,x) = -\frac{\lambda x}{2}, \tag{3.6}$$

where we denote $\lambda = (\lambda_0 + c)/\varepsilon$. Let us solve this equation directly by the method of successive approximations. Introducing new variables $\xi = x + y$, $\eta = x - y$, $G(\xi,\eta) = k(x,y)$, we convert the PDE (3.4)–(3.6) into the integral equation

$$G(\xi,\eta) = -\frac{\lambda}{4}(\xi - \eta) + \frac{\lambda}{4}\int_\eta^\xi \int_0^\eta G(\tau,s)\,ds\,d\tau. \tag{3.7}$$

Let us start with the initial guess $G^0(\xi, \eta) = 0$ and set up the recursive formula as follows:

$$G^{n+1}(\xi, \eta) = -\frac{\lambda}{4}(\xi - \eta) + \frac{\lambda}{4}\int_\eta^\xi \int_0^\eta G^n(\tau, s)\, ds\, d\tau. \tag{3.8}$$

For the difference between two consecutive terms $\Delta G^n = G^{n+1} - G^n$, we have

$$\Delta G^{n+1}(\xi, \eta) = \frac{\lambda}{4}\int_\eta^\xi \int_0^\eta \Delta G^n(\tau, s)\, ds\, d\tau. \tag{3.9}$$

The solution to (3.7) is given by

$$G(\xi, \eta) = \lim_{n\to\infty} G^n(\xi, \eta) = \sum_{n=0}^{\infty} \Delta G^n(\xi, \eta). \tag{3.10}$$

Fortunately, starting from ΔG^0 and repeatedly calculating the double integral in (3.9), we can observe a certain pattern and find the general term ΔG^n in closed form, which can be verified by induction:

$$\Delta G^n(\xi, \eta) = -\frac{(\xi - \eta)\, \xi^n \eta^n}{(n!)^2 (n+1)} \left(\frac{\lambda}{4}\right)^{n+1}. \tag{3.11}$$

Calculating the series (3.10), we get

$$G(\xi, \eta) = -\frac{\lambda(\xi - \eta)}{2} \frac{I_1(\sqrt{\lambda \xi \eta})}{\sqrt{\lambda \xi \eta}}, \tag{3.12}$$

where I_1 is a modified Bessel function of order one (see Appendix C for more details on Bessel functions). Writing (3.12) in terms of x, y gives the following solution for $k(x, y)$:

$$k(x, y) = -\lambda y \frac{I_1\left(\sqrt{\lambda(x^2 - y^2)}\right)}{\sqrt{\lambda(x^2 - y^2)}}, \tag{3.13}$$

which gives the Dirichlet controller

$$U^D(t) = -\int_0^1 \lambda y \frac{I_1\left(\sqrt{\lambda(1 - y^2)}\right)}{\sqrt{\lambda(1 - y^2)}} u(y, t)\, dy \tag{3.14}$$

or the Neumann controller

$$U^N(t) = -\frac{\lambda}{2} u(1, t) - \int_0^1 \lambda y \frac{I_2\left(\sqrt{\lambda(1 - y^2)}\right)}{1 - y^2} u(y, t)\, dy. \tag{3.15}$$

In Figure 3.1 the kernel $k(1, y)$ is plotted for several values of λ. As λ grows, the maximum of the absolute value of the gain kernel moves to the left and is approximately proportional to $\lambda^{1/4} e^{\sqrt{\lambda}}$.

Let us now derive the explicit solution to the closed-loop system (3.1), (3.2), (3.14). First, we need to find the kernel of the inverse transformation. It satisfies the

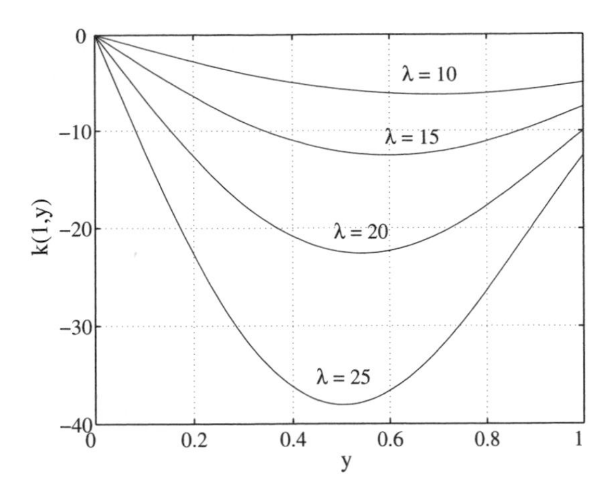

Figure 3.1 The control gain in (3.14) for different values of the parameter λ.

following PDE:

$$l_{xx}(x, y) - l_{yy}(x, y) = -\lambda l(x, y), \qquad (x, y) \in \mathcal{T}, \tag{3.16}$$

$$l(x, 0) = 0, \tag{3.17}$$

$$l(x, x) = -\frac{\lambda x}{2}. \tag{3.18}$$

Comparing (3.16)–(3.18) with (3.4)–(3.6), we can see that $l(x, y) = -k(x, y)$ when λ is replaced by $-\lambda$. Therefore, using the properties of Bessel functions (Appendix C), we get

$$l(x, y) = -\lambda y \frac{J_1\left(\sqrt{\lambda(x^2 - y^2)}\right)}{\sqrt{\lambda(x^2 - y^2)}}. \tag{3.19}$$

With both $k(x, y)$ and $l(x, y)$ given explicitly, we calculate $\phi_n(x)$ and $\psi_n(x)$ in (2.60)–(2.61) and get the following result.

Theorem 3.1. *The solution to the closed-loop system* (3.1), (3.2) *with the controller* (3.14) *is*

$$u(x, t) = \sum_{n=1}^{\infty} e^{-(c+\varepsilon\pi^2 n^2)t} \frac{2\pi n \sin\sqrt{\lambda + \pi^2 n^2}x}{\sqrt{\lambda + \pi^2 n^2}} \int_0^1 \psi_n(\xi) u_0(\xi)\, d\xi, \tag{3.20}$$

where

$$\psi_n(\xi) = \sin(\pi n \xi) + \int_\xi^1 \xi \frac{I_1\left(\sqrt{\lambda(\tau^2 - \xi^2)}\right)}{\sqrt{\lambda(\tau^2 - \xi^2)}} \sin(\pi n \tau)\, d\tau. \tag{3.21}$$

This result is a corollary of Theorem 2.7. In particular, the integral in (2.60) is solved explicitly with the help of [101].

For the case of the Neumann boundary condition at $x = 0$ for the equation (3.1), it is easy to repeat all the steps we have done for the Dirichlet case and get the following closed-form solution for the kernel:

$$k(x, y) = -\lambda x \frac{I_1\left(\sqrt{\lambda(x^2 - y^2)}\right)}{\sqrt{\lambda(x^2 - y^2)}}. \tag{3.22}$$

Note that the leading factor here is x, versus y in (3.13). The maximum of the absolute value of the kernel $k(1, y)$ is reached at $y = 0$.

3.2 A FAMILY OF PLANTS WITH SPATIALLY VARYING REACTIVITY

Consider the system

$$u_t(x, t) = u_{xx}(x, t) + \lambda_\sigma(x)u(x, t), \tag{3.23}$$

$$u(0, t) = 0, \tag{3.24}$$

$$u(1, t) = U(t), \tag{3.25}$$

where $\lambda_\sigma(x)$ is given by

$$\lambda_\sigma(x) = \frac{2\sigma^2}{\cosh^2(\sigma(x - x_0))}. \tag{3.26}$$

The function $\lambda_\sigma(x)$ parameterizes a family of "one-peak" functions. The maximum of $\lambda_\sigma(x)$ is $2\varepsilon\sigma^2$ and is achieved at $x = x_0$. The parameters σ and x_0 can be chosen to give the maximum an arbitrary value and location. Examples of $\lambda_\sigma(x)$ for different values of σ and x_0 are shown in Figure 3.2. The "sharpness" of the peak is not arbitrary and is given by $\lambda''_{\max} = -\lambda^2_{\max}/\varepsilon$. Despite the strange-looking expression for $\lambda_\sigma(x)$, the system (3.23)–(3.24) can approximate very well the linearized model of a chemical tubular reactor (see [14] and references therein) that is open-loop unstable.

Our result for stabilization of (3.23)–(3.25) is given by the following theorem.

Theorem 3.2. *The system* (3.23)–(3.25) *with the controller*

$$U(t) = -\int_0^1 \sigma e^{\sigma \tanh \sigma x_0 (1-y)}[\tanh \sigma x_0 - \tanh(\sigma(x_0 - y))]u(y, t)\, dy \tag{3.27}$$

is exponentially stable at the origin in $H^1(0, 1)$,

$$\|u(t)\|_{H^1} \leq Me^{-\pi^2 t}\|u_0\|_{H^1}. \tag{3.28}$$

Proof. The kernel PDE for (3.23)–(3.24) is

$$k_{xx}(x, y) - k_{yy}(x, y) = \lambda_\sigma(y)k(x, y), \tag{3.29}$$

$$k(x, 0) = 0, \tag{3.30}$$

$$k(x, x) = -\frac{1}{2}\int_0^x \lambda_\sigma(\tau)\, d\tau. \tag{3.31}$$

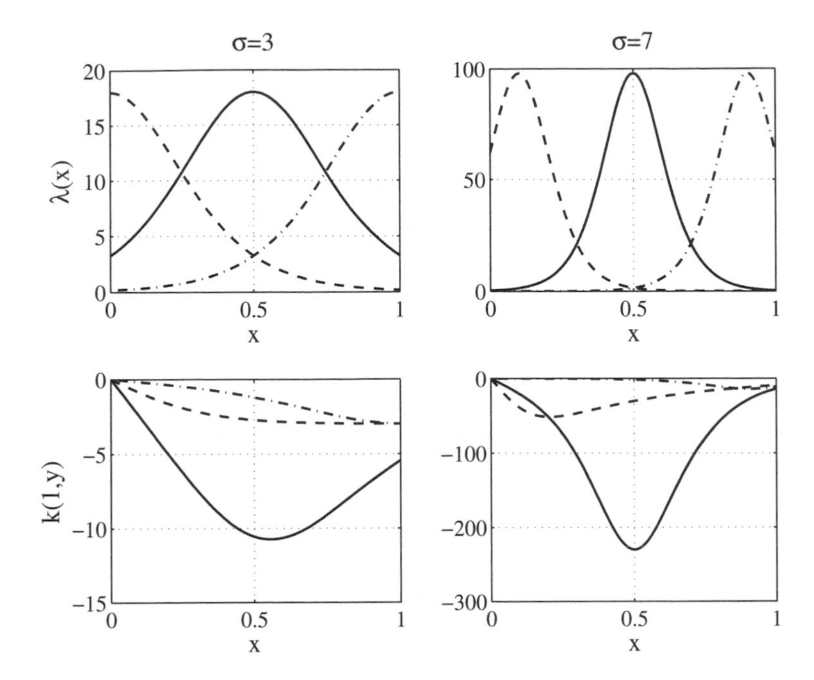

Figure 3.2 "One-peak" $\lambda_\sigma(x)$ (top) and the corresponding Dirichlet gain kernels $k(1, y)$ (bottom) for various values of σ and x_0.

Postulating $k(x, y) = X(x)Y(y)$, we have the following set of ODEs:

$$X''(x) = \mu X(x), \tag{3.32}$$

$$Y''(y) = Y(y)(\mu + 2X(y)Y'(y) + 2X'(y)Y(y)), \tag{3.33}$$

$$Y(0) = 0, \tag{3.34}$$

where μ is an arbitrary parameter. Let us choose $X(x) = e^{\sqrt{\mu}x}$ and substitute it into (3.33). We get

$$Y''(y) - 2e^{\sqrt{\mu}y}Y(y)(Y'(y) + \sqrt{\mu}Y(y)) - \mu\, Y(y) = 0. \tag{3.35}$$

With the change of variables $Y(y) = Z(y)e^{-\sqrt{\mu}y}$, we arrive at the following ODE:

$$Z''(y) - 2Z'(y)Z(y) - 2\sqrt{\mu}Z'(y) = 0, \tag{3.36}$$

$$Z(0) = 0, \tag{3.37}$$

$$Z'(0) = \mu - \sigma^2. \tag{3.38}$$

Here we have introduced an arbitrary parameter σ. The solution to the problem (3.36)–(3.38) is

$$Z(y) = -\sigma(\tanh(\sigma(y - x_0)) + \tanh\sigma x_0), \tag{3.39}$$

where $\tanh\sigma x_0 = \sqrt{\mu}/\sigma$. Now we can check that

$$\lambda_\sigma(x) = -2Z'(x) = \frac{2\sigma^2}{\cosh^2(\sigma(x - x_0))}, \tag{3.40}$$

which gives (3.26). Using (3.39), we obtain the kernel

$$k(x, y) = -\sigma e^{\sigma \tanh \sigma x_0 (x-y)} [\tanh \sigma x_0 - \tanh(\sigma(x_0 - y))]. \tag{3.41}$$

Setting $x = 1$ in (3.41) and invoking Theorem 2.6 concludes the proof. □

The gain kernels for corresponding $\lambda_\sigma(x)$ are shown in Figure 3.2. We can see that the control effort depends very much on the location of the peak of $\lambda_\sigma(x)$, which has an obvious explanation. When the peak is close to $x = 1$, the controller's influence is very high, whereas when the peak is close to $x = 0$, the boundary condition (3.24) helps to stabilize, so the worst case is a peak somewhere in the middle of the domain.

Remark 3.1. The solution (3.41) was obtained using the method of separation of variables. One can show that the most general $\lambda(x)$ for which the controller can be found using the method of separation of variables is

$$\lambda(x) = 2(\alpha^2 - \beta^2) \frac{\alpha^2 h^2(x) + (\beta^2 - \gamma^2)\sinh^2(\alpha x)}{(\alpha \cosh(\alpha x) h(x) - \sinh(\alpha x) h'(x))^2}, \tag{3.42}$$

where α, β, and γ are arbitrary constants and

$$h(x) = \cosh(\beta x) + \frac{\gamma}{\beta} \sinh(\beta x). \tag{3.43}$$

The control gain kernel for the plant with this $\lambda(x)$ is given by

$$k(x, y) = -\frac{(\alpha^2 - \beta^2) h(y) \sinh(\beta x)}{\alpha \cosh(\alpha y) h(y) - \sinh(\alpha y) h'(y)}. \tag{3.44}$$

3.3 SOLID PROPELLANT ROCKET MODEL

Consider the plant

$$u_t(x, t) = u_{xx}(x, t) + g e^{\alpha x} u(0, t), \tag{3.45}$$

$$u_x(0, t) = -q u(0, t), \tag{3.46}$$

$$u(1, t) = U(t). \tag{3.47}$$

Here, g, q, and α are arbitrary constants. This system represents a linearized model of unstable burning in solid propellant rockets (for more details, see [15] and references therein). The open-loop system is unstable for any $g + 2q > 2$, $\alpha \geq 0$.

The kernel PDE (2.20)–(2.24) takes the form

$$k_{xx}(x, y) = k_{yy}(x, y), \tag{3.48}$$

$$k_y(x, 0) = -q k(x, 0) + g e^{\alpha x} - g \int_0^x k(x, y) e^{\alpha y} \, dy, \tag{3.49}$$

$$k(x, x) = -q. \tag{3.50}$$

The general solution to (3.48) is $k(x, y) = \phi(x - y) + \psi(x + y)$, where ϕ and ψ are arbitrary functions. Using the condition (3.50), we obtain $\psi \equiv 0$ and

$\phi(0) = -q$. Substituting this solution into (3.49), we get the following integro-differential equation in one variable:

$$-\phi'(x) = -q\phi(x) + ge^{\alpha x} - g\int_0^x \phi(x-y)e^{\alpha y}\,dy, \tag{3.51}$$

$$\phi(0) = -q. \tag{3.52}$$

Let us apply the Laplace transform to both sides of (3.51). The result is

$$-s\hat{\phi}(s) - q = -q\hat{\phi}(s) + \frac{g}{s-\alpha} - \frac{g\hat{\phi}(s)}{s-\alpha}, \tag{3.53}$$

where $\hat{\phi}(s)$ is the Laplace transform of $\phi(x)$. Solving (3.53) for $\hat{\phi}(s)$, we get

$$\begin{aligned}\hat{\phi}(s) &= -\frac{q(s-\alpha)+g}{s^2-(q+\alpha)s+\alpha q-g} \\ &= \frac{2g+q(q-\alpha)}{4r(s-s_1)} - \frac{2g+q(q-\alpha)}{4r(s-s_2)} - \frac{q}{2(s-s_1)} - \frac{q}{2(s-s_2)},\end{aligned} \tag{3.54}$$

where

$$s_{1,2} = \frac{q+\alpha}{2} \mp r, \qquad r = \sqrt{g + \frac{(q-\alpha)^2}{4}}. \tag{3.55}$$

Taking the inverse Laplace transform of (3.54) and going back to $k(x, y)$, we get

$$k(x,y) = -e^{\frac{q+\alpha}{2}(x-y)}\left[q\cosh r(x-y) + \frac{2g+q(q-\alpha)}{2r}\sinh r(x-y)\right]. \tag{3.56}$$

The solution becomes particularly simple in one-parameter cases:

$$k(x,y) = -qe^{q(x-y)} \quad \text{for } g = \alpha = 0 \tag{3.57}$$

and

$$k(x,y) = -\sqrt{g}\sinh\sqrt{g}(x-y) \quad \text{for } q = \alpha = 0. \tag{3.58}$$

We arrive at the following result.

Theorem 3.3. *The system* (3.45)–(3.47) *with the controller*

$$\begin{aligned}U(t) = -\int_0^1 e^{\frac{q+\alpha}{2}(1-y)}&\left[q\cosh r(1-y)\right. \\ &\left.+\frac{2g+q(q-\alpha)}{2r}\sinh r(1-y)\right]u(y,t)\,dy\end{aligned} \tag{3.59}$$

is exponentially stable at the origin in $H^1(0, 1)$,

$$\|u(t)\|_{H^1} \le Me^{-\frac{\pi^2}{4}t}\|u_0\|_{H^1}. \tag{3.60}$$

The closed-loop solutions can also be obtained explicitly with the help of Appendix E and direct and inverse transformations. For example, for $\alpha = 0$, $q = 0$, one gets

$$u(x,t)=2\sum_{n=0}^{\infty} e^{-\mu_n^2 t}\left(\cos(\mu_n x)-\frac{g}{g+\mu_n^2}\right)$$
$$\times\int_0^1 u_0(\xi)\left[\cos(\mu_n\xi)+(-1)^n\frac{\sqrt{g}}{\mu_n}\sinh(\sqrt{g}(1-\xi))\right]d\xi, \quad (3.61)$$

and for $\alpha=0$, $g=0$, one gets

$$u(x,t)=2\sum_{n=0}^{\infty} e^{-\mu_n^2 t}\left(\mu_n\cos(\mu_n x)-q\sin(\mu_n x)\right)$$
$$\times\int_0^1 u_0(\xi)\frac{\mu_n\cos(\mu_n\xi)-q\sin(\mu_n\xi)+(-1)^n q e^{q(1-\xi)}}{\mu_n^2+q^2}\,d\xi, \quad (3.62)$$

where $\mu_n=\pi n+\pi/2$.

3.4 PLANTS WITH SPATIALLY VARYING DIFFUSIVITY

In this section we derive explicit controllers for two families of unstable plants with spatially varying diffusivity.

3.4.1 Plant with $\varepsilon(x)$

Consider the plant

$$u_t(x,t)=\varepsilon(x)u_{xx}(x,t)+\lambda u(x,t), \quad (3.63)$$
$$u(0,t)=0, \quad (3.64)$$
$$u(1,t)=U(t). \quad (3.65)$$

The boundary condition at the zero end can also be Neumann or mixed (see Remark 3.2. after Theorem 3.4). Applying the change of variables (2.5), (2.6), we get the plant

$$\bar{u}_t(\bar{x},t)=\bar{\varepsilon}\bar{u}_{\bar{x}\bar{x}}(\bar{x},t)+\left(\lambda+\frac{\varepsilon''(x)}{4}-\frac{3}{16}\frac{(\varepsilon'(x))^2}{\varepsilon(x)}\right)\bar{u}(\bar{x},t), \quad (3.66)$$
$$\bar{u}(0,t)=0, \quad (3.67)$$
$$\bar{u}(1,t)=\left(\frac{\bar{\varepsilon}}{\varepsilon(1)}\right)^{-1/4}U(t). \quad (3.68)$$

Suppose that for some constant a we have

$$\frac{\varepsilon''(x)}{4}-\frac{3}{16}\frac{(\varepsilon'(x))^2}{\varepsilon(x)}=a. \quad (3.69)$$

Then the plant (3.66)–(3.68) has constant coefficients, and we can apply the results of Section 3.1 to obtain the controller

$$U(t)=-\left(\frac{\varepsilon(1)}{\bar{\varepsilon}}\right)^{1/4}\int_0^1 \bar{y}\,\frac{\lambda+a+c}{\bar{\varepsilon}}\,\frac{I_1\left(\sqrt{\frac{\lambda+a+c}{\bar{\varepsilon}}(1-\bar{y}^2)}\right)}{\sqrt{\frac{\lambda+a+c}{\bar{\varepsilon}}(1-\bar{y}^2)}}\bar{u}(\bar{y},t)\,d\bar{y}, \quad (3.70)$$

where $c\geq 0$ is the design parameter.

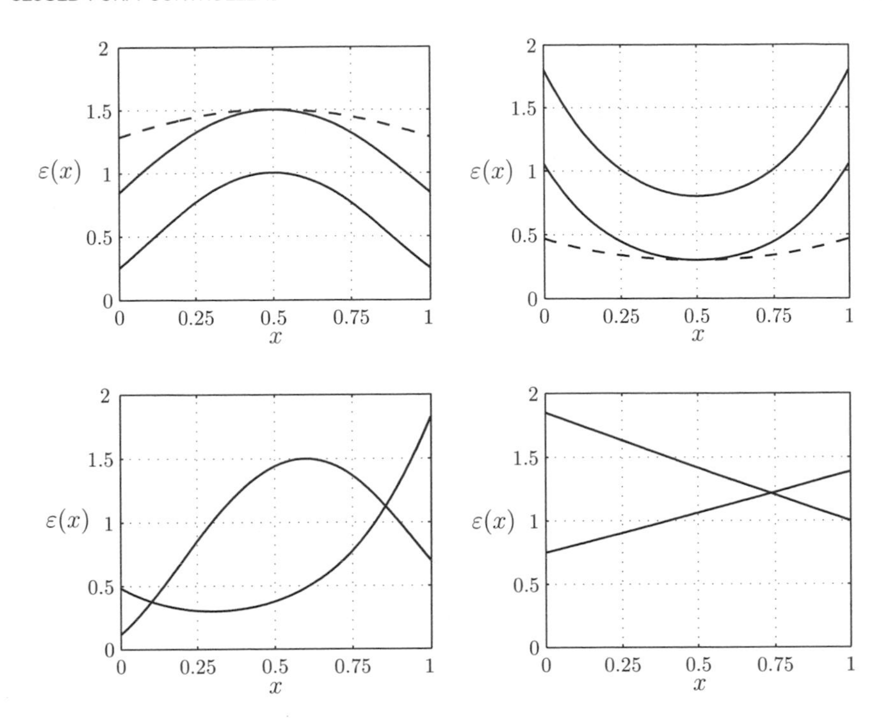

Figure 3.3 The function (3.72) for different values of ε_0, θ_0, and x_0. Top left: peak value and flatness are arbitrary. Top right: extremum can be set to max or min. Bottom left: location of extremum is arbitrary. Bottom right: linear functions matched.

There are two solutions to the ODE (3.69). The first solution is

$$\varepsilon(x) = \varepsilon_0(x - x_0)^2, \tag{3.71}$$

where $\varepsilon_0 > 0$ and $x_0 \in (-\infty, 0) \cup (1, \infty)$ are arbitrary constant parameters and $a = -\varepsilon_0/4$.

The other solution is three-parametric and thus is more interesting:

$$\varepsilon(x) = \varepsilon_0(1 + \theta_0(x - x_0)^2)^2, \tag{3.72}$$

where ε_0, θ_0, and x_0 are arbitrary constants (not violating the condition $\varepsilon(x) > 0$ on $[0, 1]$) and $a = \varepsilon_0\theta_0$. This solution gives a very good approximation on $x \in [0, 1]$ to many functions, including (3.71). Therefore, we focus our attention on the solution (3.72).

The function (3.72) always has one maximum or one minimum. The value and the location of the maximum (minimum) can be arbitrarily set by ε_0 and x_0, correspondingly. The sign of θ_0 determines whether it is a maximum or a minimum, and the value of θ_0 can set to an arbitrary "sharpness" of the extremum (Fig. 3.3). By selecting the extremum outside the region [0,1] and changing its value and sharpness, we can almost perfectly match any linear function as well (Fig. 3.3, bottom right).

Using (2.5), (2.6), and (3.72), we compute

$$\bar{\varepsilon} = \frac{\varepsilon_0 \theta_0}{\left[\operatorname{atan}\left(\sqrt{\theta_0}(1 - x_0)\right) + \operatorname{atan}\left(\sqrt{\theta_0} x_0\right)\right]^2}, \tag{3.73}$$

$$\bar{y} = \sqrt{\frac{\bar{\varepsilon}}{\varepsilon_0 \theta_0}} \left[\operatorname{atan}\left(\sqrt{\theta_0}(y - x_0)\right) + \operatorname{atan}\left(\sqrt{\theta_0} x_0\right)\right], \tag{3.74}$$

$$\bar{u}(\bar{y}, t) = \left(\frac{\bar{\varepsilon}}{\varepsilon(y)}\right)^{1/4} u(y, t). \tag{3.75}$$

Making the above change of variables in (3.70), we arrive at the following result.

THEOREM 3.4. *The plant* (3.66)–(3.68) *with* $\varepsilon(x)$ *given by (3.71) and with the controller*

$$U(t) = -\int_0^1 \phi(y) \frac{(\lambda + \varepsilon_0 \theta_0 + c)}{\sqrt{\varepsilon_0}} \frac{\varepsilon^{1/4}(1)}{\varepsilon^{3/4}(y)} \times \frac{I_1\left(\sqrt{\frac{\lambda + \varepsilon_0 \theta_0 + c}{\varepsilon_0}(\phi(1)^2 - \phi(y)^2)}\right)}{\sqrt{\frac{\lambda + \varepsilon_0 \theta_0 + c}{\varepsilon_0}(\phi(1)^2 - \phi(y)^2)}} u(y, t)\, dy, \tag{3.76}$$

where

$$\phi(y) = \frac{1}{\sqrt{\theta_0}} \left[\operatorname{atan}\left(\sqrt{\theta_0}(y - x_0)\right) + \operatorname{atan}\left(\sqrt{\theta_0} x_0\right)\right], \tag{3.77}$$

is exponentially stable at the origin in $H^1(0, 1)$ *with the decay rate* $(c + \varepsilon_0 \pi^2/\phi(1)^2)$.

Remark 3.2. If the boundary condition (3.64) is changed to $u_x(0, t) = 0$, the only change in the control gain (3.76) would be the leading factor $\phi(1)$ instead of $\phi(y)$. For the mixed boundary condition $u_x(0, t) = -qu(0, t)$ the closed-form solution can be obtained using the method described in Section 3.6.

3.4.2 Plant with Spatially Varying Thermal Diffusivity

Many problems (e.g., heat conduction in nonhomogeneous materials [21]) have a structure different from that of (3.63). The heat equation with spatially varying thermal diffusivity is usually written in the form

$$u_t(x, t) = \frac{d}{dx}\left(\varepsilon(x) \frac{d}{dx} u(x, t)\right) + \lambda u(x, t), \tag{3.78}$$

$$u(0, t) = 0, \tag{3.79}$$

$$u(1, t) = U(t). \tag{3.80}$$

With a change of variables $u = \sqrt{\varepsilon(x)} v$ we have

$$v_t(x, t) = \varepsilon(x) v_{xx}(x, t) + \left(\lambda + \frac{\varepsilon'^2(x)}{4\varepsilon(x)} - \frac{\varepsilon''(x)}{2}\right) v(x, t), \tag{3.81}$$

$$v(0, t) = 0, \tag{3.82}$$

$$v(1, t) = \frac{1}{\sqrt{\varepsilon(1)}} U(t). \tag{3.83}$$

Suppose that for some constant a_1 we have

$$\frac{\varepsilon'^2(x)}{4\varepsilon(x)} - \frac{\varepsilon''(x)}{2} = a_1, \tag{3.84}$$

then the plant (3.81)–(3.83) is in the form (3.63)–(3.65), and we can apply the results of Section 3.4.1. Therefore we only need to check whether $\varepsilon(x)$ defined by (3.71) or (3.72) also satisfies (3.84). By direct substitution we find that only (3.71) satisfies the condition (3.84) (with $a_1 = 0$).

Using (2.5), (2.6), and (3.71), we compute

$$\bar{\varepsilon} = \frac{\varepsilon_0}{\left(\ln\left|1 - \frac{1}{x_0}\right|\right)^2}, \tag{3.85}$$

$$\bar{y} = \sqrt{\frac{\bar{\varepsilon}}{\varepsilon_0}} \ln\left|1 - \frac{y}{x_0}\right|, \tag{3.86}$$

$$\bar{v}(\bar{y}, t) = \left(\frac{\bar{\varepsilon}}{\varepsilon(y)}\right)^{1/4} v(y, t). \tag{3.87}$$

Making the above change of variables in (3.70) (written in $\bar{v}$-variable) and going from v back to the original variable u, we get the following result.

THEOREM 3.5. *The plant* (3.78)–(3.80) *with* $\varepsilon(x)$ *given by* (3.71) *and with the controller*

$$U(t) = -\int_0^1 \phi(y) \frac{(\lambda + c - \varepsilon_0/4)}{\sqrt{\varepsilon_0}} \frac{|\varepsilon(1)|^{3/4}}{|\varepsilon(y)|^{5/4}} \times \frac{I_1\left(\sqrt{\frac{\lambda+c-\varepsilon_0/4}{\varepsilon_0}(\phi(1)^2 - \phi(y)^2)}\right)}{\sqrt{\frac{\lambda+c-\varepsilon_0/4}{\varepsilon_0}(\phi(1)^2 - \phi(y)^2)}} u(y, t)\, dy, \tag{3.88}$$

where

$$\phi(y) = \ln\left|1 - \frac{y}{x_0}\right|, \tag{3.89}$$

is exponentially stable at the origin in $H^1(0, 1)$ *with the decay rate* $(c + \varepsilon_0\pi^2/\phi(1)^2)$.

Note that Remark 3.2. holds here as well. Since the minimum of the function (3.71) is always zero, x_0 should be chosen outside the region [0, 1] to keep $\varepsilon(x) > 0$ for $x \in [0, 1]$. This means that $\varepsilon(x)$ given by (3.71) can approximate linear functions on [0, 1] very well. In Figure 3.4 the function $\varepsilon(x)$ and the corresponding control gains are shown for different parameter values.

3.5 THE TIME-VARYING REACTION EQUATION

Up to this point we have considered only time-invariant systems. In this section we illustrate on a simple plant how time-varying systems can be handled by the

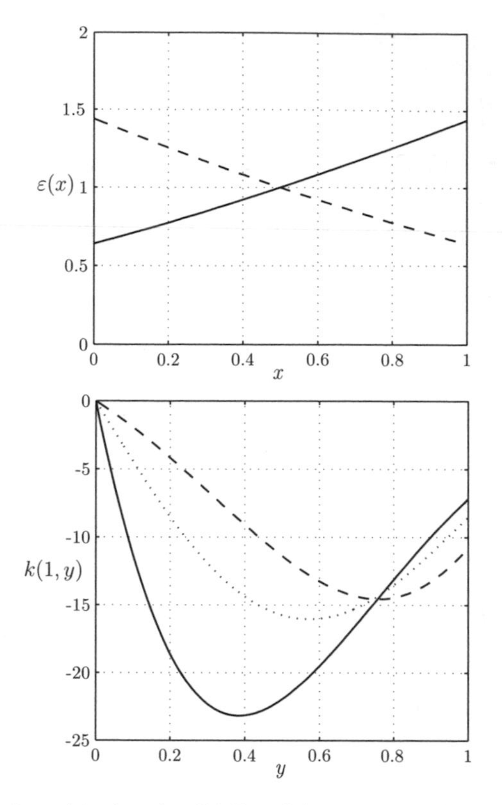

Figure 3.4 The function $\varepsilon(x)$ given by (3.71) and the corresponding kernel $k(1, y)$ for different parameter values. Dotted line shows the kernel for $\varepsilon(x) \equiv 1$.

backstepping method. Consider the plant

$$u_t(x,t) = u_{xx}(x,t) + \lambda(t)u(x,t), \tag{3.90}$$

$$u(0,t) = 0, \tag{3.91}$$

$$u(1,t) = U(t), \tag{3.92}$$

where $\lambda(t)$ is an analytic function of time. Applying the transformation

$$w(x,t) = u(x,t) - \int_0^x k(x,y,t)u(y,t)\,dy \tag{3.93}$$

along with feedback

$$U(t) = \int_0^1 k(1,y,t)u(y,t)\,dy, \tag{3.94}$$

we convert (3.90)–(3.92) into

$$w_t(x,t) = w_{xx}(x,t), \tag{3.95}$$

$$w(0,t) = 0, \tag{3.96}$$

$$w(1,t) = 0. \tag{3.97}$$

One can show that the function $k(x, y, t)$ satisfies the following PDE:

$$k_t(x, y, t) = k_{xx}(x, y, t) - k_{yy}(x, y, t) - \lambda(t)k(x, y, t), \tag{3.98}$$

$$k(x, 0, t) = 0, \tag{3.99}$$

$$k(x, x, t) = -\frac{x}{2}\lambda(t). \tag{3.100}$$

Let us make the following change of variables:

$$k(x, y, t) = -\frac{y}{2}e^{-\int_0^t \lambda(\tau)\,d\tau} f(z, t), \quad z = \sqrt{x^2 - y^2}. \tag{3.101}$$

We get the following PDE in one spatial variable for the function $f(z, t)$:

$$f_t(z, t) = f_{zz}(z, t) + 3z^{-1} f_z(z, t) \tag{3.102}$$

with boundary conditions

$$f_z(0, t) = 0, \quad f(0, t) = \lambda(t)e^{\int_0^t \lambda(\tau)\,d\tau} := F(t). \tag{3.103}$$

The $C^{2,1}_{z,t}$ solution to this problem is [99]:

$$f(z, t) = \sum_{n=0}^{\infty} \frac{1}{n!(n+1)!}\left(\frac{z}{2}\right)^{2n} F^{(n)}(t). \tag{3.104}$$

This solution is rather explicit. Since $z \leq 1$ and the product of two factorials in the denominator increases very fast with n, one can obtain very accurate approximations to $f(z, t)$ using only a few terms of the sum (3.104).

The controller is given by

$$U(t) = -\frac{1}{2}\int_0^1 y e^{-\int_0^t \lambda(\tau)d\tau}\left(\sum_{n=0}^{\infty} \frac{(1-y^2)^n F^{(n)}(t)}{4^n n!(n+1)!}\right) u(y, t)\,dy. \tag{3.105}$$

There are two cases when it is easy to compute the series (3.104) in closed form: when $F(t)$ is a combination of exponentials (since it is easy to compute the nth derivative of $F(t)$ in this case) or a polynomial (since the series is finite). Next we consider two examples that are motivated by exponentials and polynomials.

Example 3.1. (Rapid transition between two levels) Let $F(t)$ be given by

$$F(t) = e^{\lambda_0 t}\left\{\lambda_0 \cosh \omega_0(t - t_0) + \omega_0 \sinh \omega_0(t - t_0)\right\}, \tag{3.106}$$

where λ_0, ω_0 and t_0 are arbitrary constants. This $F(t)$ corresponds to the following $\lambda(t)$:

$$\lambda(t) = \lambda_0 + \omega_0 \tanh(\omega_0(t - t_0)). \tag{3.107}$$

This $\lambda(t)$ approximates a rapid change from a constant level $\lambda_0 - \omega_0$ to a constant level $\lambda_0 + \omega_0$ at $t = t_0$ (Fig. 3.5). Substituting (3.106) into (3.104) and computing the sum, we get the following control gain:

$$\begin{aligned} k(x, y, t) = &-\frac{y}{2\sqrt{x^2 - y^2}\cosh(\omega_0(t - t_0))} \\ &\times\left\{\sqrt{\lambda_0 + \omega_0}\, I_1\left(\sqrt{(\lambda_0 + \omega_0)(x^2 - y^2)}\right) e^{-\omega_0(t - t_0)}\right. \\ &\left.+\sqrt{\lambda_0 - \omega_0}\, I_1\left(\sqrt{(\lambda_0 - \omega_0)(x^2 - y^2)}\right) e^{\omega_0(t - t_0)}\right\}. \end{aligned} \tag{3.108}$$

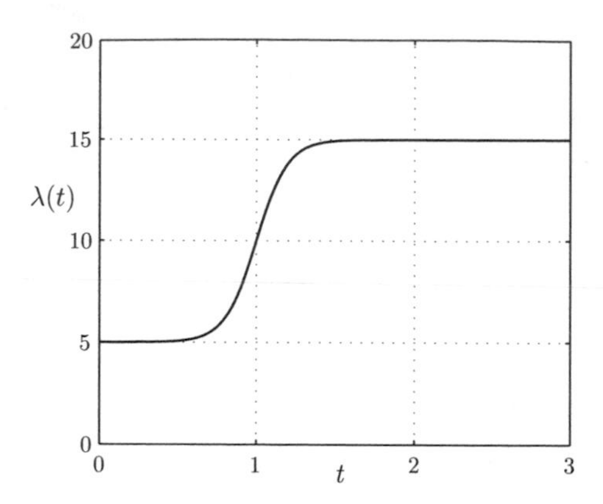

Figure 3.5 Graph of $\lambda(t)$ from (3.107) for $\lambda_0 = 10$, $\omega_0 = 5$, and $t_0 = 1$.

◇

Example 3.2. (One-peak function). Let $F(t)$ be

$$F(t) = \frac{e^{\lambda_0 t}}{a^2 + b^2}(\lambda_0((t + a)^2 + b^2) + 2(t + a)), \tag{3.109}$$

where λ_0, a and $b \neq 0$ are arbitrary constants. This $F(t)$ corresponds to the following $\lambda(t)$:

$$\lambda(t) = \lambda_0 + \frac{2(t + a)}{(t + a)^2 + b^2}. \tag{3.110}$$

This $\lambda(t)$ can approximate some one-peak functions (Fig. 3.6). Substituting (3.109) into (3.104) and computing the sum, we get the following control gain:

$$\begin{aligned} k(x, y, t) = &-\lambda_0 y \frac{I_1(\sqrt{\lambda_0} z)}{\sqrt{\lambda_0} z} - y \frac{t + a}{(t + a)^2 + b^2} I_0(\sqrt{\lambda_0} z) \\ &- \frac{y}{4\sqrt{\lambda_0}} \frac{z I_1(\sqrt{\lambda_0} z)}{(t + a)^2 + b^2}, \end{aligned} \tag{3.111}$$

where $z = \sqrt{x^2 - y^2}$. ◇

We should mention that there is a simpler solution to the problem of stabilization of (3.90)–(3.92), that is obtained using a change of variables,

$$u(x, t) = v(x, t) \exp\left\{\int_0^t \lambda(\tau)\, d\tau\right\}, \tag{3.112}$$

to convert the plant into a PDE with constant coefficients

$$v_t(x, t) = v_{xx}(x, t), \tag{3.113}$$

$$v(0, t) = 0, \tag{3.114}$$

$$v(1, t) = U(t) \exp\left\{-\int_0^t \lambda(\tau)\, d\tau\right\}. \tag{3.115}$$

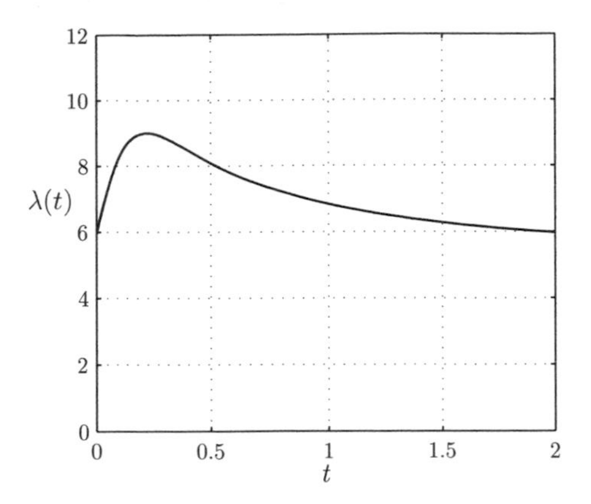

Figure 3.6 Graph of $\lambda(t)$ from (3.110) for $\lambda_0 = 5$, $a = 0.03$, and $b = 0.25$.

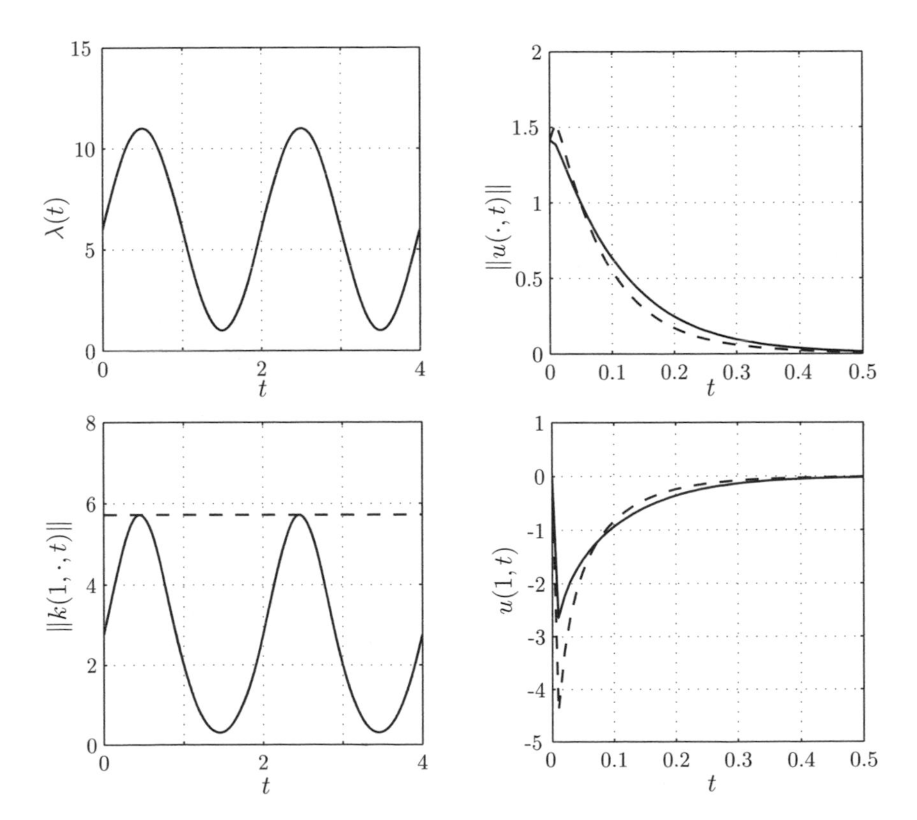

Figure 3.7 The simulation results for (3.90)–(3.92) with controllers (3.105) (solid line) and (3.116) (dashed line). Top left: $\lambda(t)$. Top right: L^2-norm of the gain kernel. Bottom left: L^2-norm of the state. Bottom right: the control effort.

This problem can then be stabilized using the results of Section 3.1 with the controller

$$U(t) = -\exp\left\{\int_0^t \lambda(\tau)\,d\tau\right\}\int_0^1 cy\frac{I_1\left(\sqrt{c(1-y^2)}\right)}{\sqrt{c(1-y^2)}}v(y,t)\,dy$$
$$= -\int_0^1 cy\frac{I_1\left(\sqrt{c(1-y^2)}\right)}{\sqrt{c(1-y^2)}}u(y,t)\,dy. \tag{3.116}$$

The decay rate of the closed-loop v-system is equal to the decay rate of the target system, that is, $e^{-(c+\pi^2)t}$. Therefore, the closed-loop stability of the u-system is guaranteed by satisfying the condition [55]

$$c > \limsup_{t\to\infty} \lambda(t) - \pi^2, \tag{3.117}$$

or $c > -\pi^2$ if $\lambda \in L^1(0,\infty) \cup L^2(0,\infty)$.

Although the controller (3.116) stabilizes (3.90)–(3.91) for any $\lambda(t)$, it is most suitable for the cases when minimum and maximum values of $\lambda(t)$ are close, for example when it is a constant plus sinusoid with a small amplitude. When $\lambda(t)$ has significant drops and rises, this method will use an unnecessarily large initial control effort (Example 3.1.) or will result in poor initial performance (Example 3.2.). In such cases the design (3.105) is advantageous, as indicated by the simulation results in Fig. 3.7.

3.6 MORE COMPLEX SYSTEMS

The solutions presented in Sections 3.1–3.5 can be combined to obtain the explicit results for even more complex systems. We give two examples that illustrate this possibility.

Example 3.3. Consider the plant

$$u_t(x,t) = \varepsilon u_{xx}(x,t) + (\lambda_\sigma(x) + \lambda_0)\,u(x,t), \tag{3.118}$$
$$u(0,t) = 0, \tag{3.119}$$
$$u(1,t) = U(t). \tag{3.120}$$

Let us denote by $k^\sigma(x,y)$ and $k^\lambda(x,y)$ the (closed-form) control gains used to obtain the controllers for the plants (3.23)–(3.25) and (3.1)–(3.3), respectively. The transformation

$$\bar{w}(x,t) = u(x,t) - \int_0^x k^\sigma(x,y)u(y,t)\,dy \tag{3.121}$$

maps (3.118)–(3.119) into the intermediate system

$$\bar{w}_t(x,t) = \varepsilon\bar{w}_{xx}(x,t) + \lambda_0\bar{w}(x,t), \tag{3.122}$$
$$\bar{w}(0,t) = 0, \tag{3.123}$$
$$\bar{w}(1,t) = U(t) - \int_0^1 k^\sigma(1,y)u(y,t)\,dy. \tag{3.124}$$

With the transformation

$$w(x,t) = \bar{w}(x,t) - \int_0^x k^\lambda(x,y)\bar{w}(y,t)\,dy, \tag{3.125}$$

we now map (3.122)–(3.123) into the system

$$w_t(x,t) = \varepsilon w_{xx}(x,t) - cw(x,t), \tag{3.126}$$

$$w(0,t) = 0, \tag{3.127}$$

$$w(1,t) = 0. \tag{3.128}$$

Using (3.125) and (3.121), we derive the transformation directly from $u(x,t)$ into $w(x,t)$:

$$\begin{aligned} w(x,t) &= u(x,t) - \int_0^x k^\sigma(x,y)u(y,t)\,dy \\ &\quad - \int_0^x k^\lambda(x,y)\left(u(y,t) - \int_0^y k^\sigma(y,\xi)u(\xi,t)\,d\xi\right)dy \\ &= u(x,t) - \int_0^x k^c(x,y)u(y,t)\,dy, \end{aligned} \tag{3.129}$$

where k^c stands for the combined kernel:

$$k^c(x,y) = k^\sigma(x,y) + k^\lambda(x,y) - \int_y^x k^\lambda(x,\xi)k^\sigma(\xi,y)\,d\xi. \tag{3.130}$$

For example, for $\lambda(x) = \lambda_0 + 2\sigma^2/\cosh^2(\sigma x)$, one gets the closed-form solution

$$k^c(x,y) = -\lambda y \frac{I_1\left(\sqrt{\lambda(x^2-y^2)}\right)}{\sqrt{\lambda(x^2-y^2)}} - \sigma\tanh(\sigma y) I_0\left(\sqrt{\lambda(x^2-y^2)}\right), \tag{3.131}$$

where $\lambda = (\lambda_0 + c)/\varepsilon$. The stabilizing controller is

$$U(t) = \int_0^1 k^c(1,y)u(y,t)\,dy. \tag{3.132}$$

Example 3.4. Consider the reaction-diffusion equation with a destabilizing boundary condition:

$$u_t(x,t) = u_{xx}(x,t) + \lambda u(x,t), \tag{3.133}$$

$$u_x(0,t) = -qu(0,t), \tag{3.134}$$

$$u(1,t) = U(t). \tag{3.135}$$

For $q = 0$, the solution to gain kernel PDE is given by (3.22), and for $\lambda = 0$ the gain kernel is given by (3.57); let us denote these solutions by $k^\lambda(x,y)$ and $k^q(x,y)$, respectively. Using the method described in Example 3.3., it is easy to show that the transformation

$$w(x,t) = u(x,t) - \int_0^x k^c(x,y)u(y,t)\,dy, \tag{3.136}$$

where

$$k^c(x,y) = k^q(x,y) + k^\lambda(x,y) - \int_y^x k^\lambda(x,\xi)k^q(\xi,y)\,d\xi \tag{3.137}$$

maps (3.133)–(3.135) into

$$w_t(x,t) = w_{xx}(x,t), \tag{3.138}$$

$$w_x(0,t) = 0, \tag{3.139}$$

$$w(1,t) = 0. \tag{3.140}$$

The controller is

$$U(t) = -\int_0^1 \left[(q+\lambda y)e^{q(1-y)} + \int_y^1 \lambda e^{q(1-\tau)} \frac{\tau I_2\left(\sqrt{\lambda(1-\tau^2)}\right)}{1-\tau^2}\,d\tau \right] u(y,t)\,dy. \tag{3.141}$$

◇

In the same fashion one can obtain explicit stabilizing controllers for even more complicated plants. For example, using the results of Sections 3.1 and 3.3, the controllers can be obtained for the following six-parameter family of plants:

$$u_t(x,t) = \varepsilon u_{xx}(x,t) + bu_x(x,t) + \lambda u(x,t) + ge^{\alpha x}u(0,t), \tag{3.142}$$

$$u_x(0,t) = -qu(0,t), \tag{3.143}$$

$$u(1,t) = U(t). \tag{3.144}$$

3.7 2D AND 3D SYSTEMS

The explicit designs presented in the previous sections can be used to stabilize higher-dimensional (2D and 3D) plants in regular geometries with control distributed along the subset of the corresponding 1D or 2D boundary. The control design for plants with constant coefficients is illustrated by the following example.

Example 3.5. Consider the reaction-diffusion equation on the rectangle with Dirichlet boundary conditions and control applied at one side of the rectangle:

$$u_t(x,y,t) = u_{xx}(x,y,t) + u_{yy}(x,y,t) + \lambda u(x,y,t), \quad (x,y) \in (0,1)\times(0,L), \tag{3.145}$$

$$u(x,0,t) = 0, \quad 0 \le x \le 1, \tag{3.146}$$

$$u(x,L,t) = 0, \quad 0 \le x \le 1, \tag{3.147}$$

$$u(0,y,t) = 0, \quad 0 \le y \le L, \tag{3.148}$$

$$u(1,y,t) = U(y,t), \quad 0 \le y \le L. \tag{3.149}$$

Using the result of Section 3.1, it is straightforward to verify that the transformation

$$w(x, y, t) = u(x, y, t) + \int_0^x \lambda\xi \frac{I_1\left(\sqrt{\lambda(x^2-\xi^2)}\right)}{\sqrt{\lambda(x^2-\xi^2)}} u(\xi, y, t)\, d\xi \tag{3.150}$$

along with the feedback

$$U(y, t) = -\int_0^1 \lambda\xi \frac{I_1\left(\sqrt{\lambda(1-\xi^2)}\right)}{\sqrt{\lambda(1-\xi^2)}} u(\xi, y, t)\, d\xi \tag{3.151}$$

maps (3.145)–(3.149) into the exponentially stable system

$$w_t(x, y, t) = w_{xx}(x, y, t) + w_{yy}(x, y, t), \quad (x, y) \in (0, 1) \times (0, L), \tag{3.152}$$

$$w(x, 0, t) = 0, \quad 0 \le x \le 1, \tag{3.153}$$

$$w(x, L, t) = 0, \quad 0 \le x \le 1, \tag{3.154}$$

$$w(0, y, t) = 0, \quad 0 \le y \le L, \tag{3.155}$$

$$w(1, y, t) = 0, \quad 0 \le y \le L. \tag{3.156}$$

◇

It is also possible to apply backstepping design to 2D and 3D plants with spatially varying coefficients if these coefficients vary only along the direction orthogonal to the control surface.

Example 3.6. Consider the unstable plant with a spatially varying coefficient in a cube:

$$\begin{aligned} u_t(x, y, z, t) = {} & u_{xx}(x, y, z, t) + u_{yy}(x, y, z, t) + u_{zz}(x, y, z, t) \\ & + g e^{\alpha x} u(0, y, z, t), \quad (x, y, z) \in (0, 1) \times (0, 1) \times (0, 1), \end{aligned} \tag{3.157}$$

$$u(x, 0, z, t) = u(x, 1, z, t) = 0, \quad 0 \le x, z \le 1, \tag{3.158}$$

$$u(x, y, 0, t) = u(x, y, 1, t) = 0, \quad 0 \le x, y \le 1, \tag{3.159}$$

$$u_x(0, y, z, t) = 0, \quad 0 \le y, z \le 1, \tag{3.160}$$

$$u(1, y, z, t) = U(y, z, t), \quad 0 \le y, z \le 1. \tag{3.161}$$

Using the result of Section 3.3, it is straightforward to check that the transformation

$$\begin{aligned} w(x, y, z, t) = {} & u(x, y, z, t) \\ & + \int_0^x g e^{\frac{\alpha}{2}(x-\xi)} \frac{\sinh\left(\sqrt{g + \frac{\alpha^2}{4}}(x-\xi)\right)}{\sqrt{g + \frac{\alpha^2}{4}}} u(\xi, y, z, t)\, d\xi \end{aligned} \tag{3.162}$$

along with the controller

$$U(y, z, t) = -\int_0^1 g e^{\frac{\alpha}{2}(1-\xi)} \frac{\sinh\left(\sqrt{g + \frac{\alpha^2}{4}}(1-\xi)\right)}{\sqrt{g + \frac{\alpha^2}{4}}} u(\xi, y, z, t)\, d\xi \tag{3.163}$$

maps (3.157)–(3.161) into the exponentially stable target system

$$w_t(x, y, z, t) = w_{xx}(x, y, z, t) + w_{yy}(x, y, z, t) + w_{zz}(x, y, z, t),$$
$$(x, y, z) \in (0, 1) \times (0, 1) \times (0, 1), \tag{3.164}$$
$$w(x, 0, z, t) = w(x, 1, z, t) = 0, \quad 0 \le x, z \le 1, \tag{3.165}$$
$$w(x, y, 0, t) = w(x, y, 1, t) = 0, \quad 0 \le x, y \le 1, \tag{3.166}$$
$$w_x(0, y, z, t) = 0, \quad 0 \le y, z \le 1, \tag{3.167}$$
$$w(1, y, z, t) = 0, \quad 0 \le y, z \le 1. \tag{3.168}$$

3.8 NOTES AND REFERENCES

Most of the closed-form solutions in this chapter have been obtained in [108] and [110]. However, Sections 3.2, 3.3, 3.6, and 3.7 contain some previously unpublished material.

Chapter Four

Observers

The measurements in distributed parameter systems are not always available across the entire domain. They are often not available even at individual points strictly inside the domain. In fact, in some of the most exciting and complex applications, such as those involving fluid flows, sensors can be placed only at the boundaries. This is the situation that we focus on here, designing observers that employ only *boundary sensing*.

The state-feedback controllers developed in Chapters 2 and 3 require information on the state at each point in the domain. The design of state observers depends on the type (Dirichlet or Neumann) and the location of measurement and actuation. We consider two setups for observer-based control design: the anti-collocated case, in which sensor and actuator are placed at the opposite ends, and the collocated case, in which sensor and actuator are placed at the same end. There is no substantial technical difference between the cases of Dirichlet and Neumann actuation, so we pick one (Dirichlet) for the anti-collocated setup and the other (Neumann) for the collocated setup.

4.1 OBSERVER DESIGN FOR THE ANTI-COLLOCATED SETUP

We consider the plant

$$u_t(x,t) = \varepsilon u_{xx}(x,t) + \lambda(x)u(x,t) + g(x)u(0,t) + \int_0^x f(x,y)u(y,t)\,dy, \quad (4.1)$$

for $x \in (0,1)$, $t > 0$, with boundary conditions

$$u_x(0,t) = -qu(0,t), \quad (4.2)$$

$$u(1,t) = U(t), \quad (4.3)$$

where input $U(t)$ can be any function of time or a feedback law.

Suppose the only available measurement of our system is at $x = 0$, the opposite end to actuation. We propose the following observer for the system (4.1)–(4.3):

$$\hat{u}_t(x,t) = \varepsilon \hat{u}_{xx}(x,t) + \lambda(x)\hat{u}(x,t) + g(x)u(0,t) + \int_0^x f(x,y)\hat{u}(y,t)\,dy + p_1(x)[u(0,t) - \hat{u}(0,t)], \quad (4.4)$$

$$\hat{u}_x(0,t) = -qu(0,t) + p_0[u(0,t) - \hat{u}(0,t)], \quad (4.5)$$

$$\hat{u}(1,t) = U(t). \quad (4.6)$$

Here $p_1(x)$ and p_0 are output injection functions (p_0 is a constant) *to be designed.* Note that we introduce output injection not only in the equation (4.4) but also at the boundary where measurements are available. Since $u(0,t)$ is measured, we also implicitly use the additional output injections "$+g(x)[u(0,t)-\hat{u}(0,t)]$" and "$-q[u(0,t)-\hat{u}(0,t)]$." These extra terms cancel the dependency on $g(x)$ and q in the error dynamics.

The observer (4.4)–(4.6) is in the standard form of "copy of the system plus injection of the output estimation error," that is, it mimics the finite-dimensional case in which observers of the form $\dot{\hat{x}} = A\hat{x} + Bu + L(y - C\hat{x})$ are used for plants $\dot{x} = Ax + Bu,\ y = Cx$. This standard form allows us to pursue duality between the observer and the controller design, that is, to find the observer gain function using the solution to the stabilization problem presented in Chapter 2, similar to the way duality is used to find the gains of a Luenberger observer based on the pole placement control algorithm or the way duality is used to construct a Kalman filter based on the LQR design.

The observer error $\tilde{u}(x,t) = u(x,t) - \hat{u}(x,t)$ satisfies the following PDE:

$$\tilde{u}_t(x,t) = \varepsilon\tilde{u}_{xx}(x,t) + \lambda(x)\tilde{u}(x,t) + \int_0^x f(x,y)\tilde{u}(y,t)\,dy - p_1(x)\tilde{u}(0,t), \tag{4.7}$$

$$\tilde{u}_x(0,t) = -p_0\tilde{u}(0,t), \tag{4.8}$$

$$\tilde{u}(1,t) = 0. \tag{4.9}$$

Observer gains $p_1(x)$ and p_0 should be designed to stabilize the system (4.7)–(4.9). We solve the problem of stabilization of (4.7)–(4.9) by the same integral transformation approach as was used for the state-feedback boundary control problem in Chapter 2. We look for a coordinate transformation

$$\tilde{u}(x,t) = \tilde{w}(x,t) - \int_0^x p(x,y)\tilde{w}(y,t)\,dy \tag{4.10}$$

that transforms (4.7)–(4.9) into the exponentially stable (for $\tilde{c} > -\varepsilon\pi^2/4$) system

$$\tilde{w}_t(x,t) = \varepsilon\tilde{w}_{xx}(x,t) - \tilde{c}\tilde{w}(x,t), \tag{4.11}$$

$$\tilde{w}_x(0,t) = 0, \tag{4.12}$$

$$\tilde{w}(1,t) = 0. \tag{4.13}$$

The free parameter $\tilde{c}$ sets the desired observer convergence speed. It is in general different from the analogous coefficient c in control design, since one usually wants the estimator to be faster than the state-feedback closed-loop dynamics.

Substituting (4.10) into (4.7)–(4.9), we obtain the following PDE for $\tilde{w}$:

$$\tilde{w}_t(x,t) = \varepsilon\tilde{w}_{xx}(x,t) - \left(\lambda(x) + 2\varepsilon\frac{d}{dx}p(x,x)\right)\tilde{w}(x,t)$$
$$+ \int_0^x \tilde{w}(y,t)\left[f(x,y) - \int_y^x f(x,\xi)p(\xi,y)\,d\xi\right]dy$$
$$- \int_0^x \left[\varepsilon p_{xx}(x,y) - \varepsilon p_{yy}(x,y) + (\lambda(x)+\tilde{c})p(x,y)\right]\tilde{w}(y,t)\,dy$$
$$- \varepsilon p(x,0)\tilde{w}_x(0,t) + (\varepsilon p_y(x,0) - p_1(x))\tilde{w}(0,t), \tag{4.14}$$

$$\tilde{w}_x(0,t) = (p(0,0) - p_0)\tilde{w}(0,t), \tag{4.15}$$

$$\tilde{w}(1,t) = \int_0^1 p(1,y)\tilde{w}(y,t)\,dy. \tag{4.16}$$

Comparing this PDE with (4.11)–(4.13), we see that $p(x,y)$ has to satisfy the PDE

$$\varepsilon p_{yy}(x,y) - \varepsilon p_{xx}(x,y) = (\lambda(x)+\tilde{c})p(x,y) - f(x,y) + \int_y^x f(x,\xi)p(\xi,y)\,d\xi \tag{4.17}$$

for $(x,y) \in \mathcal{T} = \{x, y : 0 < y < x < 1\}$, with the boundary conditions

$$\frac{d}{dx}p(x,x) = \frac{1}{2\varepsilon}(\lambda(x)+\tilde{c}), \tag{4.18}$$

$$p(1,y) = 0, \tag{4.19}$$

and the observer gains should be chosen as

$$p_1(x) = \varepsilon p_y(x,0), \qquad p_0 = p(0,0). \tag{4.20}$$

We want to establish well-posedness of the PDE (4.17)–(4.19). Once the solution to this PDE is found, the observer gains can be obtained from (4.20).

Let us make a change of variables:

$$\breve{x} = 1 - y, \quad \breve{y} = 1 - x, \tag{4.21}$$

$$\breve{\lambda}(\breve{y}) = \lambda(x), \quad \breve{f}(\breve{x},\breve{y}) = f(x,y), \quad \breve{p}(\breve{x},\breve{y}) = p(x,y). \tag{4.22}$$

In these new variables the problem (4.17)–(4.19) becomes

$$\varepsilon\breve{p}_{\breve{x}\breve{x}}(\breve{x},\breve{y}) - \varepsilon\breve{p}_{\breve{y}\breve{y}}(\breve{x},\breve{y}) = (\breve{\lambda}(\breve{y})+\tilde{c})\breve{p}(\breve{x},\breve{y}) - \breve{f}(\breve{x},\breve{y})$$
$$+ \int_{\breve{y}}^{\breve{x}} \breve{p}(\breve{x},\xi)\breve{f}(\xi,\breve{y})\,d\xi, \tag{4.23}$$

$$\breve{p}(\breve{x},0) = 0, \tag{4.24}$$

$$\breve{p}(\breve{x},\breve{x}) = -\frac{1}{2\varepsilon}\int_0^{\breve{x}} (\breve{\lambda}(\xi)+\tilde{c})\,d\xi. \tag{4.25}$$

This PDE is the same as the PDE (2.71)–(2.73) from Chapter 2 (with $g(x) = 0$ and λ, f, and c replaced by $\breve{\lambda}$, $\breve{f}$, and $\tilde{c}$, respectively). Hence, using Theorem 2.5, we obtain the following result.

THEOREM 4.1. *The equation* (4.17) *with boundary conditions* (4.18)–(4.19) *has a unique* $C^2(\mathcal{T})$ *solution. In addition, the kernel* $r(x, y)$ *of the inverse transformation*

$$\widetilde{w}(x,t) = \tilde{u}(x,t) + \int_0^x r(x,y)\tilde{u}(y,t)\,dy \tag{4.26}$$

is a unique $C^2(\mathcal{T})$ *function.*

The fact that the observer kernel in transposed and switched variables satisfies the same class of PDEs as the control kernel is reminiscent of the duality property of state-feedback and observer design problems for linear finite-dimensional systems. The difference between the equations for observer and control gain kernels is due to the fact that the observer error system does not contain $g(x)$- and q-terms (because $u(0, t)$ is measured).

The observer gains in the new variables are given by

$$p_1(x) = -\varepsilon \breve{p}_{\breve{x}}(1, 1-x), \qquad p_0 = \breve{p}(1,1). \tag{4.27}$$

Exponential stability of the target system (4.11)–(4.13) and invertibility of the transformation (4.10) (established in Theorem 4.1) imply exponential stability of (4.7)–(4.9) (see the proof of Theorem 2.3). We obtain the following result.

THEOREM 4.2. *Let* $p(x, y)$ *be the solution of the system* (4.17)–(4.19). *For any initial data* $\tilde{u}_0(x) \in H^1(0, 1)$ *compatible with boundary conditions, the system* (4.7)–(4.9) *with* $p_1(x)$ *and* p_0 *given by* (4.20) *has a unique classical solution* $\tilde{u} \in C^{2,1}([0, 1] \times (0, \infty))$ *and is exponentially stable at the origin in* $H^1(0, 1)$ *with the decay rate* $(\tilde{c} + \varepsilon\pi^2/4)$.

4.2 PLANTS WITH DIRICHLET UNCONTROLLED END AND NEUMANN MEASUREMENTS

Let us now design an observer for the plant with the Dirichlet boundary condition at the uncontrolled end:

$$u_t(x,t) = \varepsilon u_{xx}(x,t) + \lambda(x)u(x,t) + g(x)u_x(0,t) + \int_0^x f(x,y)u(y,t)\,dy, \tag{4.28}$$

$$u(0,t) = 0, \tag{4.29}$$

$$u(1,t) = U(t). \tag{4.30}$$

In this case it only makes sense to consider Neumann measurements, $u_x(0, t)$.

We propose the observer

$$\hat{u}_t(x,t) = \varepsilon\hat{u}_{xx}(x,t) + \lambda(x)\hat{u}(x,t) + g(x)u_x(0,t) + \int_0^x f(x,y)\hat{u}(y,t)\,dy + p_1(x)[u_x(0,t) - \hat{u}_x(0,t)], \tag{4.31}$$

$$\hat{u}(0,t) = p_0[u_x(0,t) - \hat{u}_x(0,t)], \tag{4.32}$$

$$\hat{u}(1,t) = U(t), \tag{4.33}$$

where $p_1(x)$ and p_0 are observer gains to be designed.

The observer error $\tilde{u}(x,t) = u(x,t) - \hat{u}(x,t)$ satisfies the following PDE:

$$\tilde{u}_t(x,t) = \varepsilon\tilde{u}_{xx}(x,t) + \lambda(x)\tilde{u}(x,t) + \int_0^x f(x,y)\tilde{u}(y,t)\,dy - p_1(x)\tilde{u}_x(0,t), \tag{4.34}$$

$$\tilde{u}(0,t) = -p_0\tilde{u}_x(0,t), \tag{4.35}$$

$$\tilde{u}(1,t) = 0. \tag{4.36}$$

Using the transformation (4.10), we map (4.34)–(4.36) into the exponentially stable (for $\tilde{c} > -\varepsilon\pi^2$) system

$$\tilde{w}_t(x,t) = \varepsilon\tilde{w}_{xx}(x,t) - \tilde{c}\tilde{w}(x,t), \tag{4.37}$$

$$\tilde{w}(0,t) = 0, \tag{4.38}$$

$$\tilde{w}(1,t) = 0. \tag{4.39}$$

As can be seen from (4.14)–(4.16), one obtains the PDE (4.17)–(4.19) for $p(x,y)$, except that the observer gains are computed in a different way:

$$p_1(x) = -\varepsilon p(x,0), \qquad p_0 = 0. \tag{4.40}$$

Repeating the arguments in the proof of Theorem 4.2, we obtain the following result.

Theorem 4.3. *Let* $p(x,y)$ *be the solution of the system* (4.17)–(4.19). *For any initial data* $u_0(x) \in H^1(0,1)$ *compatible with boundary conditions, the system* (4.34)–(4.36) *with* $p_1(x)$ *and* p_0 *given by* (4.40) *has a unique classical solution* $\tilde{u} \in C^{2,1}([0,1] \times (0,\infty))$ *and is exponentially stable at the origin in* $H^1(0,1)$ *with the decay rate* $(\tilde{c} + \varepsilon\pi^2)$.

4.3 OBSERVER DESIGN FOR THE COLLOCATED SETUP

Suppose now that the only available measurement is at the same end with actuation ($x = 1$). We will concentrate on the case when $u(1,t)$ is measured and $u_x(1,t)$ is actuated, which is the usual setting for thermal or chemical problems (temperature or concentration is available and the gradients are used for actuation). It is quite straightforward to adapt the design to the opposite setting, which usually occurs in fluid problems (shear stress is measured and velocity is a control variable).

We solve this problem with a restriction on the class (4.1)–(4.3) by setting $f(x, y) \equiv 0$, $g(x) \equiv 0$. This restriction is necessary because the observer problem in the collocated case is "upper-triangular"; thus the "lower-triangular" terms with $g(x)$ and $f(x, y)$ are not allowed.

Consider the observer

$$\hat{u}_t(x,t) = \varepsilon \hat{u}_{xx}(x,t) + \lambda(x)\hat{u}(x,t) + p_1(x)[u(1,t) - \hat{u}(1,t)], \tag{4.41}$$

$$\hat{u}_x(0,t) = -q\hat{u}(0,t), \tag{4.42}$$

$$\hat{u}_x(1,t) = p_0[u(1,t) - \hat{u}(1,t)] + U(t). \tag{4.43}$$

Here $p_1(x)$ and p_0 are output injection functions *to be designed.* The difference from the anti-collocated case (apart from injecting $u(1, t)$ instead of $u(0, t)$) is that the gain p_0 is introduced in the other boundary condition.

The observer error $\tilde{u}(x, t)$ satisfies the equation

$$\tilde{u}_t(x,t) = \varepsilon \tilde{u}_{xx}(x,t) + \lambda(x)\tilde{u}(x,t) - p_1(x)\tilde{u}(1,t), \tag{4.44}$$

$$\tilde{u}_x(0,t) = -q\tilde{u}(0,t), \tag{4.45}$$

$$\tilde{u}_x(1,t) = -p_0\tilde{u}(1,t). \tag{4.46}$$

With the transformation

$$\tilde{u}(x,t) = \widetilde{w}(x,t) - \int_x^1 p(x,y)\widetilde{w}(y,t)\,dy, \tag{4.47}$$

we map (4.44)–(4.46) into the exponentially stable target system

$$\widetilde{w}_t(x,t) = \varepsilon \widetilde{w}_{xx}(x,t) - \tilde{c}\widetilde{w}(x,t), \qquad \tilde{c} > 0, \tag{4.48}$$

$$\widetilde{w}_x(0,t) = 0, \tag{4.49}$$

$$\widetilde{w}_x(1,t) = 0. \tag{4.50}$$

Note that the transformation (4.47) is in the upper-triangular form. By substituting (4.47) into (4.44)–(4.46) we get

$$\begin{aligned} \widetilde{w}_t(x,t) = {} & \varepsilon \widetilde{w}_{xx}(x,t) - \tilde{c}\widetilde{w}(x,t) + \left(2\varepsilon \frac{d}{dx} p(x,x) + \lambda(x) + \tilde{c}\right) \widetilde{w}(x,t) \\ & - (\varepsilon p_y(x,1) + p_1(x))\widetilde{w}(1,t) + \int_x^1 (\varepsilon p_{yy}(x,y) - \varepsilon p_{xx}(x,y) \\ & - (\lambda(x) + \tilde{c})p(x,y))\widetilde{w}(y,t)\,dy, \end{aligned} \tag{4.51}$$

$$\begin{aligned} \widetilde{w}_x(0,t) = {} & -(q + p(0,0))\widetilde{w}(0,t) \\ & + \int_0^1 (p_x(0,y) + qp(0,y))\widetilde{w}(y,t)\,dy, \end{aligned} \tag{4.52}$$

$$\widetilde{w}_x(1,t) = -(p_0 + p(1,1))\widetilde{w}(1,t). \tag{4.53}$$

Comparing the above PDE with (4.48)–(4.50), we see that the observer gains should be chosen as

$$p_1(x) = -\varepsilon p_y(x,1), \qquad p_0 = -p(1,1), \tag{4.54}$$

where $p(x, y)$ satisfies the following hyperbolic PDE:

$$\varepsilon p_{yy}(x, y) - \varepsilon p_{xx}(x, y) = (\lambda(x) + \tilde{c})p(x, y), \tag{4.55}$$

with the boundary conditions

$$p_x(0, y) = -qp(0, y), \tag{4.56}$$

$$p(x, x) = -q - \frac{1}{2\varepsilon}\int_0^x (\lambda(\xi) + \tilde{c})\, d\xi. \tag{4.57}$$

Similarly to the anticollocated case, we introduce the new variables

$$\breve{x} = y, \quad \breve{y} = x, \quad \breve{p}(\breve{x}, \breve{y}) = p(x, y), \tag{4.58}$$

in which (4.55)–(4.57) becomes

$$\varepsilon \breve{p}_{\breve{x}\breve{x}}(\breve{x}, \breve{y}) - \varepsilon \breve{p}_{\breve{y}\breve{y}}(\breve{x}, \breve{y}) = (\lambda(\breve{y}) + \tilde{c})\breve{p}(\breve{x}, \breve{y}), \qquad (\breve{x}, \breve{y}) \in \mathcal{T} \tag{4.59}$$

$$\breve{p}_{\breve{y}}(\breve{x}, 0) = q\,\breve{p}(\breve{x}, 0), \tag{4.60}$$

$$\breve{p}(\breve{x}, \breve{x}) = -q - \frac{1}{2\varepsilon}\int_0^{\breve{x}} (\lambda(\xi) + \tilde{c})\, d\xi. \tag{4.61}$$

This is exactly the same PDE as (2.20), (2.21), (2.24) for $k(\breve{x}, \breve{y})$ (with $f(x, y) \equiv 0$, $g(x) \equiv 0$, and c replaced by $\tilde{c}$). Therefore, the existence and uniqueness of the solution of (4.55)–(4.57) and invertibility of the transformation (4.47) immediately follow. The duality between the observer and control design is even more evident here than in the anti-collocated case: the kernel of the observer transformation (4.47) is equal to the kernel of the control transformation (2.15) with switched variables, $p(x, y) = k(y, x)$ (for the same rate of convergence, i.e., $\tilde{c} = c$). The observer gains in the new coordinates are given by

$$p_1(x) = -\varepsilon\,\breve{p}_x(1, x), \qquad p_0 = -\breve{p}(1, 1). \tag{4.62}$$

For $\tilde{c} = c$, these gains are equal (up to a constant factor $-\varepsilon$) to the control gains.

We obtain the following result.

Theorem 4.4. *Let $p(x, y)$ be the solution of the system* (4.55)–(4.57). *For any initial data $\tilde{u}_0(x) \in H^1(0, 1)$ compatible with boundary conditions, the system* (4.44)–(4.46) *with $p_1(x)$ and p_0 given by* (4.54) *has a unique classical solution $\tilde{u} \in C^{2,1}((0, 1) \times (0, \infty))$ and is exponentially stable at the origin in $H^1(0, 1)$ with the decay rate $\tilde{c}$.*

4.4 NOTES AND REFERENCES

The observers presented in this chapter can be viewed as infinite-dimensional generalizations of Krener-Kang observers [62] for finite-dimensional nonlinear ODEs. In [62], Krener and Kang discover and exploit a triangular structure dual to that for the backstepping controller design [68]. The complexities present owing to nonlinearities in finite dimension make the Krener-Kang observer nonglobal; however, this limitation is not an issue in linear PDEs.

Other observer designs for linear PDEs include the infinite-dimensional Luenberger approach [29, 75]. A unified treatment of both interior and boundary observations and control generalized to semilinear problems can be found in [5]. Fuji [42] and Nambu [91] developed auxiliary functional observers to stabilize diffusion equations using boundary observation and feedback. For the Pritchard-Salamon class of state-space systems, several frequency domain stabilization results have been established (see, e.g., [26] and [82] for surveys). Christofides [22] developed nonlinear output-feedback controllers for parabolic PDE systems for which the eigenspectrum can be separated into a finite-dimensional slow part and an infinite-dimensional stable fast part.

Chapter Five

Output Feedback

The exponentially convergent observers developed in Chapter 4 are independent of the control input and can be used with any controller. In this chapter we combine the backstepping observers from Chapter 4 with the backstepping controllers developed in Chapter 2 to solve the output-feedback problems.

First, in Sections 5.1 and 5.2, respectively, we establish closed-loop stability results for observer-based backstepping controllers in anti-collocated and collocated configurations. These results are essentially "separation principle" results for the backstepping approach to output-feedback stabilization.

Then, in Section 5.3 we derive an explicit output-feedback law (in the state-space format) for a reaction-diffusion system with constant coefficients. Finally, in Section 5.4 we derive an explicit transfer function of an output-feedback compensator for an example of a parabolic PDE with a boundary-driven diffusion term.

5.1 ANTI-COLLOCATED SETUP

First, we establish a separation principle for an observer-based backstepping controller for the PDE class (2.9)–(2.11).

THEOREM 5.1. *For any initial data u_0, $\hat{u}_0 \in H^1(0, 1)$ compatible with boundary conditions, the system consisting of the plant*

$$u_t(x,t) = \varepsilon u_{xx}(x,t) + \lambda(x)u(x,t) + g(x)u(0,t) + \int_0^x f(x,y)u(y,t)\,dy, \tag{5.1}$$

$$u_x(0,t) = -qu(0,t), \tag{5.2}$$

$$u(1,t) = U(t), \tag{5.3}$$

the controller

$$U(t) = \int_0^1 k(1,y)\hat{u}(y,t)\,dy, \tag{5.4}$$

where $k(x, y)$ is the solution of (2.20), (2.21), (2.24), *and the observer*

$$\hat{u}_t(x,t) = \varepsilon \hat{u}_{xx}(x,t) + \lambda(x)\,\hat{u}(x,t) + g(x)\,u\,(0,t) + \int_0^x f(x,y)\hat{u}(y,t)\,dy$$
$$+ p_1(x)[u(0,t) - \hat{u}(0,t)], \tag{5.5}$$

$$\hat{u}_x(0,t) = -qu(0,t) + p_0[u(0,t) - \hat{u}(0,t)], \tag{5.6}$$

$$\hat{u}(1,t) = \int_0^1 k(1,y)\hat{u}(y,t)\,dy, \tag{5.7}$$

where $p_1(x)$, p_0 *are given by* (4.17)–(4.20), *has a unique classical solution* $u, \hat{u} \in C^{2,1}([0,1] \times (0,\infty))$ *and is exponentially stable at the origin in* $H^1(0,1)$,

$$\|u(t)\|_{H^1} + \|\hat{u}(t)\|_{H^1} \le M e^{-\left[\min(c,\tilde{c}) + \frac{\varepsilon\pi^2}{8}\right]t} \left(\|u_0\|_{H^1} + \|\hat{u}_0\|_{H^1}\right), \tag{5.8}$$

where $c, \tilde{c} \ge 0$ *and* $M > 0$ *is independent of initial conditions.*

Proof. One can show that the coordinate transformations

$$\widehat{w}(x,t) = \hat{u}(x,t) - \int_0^x k(x,y)\,\hat{u}(y,t)\,dy, \tag{5.9}$$

$$\tilde{u}(x,t) = \widetilde{w}(x,t) - \int_0^x p(x,y)\widetilde{w}(y,t)\,dy \tag{5.10}$$

map $(\hat{u}, \tilde{u})$ into the systems

$$\widehat{w}_t(x,t) = \varepsilon\widehat{w}_{xx}(x,t) - c\widehat{w}(x,t) + F(x)\widetilde{w}(0,t), \tag{5.11}$$

$$\widehat{w}_x(0,t) = (p_0 - q)\widetilde{w}(0,t), \tag{5.12}$$

$$\widehat{w}(1,t) = 0 \tag{5.13}$$

and

$$\widetilde{w}_t(x,t) = \varepsilon\widetilde{w}_{xx}(x,t) - \tilde{c}\widetilde{w}(x,t), \tag{5.14}$$

$$\widetilde{w}_x(0,t) = 0, \tag{5.15}$$

$$\widetilde{w}(1,t) = 0, \tag{5.16}$$

where

$$F(x) = p_1(x) + g(x) + \varepsilon(p_0 - q)k(x,0) - \int_0^x k(x,y)(p_1(y) + g(y))\,dy. \tag{5.17}$$

Let us introduce a new variable, $\overline{w} = \widehat{w} - x(p_0 - q)\widetilde{w}$. The purpose of this change of variables is to make boundary conditions homogeneous, so that we can use H^1-norms as Lyapunov functions. The PDE for $\bar{w}$ is

$$\begin{aligned} \overline{w}_t(x,t) = {} & \varepsilon\overline{w}_{xx}(x,t) - c\overline{w}(x,t) - x(c - \tilde{c})(p_0 - q)\widetilde{w}(x,t) \\ & + 2\varepsilon(p_0 - q)\widetilde{w}_x(x,t) + F(x)\widetilde{w}(0,t), \end{aligned} \tag{5.18}$$

$$\overline{w}_x(0,t) = 0, \tag{5.19}$$

$$\overline{w}(1,t) = 0. \tag{5.20}$$

For

$$V_1 = \frac{1}{2}\int_0^1 \overline{w}^2(x)\,dx + \frac{1}{2}\int_0^1 \overline{w}_x^2(x)\,dx, \tag{5.21}$$

we compute

$$\begin{aligned}\dot{V}_1 &= -\varepsilon(\|\overline{w}_x\|^2 + \|\overline{w}_{xx}\|^2) - c(\|\overline{w}\|^2 + \|\overline{w}_x\|^2) \\ &\quad + \int_0^1 (\overline{w}_{xx}(x) - \overline{w}(x))\big[(p_0 - q)(x(c-\tilde{c})\widetilde{w}(x) - 2\varepsilon\widetilde{w}_x(x)) - F(x)\widetilde{w}(0)\big]\, dx \\ &\le -\frac{\varepsilon}{2}(\|\overline{w}_x\|^2 + \|\overline{w}_{xx}\|^2) - 2cV_1 + A\|\widetilde{w}_x\|^2, \end{aligned} \tag{5.22}$$

where

$$A = \frac{12}{\varepsilon\pi^2}\left[\frac{4(c-\tilde{c})^2(p_0-q)^2}{\pi^2} + 4\varepsilon^2(p_0-q)^2 + \bar{F}^2\right], \quad \bar{F} = \max_{x\in[0,1]} F(x). \tag{5.23}$$

For

$$V_2 = \frac{1}{2}\int_0^1 \widetilde{w}^2(x)\, dx + \frac{1}{2}\int_0^1 \widetilde{w}_x^2(x)\, dx, \tag{5.24}$$

we already computed $\dot{V}_2$ in (2.53):

$$\dot{V}_2 \le -\varepsilon(\|\widetilde{w}_x\|^2 + \|\widetilde{w}_{xx}\|^2) - 2\tilde{c}V_2. \tag{5.25}$$

For the Lyapunov function $V = V_1 + BV_2$, where $B = 2A/\varepsilon$, we get

$$\begin{aligned}\dot{V} &\le -\frac{\varepsilon}{2}(\|\overline{w}_x\|^2 + \|\overline{w}_{xx}\|^2) - B\frac{\varepsilon}{2}(\|\widetilde{w}_x\|^2 + \|\widetilde{w}_{xx}\|^2) - 2cV_1 - 2B\tilde{c}V_2 \\ &\le -2\left[\min(c, \tilde{c}) + \frac{\varepsilon\pi^2}{8}\right]V. \end{aligned} \tag{5.26}$$

From (5.26) it follows that the system $(\overline{w}, \widetilde{w})$ is exponentially stable in $H^1(0, 1)$ with the decay rate $(\min(c, \tilde{c}) + \varepsilon\pi^2/8)$. Going through the chain of invertible transformations $(\overline{w}, \widetilde{w}) \to (\widehat{w}, \widetilde{w}) \to (\hat{u}, \tilde{u}) \to (u, \hat{u})$, we obtain (5.8). □

5.2 COLLOCATED SETUP

In this section we establish the separation principle for an observer-based backstepping controller for the PDE class (2.9)–(2.11) with $g(x) \equiv 0$, $f(x, y) \equiv 0$, namely, for reaction-advection-diffusion PDEs with nonconstant coefficients that have been converted, by a change of variable, into a reaction-diffusion class with nonconstant reactivity.

THEOREM 5.2. *For any initial data u_0, $\hat{u}_0 \in H^1(0,1)$ compatible with boundary conditions, the system consisting of the plant*

$$u_t(x,t) = \varepsilon u_{xx}(x,t) + \lambda(x)u(x,t), \tag{5.27}$$

$$u_x(0,t) = -qu(0,t), \tag{5.28}$$

$$u(1,t) = U(t), \tag{5.29}$$

the controller

$$U(t) = k(1,1)u(1,t) + \int_0^1 k_x(1,y)\hat{u}(y,t)\,dy, \tag{5.30}$$

where $k(x,y)$ is the solution of (2.20), (2.21), (2.24), *and the observer*

$$\hat{u}_t(x,t) = \varepsilon\hat{u}_{xx}(x,t) + \lambda(x)\,\hat{u}(x,t) + p_1(x)[u(1,t) - \hat{u}(1,t)], \tag{5.31}$$

$$\hat{u}_x(0,t) = -q\hat{u}(0,t), \tag{5.32}$$

$$\hat{u}_x(1,t) = k(1,1)u(1,t) + \int_0^1 k_x(1,y)\hat{u}(y,t)\,dy, \tag{5.33}$$

where $p_1(x)$, p_0 are given by (4.59)–(4.62), *has a unique classical solution u, $\hat{u} \in C^{2,1}([0,1] \times (0,\infty))$, and is exponentially stable at the origin in $H^1(0,1)$,*

$$\|u(t)\|_{H^1} + \|\hat{u}(t)\|_{H^1} \le Me^{-\min(c,\tilde{c})(1-\delta)t}\left(\|u_0\|_{H^1} + \|\hat{u}_0\|_{H^1}\right), \tag{5.34}$$

where c, $\tilde{c} > 0$, δ is an arbitrarily small positive constant, and $M > 0$ is a constant independent of initial conditions.

Proof. One can show that the coordinate transformations

$$\widehat{w}(x,t) = \hat{u}(x,t) - \int_0^x k(x,y)\,\hat{u}(y,t)\,dy, \tag{5.35}$$

$$\tilde{u}(x,t) = \widetilde{w}(x,t) - \int_x^1 p(x,y)\widetilde{w}(y,t)\,dy, \tag{5.36}$$

where $k(x,y)$ is the solution of (2.20), (2.21), (2.24) and $p(x,y)$ is the solution of (4.55)–(4.57), map $(\hat{u}, \tilde{u})$ into the systems

$$\widehat{w}_t(x,t) = \varepsilon\widehat{w}_{xx}(x,t) - c\widehat{w}(x,t) + F_1(x)\widetilde{w}(1,t), \tag{5.37}$$

$$\widehat{w}_x(0,t) = 0, \tag{5.38}$$

$$\widehat{w}_x(1,t) = p_0\widetilde{w}(1,t) \tag{5.39}$$

and

$$\widetilde{w}_t(x,t) = \varepsilon\widetilde{w}_{xx}(x,t) - \tilde{c}\widetilde{w}(x,t), \tag{5.40}$$

$$\widetilde{w}_x(0,t) = 0, \tag{5.41}$$

$$\widetilde{w}_x(1,t) = 0, \tag{5.42}$$

where $p_0 = -p(1,1)$ and

$$F_1(x) = p_1(x) - \int_0^x k(x,y)p_1(y)\,dy, \quad p_1(x) = -\varepsilon p_x(1,x). \tag{5.43}$$

The rest of the proof is similar to the proof of Theorem 5.1. □

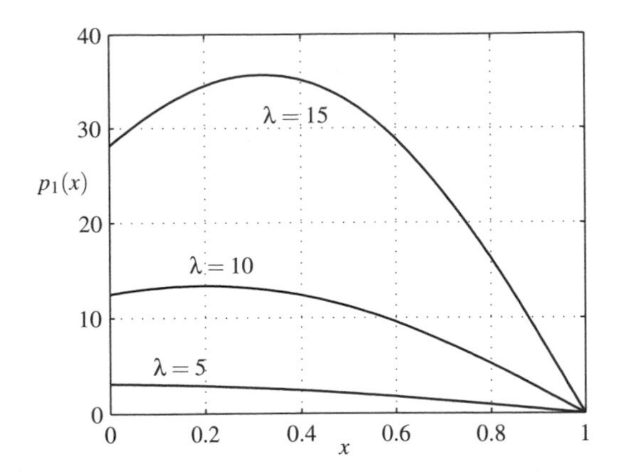

Figure 5.1 Observer gain (5.51) for different values of λ.

5.3 CLOSED-FORM COMPENSATORS

In this section we present an example of explicit output-feedback design.[1] Consider the reaction-diffusion equation

$$u_t(x,t) = \varepsilon u_{xx}(x,t) + \lambda_0 u(x,t), \tag{5.44}$$

$$u_x(0,t) = 0, \tag{5.45}$$

$$u(1,t) = U(t), \tag{5.46}$$

where only $u(0,t)$ is measured.

Observer design. The equation (4.23)–(4.25) for the observer gain takes the form

$$\breve{p}_{\breve{x}\breve{x}}(\breve{x},\breve{y}) - \breve{p}_{\breve{y}\breve{y}}(\breve{x},\breve{y}) = \lambda \breve{p}(\breve{x},\breve{y}), \tag{5.47}$$

$$\breve{p}(\breve{x},0) = 0, \tag{5.48}$$

$$\breve{p}(\breve{x},\breve{x}) = -\lambda \frac{\breve{x}}{2}, \tag{5.49}$$

where $\lambda = (\lambda_0 + c)/\varepsilon$. The solution to (5.47)–(5.49) is (see (3.4)–(3.6), (3.13))

$$\breve{p}(\breve{x},\breve{y}) = -\lambda \breve{y} \frac{I_1\left(\sqrt{\lambda(\breve{x}^2 - \breve{y}^2)}\right)}{\sqrt{\lambda(\breve{x}^2 - \breve{y}^2)}}. \tag{5.50}$$

Using (4.27) we obtain the observer gains

$$p_1(x) = \varepsilon \frac{\lambda(1-x)}{x(2-x)} I_2\left(\sqrt{\lambda x(2-x)}\right), \qquad p_0 = -\lambda/2. \tag{5.51}$$

In Figure 5.1 the observer gain $p_1(x)$ is shown for different values of the parameter λ. The exponential convergence of the observer for $\lambda = 5$ is illustrated in Figure 5.2. We can see that the observer converges to the plant even though the plant is unstable.

---

[1]For the sake of notational simplicity, we set $\tilde{c} = c$ in this section.

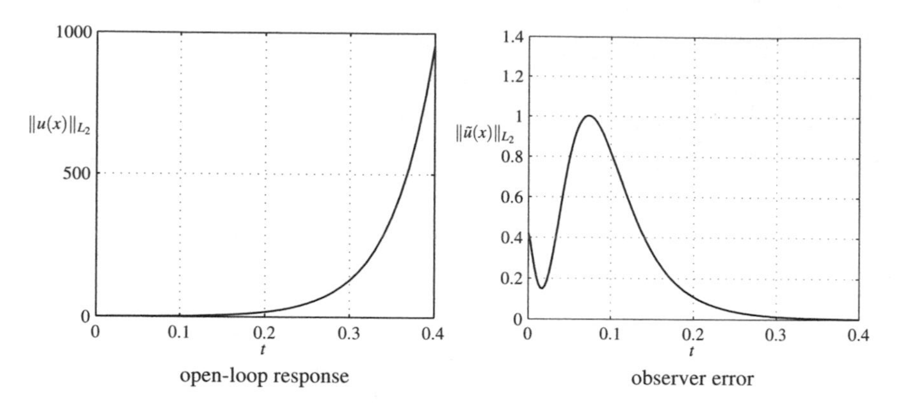

Figure 5.2 Exponential convergence of the observer for the reaction-diffusion equation (5.44)–(5.46).

Output-feedback compensator. We can now write the explicit solution to the output-feedback problem. The control gain kernel for the state-feedback problem is given by (3.22). Using (5.51) and Theorem 5.1, we get the following result.

Theorem 5.3. *The controller*

$$U(t) = -\int_0^1 \lambda \frac{I_1\left(\sqrt{\lambda(1-y^2)}\right)}{\sqrt{\lambda(1-y^2)}} \hat{u}(y,t)\,dy \tag{5.52}$$

with the observer

$$\begin{aligned} \hat{u}_t(x,t) = {} & \varepsilon \hat{u}_{xx}(x,t) + \lambda_0 \hat{u}(x,t) \\ & + \varepsilon \frac{\lambda(1-x)}{x(2-x)} I_2\left(\sqrt{\lambda x(2-x)}\right)[u(0,t) - \hat{u}(0,t)], \end{aligned} \tag{5.53}$$

$$\hat{u}_x(0,t) = -\frac{\lambda}{2}[u(0,t) - \hat{u}(0,t)], \tag{5.54}$$

$$\hat{u}(1,t) = U(t) \tag{5.55}$$

achieves exponential stability of the system (5.44)–(5.46).

The above result can easily be extended for the Neumann type of actuation.

The closed-loop system has been simulated with $\varepsilon = 1$, $\lambda_0 = 10$, $c = 5$, and $u(x,0) = 2e^{-2x}\sin(\pi x)$. With this choice of parameters the open-loop system has two unstable eigenvalues. The plant and the observer are discretized using a finite-difference method. Since designs exist where, in principle, the order of the observer can be as low as the number of unstable eigenvalues, we design the low-order compensator by taking a coarse six-point grid (keeping the fine discretization of the plant, 100 points in our case, for simulation). In Figure 5.3 the pole-zero map and Bode plots of the low-order compensator are shown. The reduced-order compensator is able to stabilize the system (Fig. 5.4).

Closed-loop solution. With every part of our design being explicit, we can even write the closed-loop solution of the system (5.44)–(5.46) together with compensator (5.52)–(5.55) explicitly, in terms of the initial conditions $u_0(x)$, $\hat{u}_0(x)$.

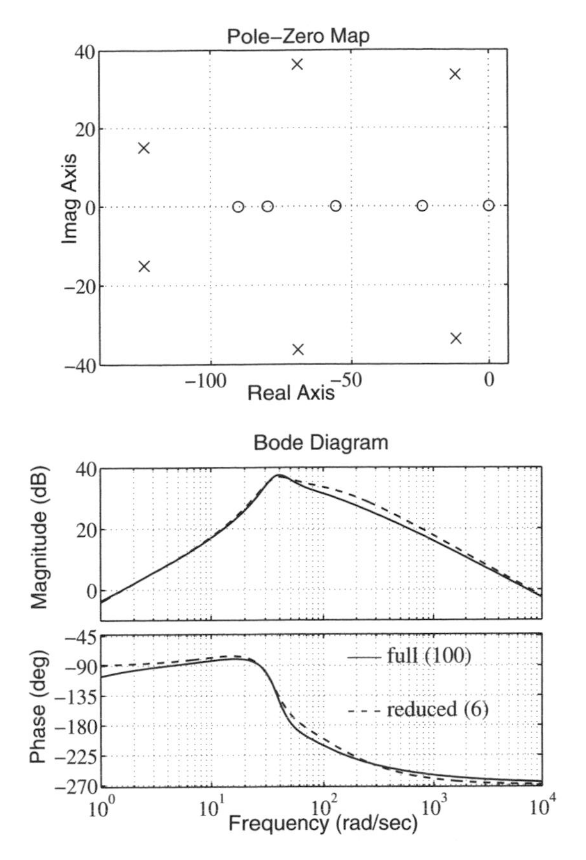

Figure 5.3 Pole-zero map and Bode plot of the compensator for a reaction-diffusion equation.

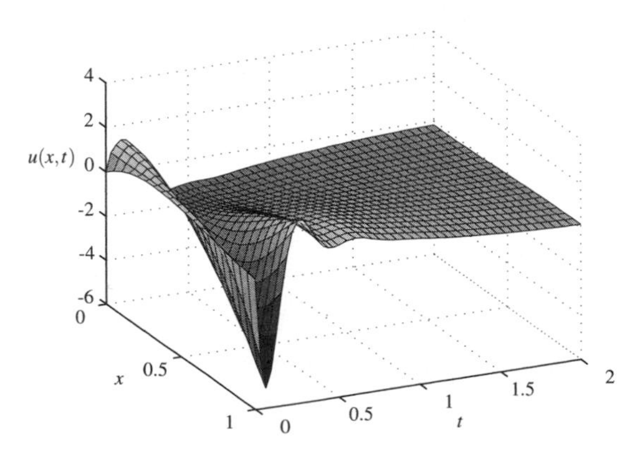

Figure 5.4 Closed-loop response with a low-order compensator.

THEOREM 5.4. *The solution to the closed-loop system (5.44)–(5.46), (5.52)–(5.55) is*

$$u(x,t) = 2\sum_{n=0}^{\infty} e^{-(c+\varepsilon\mu_n^2)t} \cos\sqrt{\lambda+\mu_n^2}x \left\{ \int_0^1 \psi_n(\xi)u_0(\xi)\,d\xi + \varepsilon\mu_n(-1)^n C_n t \right.$$

$$\left. +\mu_n(-1)^n \sum_{m\neq n}^{\infty} C_m \frac{1-e^{\varepsilon(\mu_n^2-\mu_m^2)t}}{\mu_n^2-\mu_m^2} \right\}, \tag{5.56}$$

where $\mu_n = \pi(n+1/2)$,

$$C_n = 2\left(\int_0^1 \lambda \frac{I_1\left(\sqrt{\lambda\xi(2-\xi)}\right)}{\sqrt{\lambda\xi(2-\xi)}}\psi_n'(\xi)\,d\xi\right)$$

$$\times\left(\int_0^1 \frac{\sin\sqrt{\lambda+\mu_n^2}(1-\xi)}{\sqrt{\lambda+\mu_n^2}}(u_0(\xi)-\hat{u}_0(\xi))\,d\xi\right), \tag{5.57}$$

$$\psi_n(x) = \cos(\mu_n x) + \int_0^x \lambda\xi \frac{I_1\left(\sqrt{\lambda(x^2-\xi^2)}\right)}{\sqrt{\lambda(x^2-\xi^2)}}\cos(\mu_n\xi)\,d\xi\,. \tag{5.58}$$

Proof. We start by solving the damped heat equation (5.14)–(5.16):

$$\widetilde{w}(x,t) = 2\sum_{n=0}^{\infty} e^{-(c+\varepsilon\mu_n^2)t}\cos(\mu_n x)\int_0^1 \widetilde{w}_0(\xi)\cos(\mu_n\xi)\,d\xi, \tag{5.59}$$

where $\mu_n = \pi n + \pi/2$. The initial condition $\widetilde{w}_0$ can be calculated from $\tilde{u}_0$ via the inverse of (5.10). Substituting (5.59) into (5.10), changing the order of integration, and calculating some of the integrals we obtain

$$\tilde{u}(x,t) = 2\sum_{n=0}^{\infty} e^{-(c+\varepsilon\mu_n^2)t}\frac{\mu_n}{\lambda+\mu_n^2}\sin\left(\sqrt{\lambda+\mu_n^2}(1-x)\right)$$

$$\times\int_0^1\left[\sin(\mu_n\xi) + \int_\xi^1 \lambda\xi\frac{I_1\left(\sqrt{y^2-\xi^2}\right)}{\sqrt{y^2-\xi^2}}\sin(\mu_n y)\,dy\right]\tilde{u}_0(1-\xi)\,d\xi. \tag{5.60}$$

The transformation (2.15) with the kernel (3.22) maps the plant into

$$w_t(x,t) = w_{xx}(x,t) - cw(x,t), \tag{5.61}$$

$$w_x(0,t) = 0, \tag{5.62}$$

$$w(1,t) = -\int_0^1 k(1,y)\tilde{u}(y,t)\,dy \equiv d(t). \tag{5.63}$$

The solution of this PDE is

$$w(x,t) = 2\sum_{n=0}^{\infty} e^{-(c+\varepsilon\mu_n^2)t}\cos(\mu_n x)\left(\int_0^1 w_0(\xi)\cos(\mu_n\xi)\,d\xi\right.$$

$$\left. +(-1)^n\varepsilon\mu_n\int_0^t e^{(c+\varepsilon\mu_n^2)\tau}d(\tau)\,d\tau\right). \tag{5.64}$$

The initial condition w_0 can be calculated explicitly from u_0 via (2.15). Substituting (5.64) into the inverse transformation (2.46) and calculating the integrals, we obtain (5.56)–(5.58). □

5.4 FREQUENCY DOMAIN COMPENSATOR

The solutions obtained in previous sections can be used to get explicit compensator transfer functions (treating $u(0,t)$ or $u(1,t)$ as an input and $u(1,t)$ or $u_x(1,t)$ as an output). We illustrate this point with the following system, inspired by a solid propellant rocket model [16]:

$$u_t(x,t) = u_{xx}(x,t) + gu(0,t), \tag{5.65}$$

$$u_x(0,t) = 0, \tag{5.66}$$

$$u(1,t) = U(t). \tag{5.67}$$

The observer

$$\hat{u}_t(x,t) = \hat{u}_{xx}(x,t) + gu(0,t), \tag{5.68}$$

$$\hat{u}_x(0,t) = 0, \tag{5.69}$$

$$\hat{u}(1,t) = U(t), \tag{5.70}$$

with direct injection of the reaction term $gu(0,t)$, is exponentially convergent. The stabilizing controller, whose state-feedback version was found in Chapter 3, is

$$u(1,t) = U(t) = -\sqrt{g}\int_0^1 \sinh(\sqrt{g}(1-y))\hat{u}(y,t)\,dy. \tag{5.71}$$

We want to find a transfer function from the input $u(0,t)$ to the output $u(1,t)$, that is, $u(1,s) = -C(s)u(0,s)$. Taking the Laplace transform of (5.68)–(5.70) and setting the initial condition to zero, $\hat{u}(x,0) = 0$, we have (for simplicity of notation we denote by $\hat{u}(x,s)$ and $u(0,s)$ the Laplace transforms of $\hat{u}(x,t)$ and $u(0,t)$, respectively):

$$s\hat{u}(x,s) = \hat{u}_{xx}(x,s) + gu(0,s), \tag{5.72}$$

$$\hat{u}_x(0,s) = 0, \tag{5.73}$$

$$\hat{u}(1,s) = -\sqrt{g}\int_0^1 \sinh(\sqrt{g}(1-y))\hat{u}(y,s)\,dy. \tag{5.74}$$

The equation (5.72) with boundary conditions (5.73)–(5.74) is a second-order ODE with respect to x (we regard s as a parameter). The solution of (5.72) satisfying (5.73) is:

$$\hat{u}(x,s) = \hat{u}(0,s)\cosh(\sqrt{s}x) + \frac{g}{s}(1-\cosh(\sqrt{s}x))u(0,s). \tag{5.75}$$

Using boundary condition (5.74), we obtain $\hat{u}(0,s)$:

$$\hat{u}(0,s) = \frac{\cosh(\sqrt{s}) - \cosh(\sqrt{g})}{s\cosh(\sqrt{s}) - g\cosh(\sqrt{g})}\,gu(0,s). \tag{5.76}$$

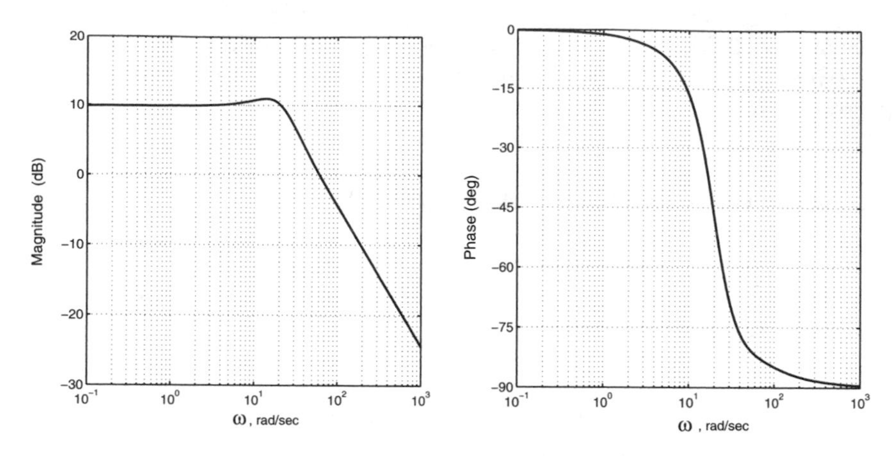

Figure 5.5 Bode plot of $C(j\omega)$ for $g = 8$.

Substituting now (5.76) into (5.75) with $x = 1$, we obtain the following result.

THEOREM 5.5. *The transfer function of the system* (5.68)–(5.71) *with* $u(0, t)$ *as an input and* $u(1, t)$ *as an output is*

$$C(s) = \frac{g}{s}\left(-1 + \frac{(s-g)\cosh(\sqrt{s})\cosh(\sqrt{g})}{s\cosh(\sqrt{s}) - g\cosh(\sqrt{g})}\right). \tag{5.77}$$

Note that $s = 0$ is not a pole:

$$C(0) = \frac{g}{2} + \frac{1}{\cosh(\sqrt{g})} - 1. \tag{5.78}$$

The transfer function (5.77) has infinitely many poles; all of them are real and negative. The Bode plots of $C(s)$ for $g = 8$ are presented in Figure 5.5. It is evident from the Bode plots that $C(s)$ can be approximated by a second-order, relative degree one transfer function. For example, a pretty good estimate would be

$$C(s) \approx 60\frac{s + 17}{s^2 + 25s + 320}. \tag{5.79}$$

The relative degree one nature of the compensator is the result of employing a full-order (rather than a reduced-order) observer.

5.5 NOTES AND REFERENCES

The material presented in this section is based on [109]. The validation of application of the procedure in Section 5.4 for linear parabolic PDEs (which proves that $C(s)$ is indeed a transfer function) can be found in [29, Chap. 4].

Chapter Six

Control of Complex-Valued PDEs

In this chapter we extend the designs developed in Chapters 2–5 to the case of plants with a complex-valued state. Such plants can also be viewed as coupled parabolic PDEs. We consider two classes of complex-valued PDEs, the Ginzburg-Landau equation and its special case, the Schrödinger equation. For the Schrödinger equation, which we treat as a single complex-valued equation, the controllers and observers are obtained in closed form. The Ginzburg-Landau equation is treated as two coupled PDEs and serves as an example of the extension of the backstepping designs to (semi-)infinite domains. This is not trivial, since control and observer gains grow exponentially with the size of the domain. However, it is possible to choose the desired behavior of the closed-loop system in such a way that the gains have compact support.

In the sequel, "j" denotes the imaginary unit, and $\mathrm{Re}\{\cdot\}$, $\mathrm{Im}\{\cdot\}$ stand for real and imaginary parts, respectively.

6.1 STATE-FEEDBACK DESIGN FOR THE SCHRÖDINGER EQUATION

The linear Schrödinger equation is

$$\psi_t(x,t) = -j\psi_{xx}(x,t), \quad 0 < x < 1, \tag{6.1}$$

$$\psi_x(0,t) = 0, \tag{6.2}$$

$$\psi(1,t) = U(t), \tag{6.3}$$

where $U(t)$ is the control input and $\psi(x,t)$ is a complex-valued state. Without control, this system displays oscillatory behavior and is not asymptotically stable.

Treating (6.1)–(6.3) formally as a heat equation with imaginary diffusivity allows us to solve the stabilization problem using the approach presented in Chapter 3 (Section 3.1). Consider the transformation

$$v(x,t) = \psi(x,t) - \int_0^x k(x,y)\psi(y,t)\,dy, \tag{6.4}$$

where $k(x,y)$ is a complex-valued function that satisfies the PDE

$$k_{xx}(x,y) - k_{yy}(x,y) = cjk(x,y), \tag{6.5}$$

$$k_y(x,0) = 0, \tag{6.6}$$

$$k(x,x) = -\frac{cj}{2}, \tag{6.7}$$

with $c > 0$. As shown in Section 3.1, (6.4)–(6.7) with the control law

$$U(t) = \int_0^1 k(1, y)\psi(y, t)\, dy \tag{6.8}$$

maps (6.1)–(6.3) into the system

$$v_t(x, t) = -jv_{xx}(x, t) - cv(x, t), \tag{6.9}$$

$$v_x(0, t) = 0, \tag{6.10}$$

$$v(1, t) = 0. \tag{6.11}$$

The eigenvalues of this system are

$$\sigma = -c + j\frac{\pi^2(2n + 1)^2}{4}, \quad n = 0, 1, 2, \ldots; \tag{6.12}$$

therefore, the design parameter c allows moving them arbitrarily to the left in the complex plane.

The solution to the PDE (6.5)–(6.7) is (see (3.22)):

$$\begin{aligned} k(x, y) &= -cjx\frac{I_1\left(\sqrt{cj(x^2 - y^2)}\right)}{\sqrt{cj(x^2 - y^2)}} \\ &= x\sqrt{\frac{c}{2(x^2 - y^2)}}\left[(j - 1)\mathrm{ber}_1\left(\sqrt{c(x^2 - y^2)}\right)\right. \\ &\quad \left. - (1 + j)\mathrm{bei}_1\left(\sqrt{c(x^2 - y^2)}\right)\right]. \end{aligned} \tag{6.13}$$

Here, $\mathrm{ber}_1(\cdot)$ and $\mathrm{bei}_1(\cdot)$ are the Kelvin functions, which are defined in terms of I_1 as

$$\mathrm{ber}_1(x) = -\mathrm{Im}\left\{I_1\left(\frac{1 + j}{\sqrt{2}}x\right)\right\}, \qquad \mathrm{bei}_1(x) = \mathrm{Re}\left\{I_1\left(\frac{1 + j}{\sqrt{2}}x\right)\right\}. \tag{6.14}$$

The precise statement of stability of the closed-loop system (6.1), (6.3)–(6.8) is given by the following theorem.

Theorem 6.1. *Consider the system*

$$\frac{d}{dt}\psi(\cdot, t) = A\psi(\cdot, t) \tag{6.15}$$

on $H = L^2(0, 1)$, *where the operator* A *is defined as*

$$A\varphi(x) = -j\varphi''(x), \quad \varphi \in D(A), \tag{6.16}$$

$$D(A) = \left\{\varphi \mid \varphi'(0) = 0, \varphi(1) = \int_0^1 k(1, y)\varphi(y)\, dy\right\} \tag{6.17}$$

and k is given by (6.13). *Then A is a Riesz spectral operator in H, and the following statements hold:*

(i) *The eigenvalues of A are given by* (6.12).

(ii) *The spectrum-determined growth condition holds true for the semigroup* e^{At}.

(iii) *The semigroup* e^{At} *is exponentially stable in the sense*

$$\|e^{At}\| \leq Me^{-ct}, \quad \forall\, t \geq 0,$$

where $M > 0$ *and c is an arbitrary positive design parameter.*

Proof. The main idea of the proof is first to establish well-posedness and stability of the target system (6.9)–(6.11) and then to use the fact that the transformation (6.4) is invertible to get well-posedness and stability of the closed-loop system. One can write (6.9)–(6.11) as

$$\frac{d}{dt}v(\cdot, t) = Bv(\cdot, t) \tag{6.18}$$

on H, where the operator B is defined by

$$\begin{aligned} B\varphi(x) &= -j\varphi''(x) - c\varphi(x), \quad \varphi \in D(B), \\ D(B) &= \{\varphi |\, \varphi'(0) = \varphi(1) = 0\}. \end{aligned} \tag{6.19}$$

A simple computation shows that B is a self-adjoint operator in H with eigenvalues given by (6.12) and eigenfunctions

$$\phi_n(x) = \cos\left(n + \frac{1}{2}\right)\pi n, \quad n = 0, 1, 2, \ldots, \tag{6.20}$$

which form an orthonormal basis. This shows that the spectrum-determined growth condition holds for (6.18). Therefore, there exists a constant $M_1 > 0$ such that

$$\|e^{Bt}\| \leq M_1 e^{-ct}, \quad \forall\, t \geq 0. \tag{6.21}$$

One can show by direct substitution that the transformation

$$\psi(x, t) = v(x, t) + \int_0^x l(x, y)v(y, t)\,dy \tag{6.22}$$

with

$$\begin{aligned} l(x, y) &= -cjx\frac{J_1\left(\sqrt{cj(x^2 - y^2)}\right)}{\sqrt{cj(x^2 - y^2)}} \\ &= x\sqrt{\frac{c}{2(x^2 - y^2)}}\left[(j + 1)\mathrm{ber}_1\left(\sqrt{c(x^2 - y^2)}\right)\right. \\ &\quad \left. + (1 - j)\mathrm{bei}_1\left(\sqrt{c(x^2 - y^2)}\right)\right] \end{aligned} \tag{6.23}$$

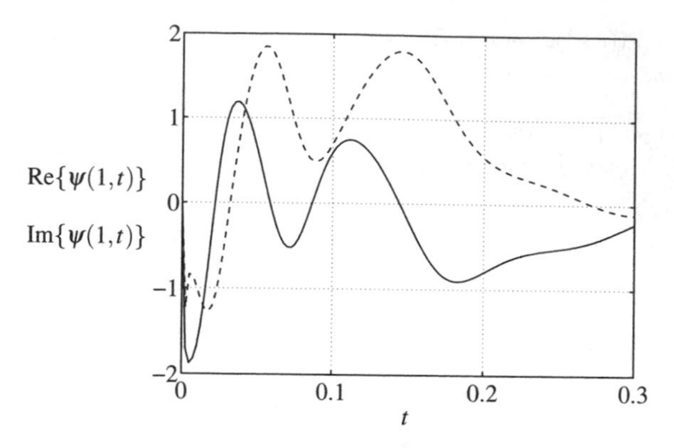

Figure 6.1 Real (solid line) and imaginary (dashed line) parts of the control effort for the Schrödinger equation.

is inverse to the transformation (6.4). The eigenfunctions of the closed-loop system are given by

$$\phi^c(x) = \phi(x) + \int_0^x l(x, y)\phi(y)\,dy. \tag{6.24}$$

It is easy to see that $\phi^c \in D(A)$ if and only if $\phi \in D(B)$, and (μ, ϕ^c) is an eigenpair of A if and only if (μ, ϕ) is an eigenpair of B. These facts together with (6.22) and (6.12) give the properties of A stated in the theorem. □

The results of numerical simulation of the above design are presented in Figures 6.2 and 6.1. The control effort is shown in Figure 6.1 and the open-loop and closed-loop behavior of the Schrödinger equation are shown (only for the real part of the state; the imaginary part is similar) in Figure 6.2.

6.2 OBSERVER DESIGN FOR THE SCHRÖDINGER EQUATION

The controller (6.8) relies on full-state measurements. Let us assume now that only boundary measurements are available (so that $\psi_x(1)$ is measured and $\psi(1)$ is actuated). We denote the estimate of the state by $\widehat{\psi}$ and use the observer

$$\widehat{\psi}_t(x,t) = -j\widehat{\psi}_{xx}(x,t) + p_1(x)[\psi_x(1,t) - \widehat{\psi}_x(1,t)], \tag{6.25}$$

$$\widehat{\psi}_x(0,t) = 0, \tag{6.26}$$

$$\widehat{\psi}(1,t) = \psi(1,t), \tag{6.27}$$

which is in the familiar form of the copy of the plant (6.1)–(6.3) plus output injection, with the observer gain $p_1(x)$ to be designed.

The observer error $\widetilde{\psi} = \psi - \widehat{\psi}$ satisfies the PDE

$$\widetilde{\psi}_t(x,t) = -j\widetilde{\psi}_{xx}(x,t) - p_1(x)\widetilde{\psi}_x(1,t), \tag{6.28}$$

$$\widetilde{\psi}_x(0,t) = 0, \tag{6.29}$$

$$\widetilde{\psi}(1,t) = 0. \tag{6.30}$$

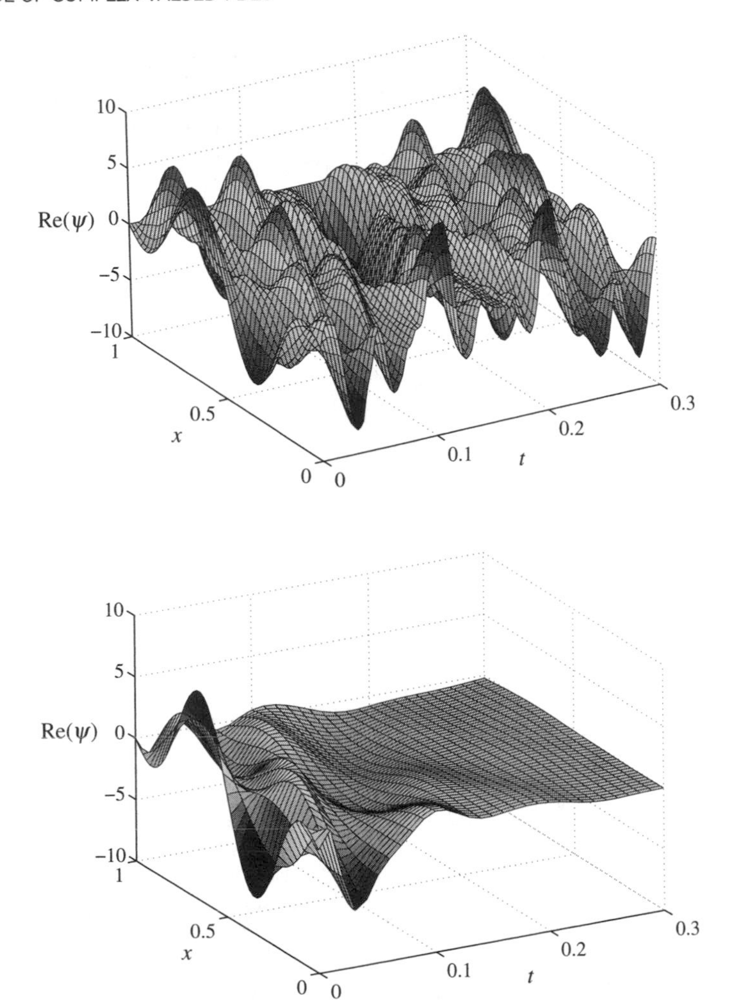

Figure 6.2 The open-loop (top) and closed-loop (bottom) response of the Schrödinger equation.

Consider the transformation

$$\widetilde{\psi}(x,t) = \widetilde{w}(x,t) + \int_x^1 p(x,y)\widetilde{w}(y,t)\,dy. \tag{6.31}$$

Note that the integral runs from x to 1 here, which is the consequence of the collocated input-output architecture. Let us select the following target system for the observer error:

$$\widetilde{w}_t(x,t) = -j\widetilde{w}_{xx}(x,t) - \tilde{c}\widetilde{w}(x,t), \tag{6.32}$$

$$\widetilde{w}_x(0,t) = 0, \tag{6.33}$$

$$\widetilde{w}(1,t) = 0. \tag{6.34}$$

The design parameter $\tilde{c}$ sets the desired observer convergence rate (which usually needs to be faster than the closed-loop system decay rate). Substituting the transformation (6.31) into (6.28)–(6.30) and matching the terms, we get the PDE for $p(x, y)$,

$$p_{xx}(x, y) - p_{yy}(x, y) = -j\tilde{c}p(x, y), \tag{6.35}$$

$$p_x(0, y) = 0, \tag{6.36}$$

$$p(x, x) = j\frac{\tilde{c}}{2}x, \tag{6.37}$$

and the condition for observer gain $p_1(x)$,

$$p_1(x) = jp(x, 1). \tag{6.38}$$

With a change of variables $\bar{x} = y$, $\bar{y} = x$, $\bar{p}(\bar{x}, \bar{y}) = p(x, y)$ we convert (6.35)–(6.37) into

$$\bar{p}_{\bar{x}\bar{x}}(\bar{x}, \bar{y}) - \bar{p}_{\bar{y}\bar{y}}(\bar{x}, \bar{y}) = j\tilde{c}\bar{p}(\bar{x}, \bar{y}), \tag{6.39}$$

$$\bar{p}_{\bar{y}}(\bar{x}, 0) = 0, \tag{6.40}$$

$$\bar{p}(\bar{x}, \bar{x}) = j\frac{\tilde{c}}{2}\bar{x}. \tag{6.41}$$

The PDE (6.39)–(6.41) has the solution (see (3.22))

$$\begin{aligned}\bar{p}(\bar{x}, \bar{y}) &= j\tilde{c}\bar{x}\frac{I_1\left(\sqrt{\tilde{c}j(\bar{x}^2 - \bar{y}^2)}\right)}{\sqrt{\tilde{c}j(\bar{x}^2 - \bar{y}^2)}} \\ &= \tilde{c}\bar{x}\sqrt{\frac{\tilde{c}}{2(\bar{x}^2 - \bar{y}^2)}}\left[(1 - j)\mathrm{ber}_1\left(\sqrt{\tilde{c}(\bar{x}^2 - \bar{y}^2)}\right)\right. \\ &\quad \left. + (1 + j)\mathrm{bei}_1\left(\sqrt{\tilde{c}(\bar{x}^2 - \bar{y}^2)}\right)\right].\end{aligned} \tag{6.42}$$

Using (6.38), we get the observer gain

$$\begin{aligned}p_1(x) = j\bar{p}(1, x) &= \sqrt{\frac{\tilde{c}}{2(1 - x^2)}}\left[(1 + j)\mathrm{ber}_1\left(\sqrt{\tilde{c}(1 - x^2)}\right)\right. \\ &\quad \left. + (j - 1)\mathrm{bei}_1\left(\sqrt{\tilde{c}(1 - x^2)}\right)\right].\end{aligned} \tag{6.43}$$

Note that when $\tilde{c} = c$, we have $p_1(x) = jk(1, x)$, which is the well-known duality property between observer and control gains.

The result of this section is summarized in the following theorem.

Theorem 6.2. *Consider the system*

$$\frac{d}{dt}\tilde{\psi}(\cdot, t) = \tilde{A}\tilde{\psi}(\cdot, t) \tag{6.44}$$

on $H = L^2(0,1)$, where the operator $\tilde{A}$ is defined as

$$\tilde{A}\varphi(x) = -j\varphi''(x) - p_1(x)\varphi'(1), \quad \varphi \in D(\tilde{A}), \tag{6.45}$$
$$D(\tilde{A}) = \left\{\varphi \mid \varphi'(0) = 0,\ \varphi(1) = 0\right\}$$

and $p_1(x)$ is given by (6.43). Then $\tilde{A}$ is a Riesz spectral operator in H, and the following statements hold:

(i) The eigenvalues of $\tilde{A}$ are

$$\sigma = -\tilde{c} + j\frac{\pi^2(2n+1)^2}{4}, \qquad n = 0, 1, 2, \ldots. \tag{6.46}$$

(ii) The spectrum-determined growth condition holds true for the semigroup $e^{\tilde{A}t}$.

(iii) The semigroup $e^{\tilde{A}t}$ is exponentially stable in the sense

$$\|e^{\tilde{A}t}\| \le M_2 e^{-\tilde{c}t}, \qquad \forall\, t \ge 0,$$

where $M_2 > 0$ and $\tilde{c}$ is an arbitrary positive design parameter.

Proof. The proof closely follows the proof of Theorem 6.1. We define the operator $\tilde{B}$ as

$$\tilde{B}\varphi(x) = -j\varphi''(x) - \tilde{c}\varphi(x), \quad \varphi \in D(\tilde{B}),$$
$$D(\tilde{B}) = \{\varphi \mid \varphi'(0) = \varphi(1) = 0\}, \tag{6.47}$$

and write the system (6.32)–(6.34) on H as

$$\frac{d}{dt}\widetilde{w}(\cdot,t) = \tilde{B}\widetilde{w}(\cdot,t). \tag{6.48}$$

Using the same argument as in the proof of Theorem 6.1, we get the result that for any $\widetilde{w}(\cdot,0) \in D(\tilde{B})$, there exists a unique C_0-semigroup solution $\widetilde{w}(\cdot,t) = e^{\tilde{B}t}\widetilde{w}(\cdot,0) \in D(\tilde{B})$ such that

$$\|\widetilde{w}(\cdot,t)\| = \|e^{\tilde{B}t}\widetilde{w}(\cdot,0)\| \le \tilde{M}_1 e^{-\tilde{c}t}\|\widetilde{w}(\cdot,0)\|. \tag{6.49}$$

The statements of the theorem follow from (6.49) and the transformation (6.31). □

6.3 OUTPUT-FEEDBACK COMPENSATOR FOR THE SCHRÖDINGER EQUATION

The combination of the observer (6.25)–(6.27) with the state-feedback controller developed in Section 6.1 gives the output-feedback controller

$$U(t) = \int_0^1 k(1,y)\widehat{\psi}(y,t)\,dy. \tag{6.50}$$

The following theorem establishes stability of the closed-loop system.

THEOREM 6.3. *For any $\widetilde{w}(\cdot,0) \in D(\tilde{B})$, $v(\cdot,0) \in H$, there exists a unique solution to* (6.58)–(6.62) *such that $(\widetilde{w}, v) \in C((0,\infty); D(\tilde{B}) \times H)$, and*

$$\|(\widetilde{w}(\cdot,t), v(\cdot,t))\|_{H\times H} \leq K \begin{cases} e^{-(\min\{c,\tilde{c}\})t} \; if\ c \neq \tilde{c}, \\ (1+t)e^{-ct} \; if\ c = \tilde{c} \end{cases} \|(\tilde{B}\widetilde{w}(\cdot,0), v(\cdot,0))\|_{H\times H} \tag{6.51}$$

for some constant $K > 0$.

Proof. Under the feedback (6.50), the plant (6.1)–(6.3) becomes

$$\psi_t(x,t) = -j\psi_{xx}(x,t), \tag{6.52}$$

$$\psi_x(0,t) = 0, \tag{6.53}$$

$$\psi(1,t) = \int_0^1 k(1,y)\psi(y,t)\,dy - \int_0^1 k(1,y)\widetilde{\psi}(y,t)\,dy, \tag{6.54}$$

where $\widetilde{\psi}$ is the solution of (6.28)–(6.30). Applying the transformation (6.4), we obtain

$$v_t(x,t) = -jv_{xx}(x,t) - cv(x,t), \tag{6.55}$$

$$v_x(0,t) = 0, \tag{6.56}$$

$$v(1,t) = -\int_0^1 k(1,y)\widetilde{\psi}(y,t)\,dy. \tag{6.57}$$

Now $\widetilde{\psi}$ in (6.57) can be expressed through $\tilde{w}$ using the transformation (6.31), and we get the following system $(v, \tilde{w})$ in $H \times H$, equivalent to the closed-loop system $(\psi, \widehat{\psi})$:

$$v_t(x,t) = -jv_{xx}(x,t) - cv(x,t), \tag{6.58}$$

$$v_x(0,t) = 0, \tag{6.59}$$

$$v(1,t) = \int_0^1 L(x)\widetilde{w}(x,t)\,dx \triangleq f(t), \tag{6.60}$$

$$\widetilde{w}_t(x,t) = -j\widetilde{w}_{xx}(x,t) - \tilde{c}\widetilde{w}(x,t), \tag{6.61}$$

$$\widetilde{w}_x(0,t) = \widetilde{w}(1,t) = 0, \tag{6.62}$$

where $L(x) = -k(1,x) - \int_0^x k(1,s)p(s,x)ds \in C^2(0,1)$. From (6.48) and (6.49) we get (note that $\tilde{B}$ commutes with $e^{\tilde{B}t}$):

$$\|\widetilde{w}_t(\cdot,t)\| = \|\tilde{B}e^{\tilde{B}t}\widetilde{w}(\cdot,0)\| \leq \tilde{M}_1 e^{-\tilde{c}t}\|\tilde{B}\widetilde{w}(\cdot,0)\|. \tag{6.63}$$

Therefore, $f'(t) \in C(0,\infty)$ and

$$|f(t)| \leq C_1 e^{-\tilde{c}t}\|\widetilde{w}(\cdot,0)\|, \qquad |f'(t)| \leq C_2 e^{-\tilde{c}t}\|\tilde{B}\widetilde{w}(\cdot,0)\| \tag{6.64}$$

for some positive constants C_1, C_2. Let

$$w(x,t) = v(x,t) - f(t). \tag{6.65}$$

Then w satisfies

$$w_t(x,t) = -jw_{xx}(x,t) - cw(x,t) - f'(t) - cf(t), \tag{6.66}$$

$$w_x(0,t) = w(1,t) = 0. \tag{6.67}$$

Since the inhomogeneous term $f' + cf \in C(0,\infty)$, for any $w(\cdot,0) \in H$ there exists a unique solution to (6.66)–(6.67) such that

$$w(\cdot,t) = e^{Bt}w(\cdot,0) - \int_0^t e^{B(t-s)}I(\cdot)[f'(s) + cf(s)]\,ds, \tag{6.68}$$

where $I(x) = 1$ for all $x \in [0,1]$. Using (6.21) and (6.64), we have

$$\begin{aligned}\|w(\cdot,t)\| &\le M_1 e^{-ct}\|w(\cdot,0)\| + M_1 \int_0^t e^{-c(t-s)}e^{-\tilde{c}s}[C_2\|\tilde{B}\widetilde{w}(\cdot,0)\| \\ &\quad + cC_1\|\widetilde{w}(\cdot,0)\|]\,ds \le M_1 e^{-ct}\|w(\cdot,0)\| + M_1[C_2\|\tilde{B}\widetilde{w}(\cdot,0)\| \\ &\quad + cC_1\|\widetilde{w}(\cdot,0)\|]\cdot \begin{cases} te^{-ct} & \text{if } c = \tilde{c}, \\ \dfrac{1}{c-\tilde{c}}[e^{-\tilde{c}t} - e^{-ct}] & \text{if } c \neq \tilde{c}. \end{cases}\end{aligned} \tag{6.69}$$

Since $\|v(\cdot,t)\| \le \|w(\cdot,t)\| + |f(t)|$ and $\|w(\cdot,0)\| \le \|v(\cdot,0)\| + |f(0)|$, the estimate (6.51) follows from (6.69) and (6.64). □

6.4 THE GINZBURG-LANDAU EQUATION

In flows past submerged obstacles, the phenomenon of vortex shedding occurs when the Reynolds number is sufficiently large. A popular prototype model flow for studying vortex shedding is the flow past a 2D circular cylinder, shown in Figure 6.3. The vortices, which are alternately shed from the upper and lower sides of the cylinder, induce an undesirable periodic force that acts on the cylinder. The dynamics of the cylinder wake, often referred to as the von Kármán vortex street, is governed by the Navier-Stokes equation; however, a simplified model exists in the form of the Ginzburg-Landau equation [50, 103],

$$\frac{\partial A}{\partial t} = a_1\frac{\partial^2 A}{\partial \check{x}^2} + a_2(\check{x})\frac{\partial A}{\partial \check{x}} + a_3(\check{x})A + a_4|A|^2 A + \delta(\check{x} - 1)U, \tag{6.70}$$

where A is a complex-valued function of the spatial variable, $\check{x} \in \mathbb{R}$, and time. The boundary conditions are $A(\pm\infty, t) = 0$. The control input, denoted by U, is in the form of point actuation at the location of the cylinder (δ denotes the Dirac distribution), and the coefficients a_i, $i = 1, \ldots, 4$, are fitted to data from laboratory experiments [103]. The state $A(\check{x}, t)$ may represent any physical variable (velocities

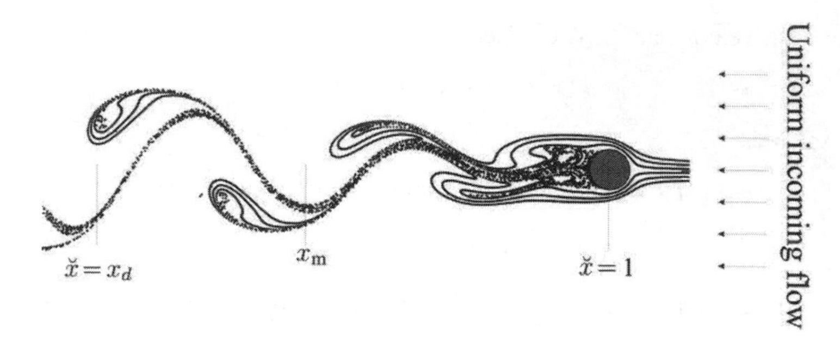

Figure 6.3 Vortex shedding in the 2D flow past a cylinder.

or pressure) or derivatives thereof, along the centerline of the 2D cylinder flow. Associating A with the transverse fluctuating velocity $v(\breve{x}, \breve{y} = 0, t)$ seems to be a particularly good choice [88]. As pointed out in [78], the model is derived for Reynolds numbers close to the critical Reynolds number for the onset of vortex shedding but has been shown to remain accurate far outside this vicinity for a wide variety of flows.

We consider here a simplification of (6.70). We linearize around the zero solution, discard the upstream subsystem by replacing the local forcing at $\breve{x} = 1$ with boundary input at this location, and truncate the downstream subsystem at some $x_d \in (-\infty, 1)$. Note that the fluid flows in the negative $\breve{x}$ direction to fit with our notation in the other chapters of the book, where the control input is at the right boundary. We justify the truncation of the system by noting that the upstream subsystem is approximately uniform flow, whereas the downstream subsystem can be approximated to any desired level of accuracy by selecting x_d sufficiently far from the cylinder.[1] The resulting system is given by

$$\frac{\partial A}{\partial t} = a_1 \frac{\partial^2 A}{\partial \breve{x}^2} + a_2(\breve{x}) \frac{\partial A}{\partial \breve{x}} + a_3(\breve{x}) A \tag{6.71}$$

for $\breve{x} \in (x_d, 1)$, with boundary conditions

$$A(x_d, t) = 0, \tag{6.72}$$

$$A(1, t) = U(t) \quad \text{or} \quad \frac{\partial A}{\partial \breve{x}}(1, t) = U(t), \tag{6.73}$$

where $A\colon [x_d, 1] \times \mathbb{R}_+ \to \mathbb{C}$, $a_2 \in C^2([x_d, 1]; \mathbb{C})$, $a_3 \in C^1([x_d, 1]; \mathbb{C})$, $a_1 \in \mathbb{C}$, and U is the control input. The coefficient a_1 is assumed to have a strictly positive real part.

We now rewrite the model in real variables by defining

$$\rho(x, t) = \mathrm{Re}\{B(x, t)\} = \frac{B(x, t) + B^*(x, t)}{2}, \tag{6.74}$$

$$\iota(x, t) = \mathrm{Im}\{B(x, t)\} = \frac{B(x, t) - B^*(x, t)}{2j}, \tag{6.75}$$

[1] This claim is postulated from the observation that the local damping effect in (6.70) increases with increasing distance from the cylinder, which follows from the coefficients reported in [103].

where

$$x = \frac{\breve{x} - x_d}{1 - x_d}, \qquad B(x,t) = A(\breve{x},t)\exp\left(\frac{1}{2a_1}\int_{x_d}^{\breve{x}} a_2(\tau)\,d\tau\right), \tag{6.76}$$

and B^* is the complex conjugate of B. Equation (6.71) gives

$$\rho_t = a_R\rho_{xx} + b_R(x)\rho - a_I\iota_{xx} - b_I(x)\iota, \tag{6.77}$$

$$\iota_t = a_I\rho_{xx} + b_I(x)\rho + a_R\iota_{xx} + b_R(x)\iota, \tag{6.78}$$

for $x \in (0,1)$, with boundary conditions

$$\rho(0,t) = 0, \qquad \iota(0,t) = 0, \tag{6.79}$$

and either

$$\rho(1,t) = U_R(t), \qquad \iota(1,t) = U_I(t), \tag{6.80}$$

or

$$\rho_x(1,t) = U_R(t), \qquad \iota_x(1,t) = U_I(t), \tag{6.81}$$

where

$$a_R = \frac{1}{(1-x_d)^2}\mathrm{Re}\{a_1\}, \qquad a_I = \frac{1}{(1-x_d)^2}\mathrm{Im}\{a_1\} \tag{6.82}$$

and

$$b_R(x) = \mathrm{Re}\left\{a_3(\breve{x}) - \frac{1}{2}a_2'(\breve{x}) - \frac{1}{4a_1}a_2^2(\breve{x})\right\}, \tag{6.83}$$

$$b_I(x) = \mathrm{Im}\left\{a_3(\breve{x}) - \frac{1}{2}a_2'(\breve{x}) - \frac{1}{4a_1}a_2^2(\breve{x})\right\}. \tag{6.84}$$

6.5 STATE FEEDBACK FOR THE GINZBURG-LANDAU EQUATION

The main result of this section is stated in the following theorem.

Theorem 6.4. *There exist feedback gain kernels, $k(1,\cdot), k_c(1,\cdot) \in C^2(0,1)$, such that for arbitrary initial data $\rho_0, \iota_0 \in H^1(0,1)$ compatible with the boundary conditions, the closed-loop system* (6.77)–(6.80) *with the feedbacks*

$$U_R(t) = \int_0^1 [k(1,y)\rho(y,t) + k_c(1,y)\iota(y,t)]\,dy, \tag{6.85}$$

$$U_I(t) = \int_0^1 [-k_c(1,y)\rho(y,t) + k(1,y)\iota(y,t)]\,dy \tag{6.86}$$

has a unique classical solution $\rho, \iota \in C^{2,1}([0,1]\times(0,\infty))$. The solution satisfies

$$\|(\rho,\iota)\|_{H^1} \le M\,\|(\rho_0,\iota_0)\|_{H^1}\, e^{-\frac{c}{2}t}, \tag{6.87}$$

where $M > 0$ and c is an arbitrary positive constant.

We give the proof of this theorem in Section 6.5.4 after deriving the PDEs for the gain kernels and proving their well-posedness.

6.5.1 PDEs for the Gain Kernels

Following the backstepping approach, we use a coordinate transformation

$$\sigma(x,t) = \rho(x,t) - \int_0^x [k(x,y)\rho(y,t) + k_c(x,y)\iota(y,t)]\,dy, \tag{6.88}$$

$$\kappa(x,t) = \iota(x,t) - \int_0^x [-k_c(x,y)\rho(y,t) + k(x,y)\iota(y,t)]\,dy, \tag{6.89}$$

to map (6.77)–(6.80), (6.85), (6.86) into the system

$$\sigma_t = a_R\sigma_{xx} + f_R(x)\sigma - a_I\kappa_{xx} - f_I(x)\kappa, \tag{6.90}$$

$$\kappa_t = a_I\sigma_{xx} + f_I(x)\sigma + a_R\kappa + f_R(x)\kappa, \tag{6.91}$$

for $x \in (0, 1)$, with boundary conditions

$$\sigma(0,t) = \kappa(0,t) = 0, \tag{6.92}$$

$$\sigma(1,t) = \kappa(1,t) = 0, \tag{6.93}$$

where $f_R, f_I \in C^1([0,1])$. The choice of the target system (6.90)–(6.93) is motivated by the fact that, under appropriate conditions on $f_R(x)$ and $f_I(x)$, it is exponentially stable at the origin (see stability analysis in Section 6.5.3).

The skew-symmetric form of the transformation (6.88), (6.89) is postulated from the skew-symmetric form of (6.77), (6.78).

Lemma 6.1. *If the pair of kernels, $k(x,y)$ and $k_c(x,y)$, satisfies the partial differential equations*

$$k_{xx} = k_{yy} + \beta(x,y)k + \beta_c(x,y)k_c, \tag{6.94}$$

$$k_{c,xx} = k_{c,yy} - \beta_c(x,y)k + \beta(x,y)k_c, \tag{6.95}$$

for $(x,y) \in \mathcal{T} = \{x, y : 0 < y < x < 1\}$, with boundary conditions

$$k(x,x) = -\frac{1}{2}\int_0^x \beta(\gamma,\gamma)\,d\gamma, \tag{6.96}$$

$$k_c(x,x) = \frac{1}{2}\int_0^x \beta_c(\gamma,\gamma)\,d\gamma, \tag{6.97}$$

$$k(x,0) = 0, \tag{6.98}$$

$$k_c(x,0) = 0, \tag{6.99}$$

where

$$\beta(x,y) = \frac{1}{a_R^2 + a_I^2}[a_R(b_R(y) - f_R(x)) + a_I(b_I(y) - f_I(x))], \tag{6.100}$$

$$\beta_c(x,y) = \frac{1}{a_R^2 + a_I^2}[a_R(b_I(y) - f_I(x)) - a_I(b_R(y) - f_R(x))], \tag{6.101}$$

and if (ρ, ι) *satisfies* (6.77)–(6.80), (6.85), (6.86), *then* (σ, κ) *satisfies* (6.90)–(6.93).

Proof. The boundary conditions (6.92) and (6.93) are verified by setting $x = 0$ and $x = 1$ in (6.88), (6.89) and using (6.79), (6.85)–(6.86). Differentiating (6.88) with respect to time and using (6.77), (6.78), we have

$$\begin{aligned}\sigma_t(x,t) = &-\int_0^x k(x,y)(a_R\rho_{yy}(y,t) + b_R(y)\rho(y,t) - a_I\iota_{yy}(y,t)\\ &- b_I(y)\iota(y,t))\,dy + \int_0^x k_c(x,y)(a_I\rho_{yy}(y,t) + b_I(y)\rho(y,t)\\ &+ a_R\iota_{yy}(y,t) + b_R(y)\iota(y,t))\,dy + a_R\rho_{xx}(x,t) + b_R(x)\rho(x,t)\\ &- a_I\iota_{xx}(x,t) - b_I(x)\iota(x,t).\end{aligned} \tag{6.102}$$

Integrating the terms with ρ_{yy} and ι_{yy} in (6.102) by parts and using the conditions (6.79), (6.88), (6.89), (6.98), and (6.99) gives

$$\begin{aligned}\sigma_t(x,t) = &\,a_R\sigma_{xx}(x,t) - a_I\kappa_{xx}(x,t) + f_R(x)\sigma(x,t) - f_I(x)\kappa(x,t)\\ &+ \left(2a_R\frac{dk(x,x)}{dx}\sigma(x,t) + 2a_I\frac{dk_c(x,x)}{dx} + b_R(x) - f_R(x)\right)\rho(x,t)\\ &+ \left(2a_R\frac{dk_c(x,x)}{dx}\sigma(x,t) - 2a_I\frac{dk(x,x)}{dx} - b_I(x) + f_I(x)\right)\iota(x,t)\\ &+ \int_0^x R(x,y)\rho(y,t)\,dy + \int_0^x I(x,y)\iota(y,t)\,dy,\end{aligned} \tag{6.103}$$

where

$$\begin{aligned}R(x,y) = &\,a_R(k_{xx}(x,y) - k_{yy}(x,y)) + a_I(k_{c,xx}(x,y) - k_{c,yy}(x,y))\\ &+ (f_R(x) - b_R(y))k(x,y) + (f_I(x) - b_I(y))k_c(x,y)\end{aligned} \tag{6.104}$$

and

$$\begin{aligned}I(x,y) = &-a_I(k_{xx}(x,y) - k_{yy}(x,y)) + a_R(k_{c,xx}(x,y) - k_{c,yy}(x,y))\\ &+ (b_I(y) - f_I(x))k(x,y) + (f_R(x) - b_R(y))k_c(x,y).\end{aligned} \tag{6.105}$$

Using equations (6.94)–(6.101), we see that the last three lines in (6.103) become zero, which gives (6.90). Equation (6.91) follows similarly by starting from the time derivative of (6.89). □

6.5.2 Well-Posedness of Kernel PDEs

To establish the existence and uniqueness of the solutions of (6.94)–(6.99), we first convert these PDEs into integral equations suitable for analysis by a fixed point method.

Lemma 6.2. *Any pair of kernels, $k(x, y)$ and $k_c(x, y)$, satisfying (6.94)–(6.99) also satisfies the integral equation*

$$G(\xi,\eta) = -\frac{1}{4}\int_\eta^\xi b(\tau,0)\,d\tau + \frac{1}{4}\int_\eta^\xi \int_0^\eta b(\tau,s)G(\tau,s)\,ds\,d\tau + \frac{1}{4}\int_\eta^\xi \int_0^\eta b_c(\tau,s)G_c(\tau,s)\,ds\,d\tau, \tag{6.106}$$

$$G_c(\xi,\eta) = \frac{1}{4}\int_\eta^\xi b_c(\tau,0)\,d\tau - \frac{1}{4}\int_\eta^\xi \int_0^\eta b_c(\tau,s)G(\tau,s)\,ds\,d\tau + \frac{1}{4}\int_\eta^\xi \int_0^\eta b(\tau,s)G_c(\tau,s)\,ds\,d\tau, \tag{6.107}$$

where

$$\xi = x + y, \qquad \eta = x - y, \tag{6.108}$$

$$G(\xi,\eta) = k\left(\frac{\xi+\eta}{2}, \frac{\xi-\eta}{2}\right), \qquad G_c(\xi,\eta) = k_c\left(\frac{\xi+\eta}{2}, \frac{\xi-\eta}{2}\right), \tag{6.109}$$

$$b(\xi,\eta) = \beta\left(\frac{\xi+\eta}{2}, \frac{\xi-\eta}{2}\right), \qquad b_c(\xi,\eta) = \beta_c\left(\frac{\xi+\eta}{2}, \frac{\xi-\eta}{2}\right). \tag{6.110}$$

Proof. In variables (ξ, η), the PDEs (6.94)–(6.99) become

$$G_{\xi\eta}(\xi,\eta) = \frac{1}{4}\left[b(\xi,\eta)G(\xi,\eta) + b_c(\xi,\eta)G_c(\xi,\eta)\right], \tag{6.111}$$

$$G_{c,\xi\eta}(\xi,\eta) = \frac{1}{4}\left[-b_c(\xi,\eta)G(\xi,\eta) + b(\xi,\eta)G_c(\xi,\eta)\right], \tag{6.112}$$

$$G(\xi,0) = -\frac{1}{4}\int_0^\xi b(\tau,0)\,d\tau, \tag{6.113}$$

$$G_c(\xi,0) = \frac{1}{4}\int_0^\xi b_c(\tau,0)\,d\tau, \tag{6.114}$$

$$G(\xi,\xi) = 0, \tag{6.115}$$

$$G_c(\xi,\xi) = 0. \tag{6.116}$$

Integrating (6.111) and (6.112) with respect to η from 0 to η, we obtain

$$G_\xi(\xi,\eta) = G_\xi(\xi,0) + \frac{1}{4}\int_0^\eta \left[b(\xi,s)G(\xi,s) + b_c(\xi,s)G_c(\xi,s)\right]ds, \tag{6.117}$$

$$G_{c,\xi}(\xi,\eta) = G_{c,\xi}(\xi,0) - \frac{1}{4}\int_0^\eta \left[b_c(\xi,s)G(\xi,s) + b(\xi,s)G_c(\xi,s)\right]ds. \tag{6.118}$$

Integrating (6.117) and (6.118) with respect to ξ from η to ξ and using (6.113)–(6.116), we obtain (6.106), (6.107). □

To find the solution of (6.106), (6.107), we use the method of successive approximations.

THEOREM 6.5. *The equations* (6.94)–(6.99) *have a unique* $C^2(\overline{\mathcal{T}})$ *solution satisfying*

$$|k(x, y)| \leq Me^{2Mx}, \qquad |k_c(x, y)| \leq Me^{2Mx}, \tag{6.119}$$

where M depends only on a_1, $a_2(\cdot)$, $a_3(\cdot)$, $f_R(\cdot)$, $f_I(\cdot)$ *and is given in* (6.123).

Proof. Set

$$\Delta G^0(\xi,\eta) = -\frac{1}{4}\int_\eta^\xi b(\tau,0)\,d\tau, \qquad \Delta G_c^0(\xi,\eta) = \frac{1}{4}\int_\eta^\xi b_c(\tau,0)\,d\tau \tag{6.120}$$

and

$$\Delta G^{n+1}(\xi,\eta) = \frac{1}{4}\int_\eta^\xi\int_0^\eta \big[b(\tau,s)\Delta G^n(\tau,s) + b_c(\tau,s)\Delta G_c^n(\tau,s)\big]\,ds\,d\tau, \tag{6.121}$$

$$\Delta G_c^{n+1}(\xi,\eta) = \frac{1}{4}\int_\eta^\xi\int_0^\eta \big[b(\tau,s)\Delta G_c^n(\tau,s) - b_c(\tau,s)\Delta G^n(\tau,s)\big]\,ds\,d\tau. \tag{6.122}$$

Denote

$$B = \sup_{(\xi,\eta)\in\mathcal{T}_1} |b(\xi,\eta)|, \quad B_c = \sup_{(\xi,\eta)\in\mathcal{T}_1} |b_c(\xi,\eta)|, \quad M = \max\{B, B_c\}, \tag{6.123}$$

where $\mathcal{T}_1 \triangleq \{\xi,\eta : 0 < \xi < 2, 0 < \eta < \min(\xi, 2-\xi)\}$. For ΔG^0 and ΔG_c^0 we have

$$|\Delta G^0(\xi,\eta)| \leq \frac{1}{4}\int_\eta^\xi |b(\tau,0)|\,d\tau \leq \frac{1}{4}B(\xi-\eta) \leq \frac{B}{2}, \tag{6.124}$$

$$|\Delta G_c^0(\xi,\eta)| \leq \frac{1}{4}\int_\eta^\xi |b_c(\tau,0)|\,d\tau \leq \frac{1}{4}B_c(\xi-\eta) \leq \frac{B_c}{2}, \tag{6.125}$$

where we have used the fact that $0 \leq \xi - \eta \leq 2$. Suppose that

$$|\Delta G^n(\xi,\eta)| \leq MK^n\frac{(\xi+\eta)^n}{n!}, \qquad |\Delta G_c^n(\xi,\eta)| \leq MK^n\frac{(\xi+\eta)^n}{n!}, \tag{6.126}$$

where $K > 0$ is a constant that will be determined later. Clearly, (6.126) holds for $n = 0$. Noting that

$$\begin{aligned}\int_\eta^\xi\int_0^\eta \big|\Delta G^n(\tau,s)\big|\,ds d\tau &\leq \frac{MK^n}{n!}\int_\eta^\xi\int_0^\eta (\tau+s)^n ds\,d\tau \\ &\leq \frac{MK^n}{(n+1)!}\int_\eta^\xi (\tau+\eta)^{n+1}d\tau \\ &\leq 2MK^n\frac{(\xi+\eta)^{n+1}}{(n+1)!},\end{aligned} \tag{6.127}$$

we obtain from (6.121) that

$$\left|\Delta G^{n+1}(\xi,\eta)\right| \leq \frac{1}{2}M(B+B_c)K^n\frac{(\xi+\eta)^{n+1}}{(n+1)!} \tag{6.128}$$

and from (6.122) that $|\Delta G_c^{n+1}(\xi,\eta)|$ satisfies the same bound (6.128). Therefore, setting $K = M$, we obtain

$$\left|\Delta G^{n+1}(\xi,\eta)\right| \leq MK^{n+1}\frac{(\xi+\eta)^{n+1}}{(n+1)!}, \tag{6.129}$$

$$\left|\Delta G_c^{n+1}(\xi,\eta)\right| \leq MK^{n+1}\frac{(\xi+\eta)^{n+1}}{(n+1)!}. \tag{6.130}$$

Thus, bounds (6.126) are proved by induction, and

$$G(\xi,\eta) = \sum_{n=0}^{\infty}\Delta G^n(\xi,\eta), \qquad G_c(\xi,\eta) = \sum_{n=0}^{\infty}\Delta G_c^n(\xi,\eta) \tag{6.131}$$

is a solution of (6.106), (6.107). This solution is $C^2(\overline{\mathcal{T}_1})$, since b and b_c are $C^1(\overline{\mathcal{T}_1})$. The bounds (6.119) follow from (6.126), (6.131) and the fact that $K = M$. The uniqueness of this solution can be proved by the following argument. Suppose G', G_c' and G'', G_c'' are two different solutions of (6.106), (6.107). Then $\delta G = G' - G''$, $\delta G_c = G_c' - G_c''$ satisfy the homogeneous integral equations (6.121), (6.122) in which ΔG^n, ΔG_c^n and ΔG^{n+1}, ΔG_c^{n+1} are replaced by δG, δG_c, respectively. Using the estimates above, we get that δG satisfies for all n

$$|\delta G(\xi,\eta)| \leq (B+B_c)M^{n+1}e^{2M}\frac{(\xi+\eta)^n}{n!} \to 0 \text{ as } n\to\infty, \tag{6.132}$$

and δG_c satisfies the same bound. Thus, $\delta G \equiv 0$, $\delta G_c \equiv 0$, which means that (6.131) is a unique solution of (6.106), (6.107). By direct substitution we can check that it is also a unique (by Lemma 2.1) solution of PDE (6.94)–(6.99). □

6.5.3 Stability Analysis

In this section we study the stability of the target system (6.90)–(6.93).

THEOREM 6.6. *Suppose $c > 0$, and select $f_R(x)$ and $f_I(x)$ such that*

$$\sup_{x\in[0,1]}\left(f_R(x) + \frac{1}{2}\left|f_I'(x)\right|\right) \leq -\frac{1}{2}c. \tag{6.133}$$

Then the solution $(\sigma,\kappa) \equiv (0,0)$ of system (6.90)–(6.93) is exponentially stable in the $L^2(0,1)$ and $H^1(0,1)$ norms.

Proof. Consider the function

$$E(t) = \frac{1}{2}\int_0^1 \left(\sigma(x,t)^2 + \kappa(x,t)^2\right)dx. \tag{6.134}$$

Its time derivative along the solutions of system (6.90)–(6.93) is

$$\begin{aligned}
\dot{E}(t) &= \int_0^1 [\sigma(a_R\sigma_{xx} + f_R(x)\sigma - a_I\kappa_{xx} - f_I(x)\kappa) \\
&\quad + \kappa(a_I\sigma_{xx} + f_I(x)\sigma + a_R\kappa_{xx} + f_R(x)\kappa)]\,dx \\
&= \int_0^1 (\sigma(a_R\sigma_{xx} + f_R(x)\sigma - a_I\kappa_{xx}) + \kappa(a_I\sigma_{xx} + a_R\kappa_{xx} + f_R(x)\kappa))\,dx \\
&= -\int_0^1 a_R\left(\sigma_x^2 + \kappa_x^2\right)dx + \int_0^1 f_R(x)(\sigma^2 + \kappa^2)dx + a_I\int_0^1 (\sigma_x\kappa_x - \kappa_x\sigma_x)dx \\
&\le \int_0^1 f_R(x)\left(\sigma^2 + \kappa^2\right)dx. \qquad (6.135)
\end{aligned}$$

Using (6.133) and the comparison principle, we have

$$E(t) \le E(0)e^{-ct}. \qquad (6.136)$$

Set

$$V(t) = \frac{1}{2}\int_0^1 \left(\sigma_x^2(x,t) + \kappa_x^2(x,t)\right)dx. \qquad (6.137)$$

The time derivative of $V(t)$ along the solutions of system (6.90)–(6.93) is

$$\begin{aligned}
\dot{V}(t) &= \int_0^1 (\sigma_x\sigma_{xt} + \kappa_x\kappa_{xt})\,dx \\
&= -\int_0^1 (\sigma_{xx}\sigma_t + \kappa_{xx}\kappa_t)\,dx \\
&= -\int_0^1 [\sigma_{xx}(a_R\sigma_{xx} + f_R(x)\sigma - a_I\kappa_{xx} - f_I(x)\kappa) \\
&\quad + \kappa_{xx}(a_I\sigma_{xx} + f_I(x)\sigma + a_R\kappa_{xx} + f_R(x)\kappa)]\,dx \\
&= -a_R\int_0^1 \left(\sigma_{xx}^2 + \kappa_{xx}^2\right)dx + \int_0^1 f_R(x)\left(\sigma_x^2 + \kappa_x^2\right)dx \\
&\quad + \int_0^1 f_I'(x)\,(\kappa_x\sigma - \sigma_x\kappa)\,dx - \frac{1}{2}\int_0^1 f_R''(x)(\sigma^2 + \kappa^2)\,dx \\
&\le \int_0^1 \left(f_R(x) + \frac{1}{2}\left|f_I'(x)\right|\right)\left(\sigma_x^2 + \kappa_x^2\right)dx \\
&\quad + \frac{1}{2}\int_0^1 \left(\left|f_I'(x)\right| - f_R''(x)\right)(\sigma^2 + \kappa^2)\,dx \\
&\le -cV(t) + c_2E(0)e^{-ct}, \qquad (6.138)
\end{aligned}$$

where we have used (6.133) and defined

$$c_2 \triangleq \max\left\{\sup_{x\in[0,1]}\left(\left|f_I'(x)\right| - f_R''(x)\right), 0\right\}. \tag{6.139}$$

From the comparison principle, we get

$$V(t) \leq \left(V(0) + \frac{c_2}{c}E(0)\right)e^{-ct} - \frac{c_2}{c}E(0)e^{-ct}, \tag{6.140}$$

so we obtain

$$V(t) \leq \left(V(0) + \frac{c_2}{c}E(0)\right)e^{-ct}. \tag{6.141}$$

Using the Poincaré inequality, we get

$$V(t) \leq \left(1 + \frac{c_2}{c}\right)V(0)e^{-ct}. \tag{6.142}$$

□

6.5.4 Proof of Theorem 6.4

To infer stability of the closed-loop system (6.77)–(6.80), (6.85), (6.86) from stability of the target system, we need to establish the equivalence of H^1-norms of (ρ, ι) and (σ, κ). The inverse transformation to (6.88), (6.89) has the form

$$\rho(x,t) = \sigma(x,t) - \int_0^x [l(x,y)\sigma(y,t) + l_c(x,y)\kappa(y,t)]\,dy, \tag{6.143}$$

$$\iota(x,t) = \kappa(x,t) - \int_0^x [-l_c(x,y)\sigma(y,t) + l(x,y)\kappa(y,t)]\,dy. \tag{6.144}$$

The following theorem establishes properties of the kernels $l(x, y)$ and $l_c(x, y)$.

Theorem 6.7. *The pair of kernels, $l(x, y)$ and $l_c(x, y)$, satisfies the PDEs*

$$l_{xx} = l_{yy} - \beta(y,x)l - \beta_c(y,x)l_c, \tag{6.145}$$

$$l_{c,xx} = l_{c,yy} + \beta_c(y,x)l - \beta(y,x)l_c, \tag{6.146}$$

with boundary conditions

$$l(x,x) = \frac{1}{2}\int_0^x \beta(\gamma,\gamma)\,d\gamma, \tag{6.147}$$

$$l_c(x,x) = -\frac{1}{2}\int_0^x \beta_c(\gamma,\gamma)\,d\gamma, \tag{6.148}$$

$$l(x,0) = 0, \tag{6.149}$$

$$l_c(x,0) = 0. \tag{6.150}$$

These PDEs have a unique $C^2(\overline{\mathcal{T}})$ solution satisfying

$$|l(x,y)| \leq Me^{2Mx}, \qquad |l_c(x,y)| \leq Me^{2Mx}, \tag{6.151}$$

where M is given in (6.123).

Proof. The proof is similar to those of Lemmas 6.1 and 6.2 and Theorem 6.5. □

From Theorems 6.5, 6.6, and 6.7 we conclude that the closed-loop system (6.77)–(6.80), (6.85), (6.86) is exponentially stable at the origin in the L^2- and H^1-norms. From standard results for uniformly parabolic equations (see, e.g., [41]), it follows that the target system (6.90)–(6.93) with initial data $\sigma_0, \kappa_0 \in H^1(0,1)$ has a unique classical solution $\sigma, \kappa \in C^{2,1}([0,1] \times (0,\infty))$. The smoothness properties of k, k_c, l, and l_c stated in Theorems 6.5 and 6.7 then provide well-posedness of the closed-loop system.

6.5.5 Neumann Actuation

For the case of Neumann actuation, we have the following result.

THEOREM 6.8. *For arbitrary initial data $\rho_0, \iota_0 \in H^1(0,1)$ compatible with the boundary conditions, the closed-loop system* (6.77)–(6.79), (6.81) *with the feedbacks*

$$U_R(t) = k(1,1)\rho(1,t) + k_c(1,1)\iota(1,t) + \int_0^1 [k_x(1,y)\rho(y,t) + k_{c,x}(1,y)\iota(y,t)]\,dy, \tag{6.152}$$

$$U_I(t) = k(1,1)\iota(1,t) - k_c(1,1)\rho(1,t) + \int_0^1 [-k_{c,x}(1,y)\rho(y,t) + k_x(1,y)\iota(y,t)]\,dy, \tag{6.153}$$

where $k(x,y)$, $k_c(x,y)$ are the solutions of (6.94)–(6.99), *has a unique classical solution $\rho, \iota \in C^{2,1}([0,1] \times (0,\infty))$. The solution satisfies*

$$\|(\rho,\iota)\|_{H^1} \le M\|(\rho_0,\iota_0)\|_{H^1} e^{-ct}, \tag{6.154}$$

where $M > 0$ and c is an arbitrary positive constant.

The transformation (6.88), (6.89) is the same as in the Dirichlet actuation case; therefore, the proof of this theorem is very similar to the proof of Theorem 6.4. The only difference stems from the fact that the target system is different compared to (6.90)–(6.93):

$$\sigma_t = a_R\sigma_{xx} + f_R(x)\sigma - a_I\kappa_{xx} - f_I(x)\kappa, \tag{6.155}$$

$$\kappa_t = a_I\sigma_{xx} + f_I(x)\sigma + a_R\kappa + f_R(x)\kappa, \tag{6.156}$$

$$\sigma(0,t) = \kappa(0,t) = 0, \tag{6.157}$$

$$\sigma_x(1,t) = \kappa_x(1,t) = 0, \tag{6.158}$$

The stability of the above system is established by the following theorem.

THEOREM 6.9. *Suppose $c > 0$, and select $f_R(x)$, $f_I(x)$ such that $f_R'(1) = 0$ and*

$$\sup_{x\in[0,1]} \left(f_R(x) + \frac{1}{2}|f_I'(x)| \right) \le -\frac{1}{2}c. \tag{6.159}$$

Then the solution $(\sigma,\kappa) \equiv (0,0)$ of system (6.155)–(6.158) *is exponentially stable in the $L^2(0,1)$ and $H^1(0,1)$ norms.*

Proof. Using the Lyapunov function (6.134), we show that (6.136) holds (the calculation is exactly the same). To show H^1 stability, we use the Lyapunov function (6.137). Under the additional conditions $f_R'(1) = 0$, the calculation (6.138) holds for the system (6.155)–(6.158). Therefore, we obtain (6.142), which concludes the proof. □

It is well known that the target system (6.155)–(6.158) with initial data in $H^1(0,1)$ (compatible with boundary conditions) has a unique classical solution [41]. Stability and well-posedness of the closed-loop system (6.77)–(6.79), (6.81), (6.152), (6.153) follow from stability and well-posedness of the target system (6.155)–(6.158).

6.5.6 Minimizing the Effect of Truncating the Domain

The control law in the original variables is given by

$$U(t) = \int_{x_d}^{1} k_u(\check{x}) A(\check{x}, t)\, d\check{x}, \tag{6.160}$$

where

$$\begin{aligned} k_u(\check{x}) = {} & \frac{1}{1 - x_d}\left(k\left(1, \frac{\check{x} - x_d}{1 - x_d}\right) - jk_c\left(1, \frac{\check{x} - x_d}{1 - x_d}\right)\right) \\ & \times \exp\left(\frac{1}{2a_1}\int_{\check{x}}^{1} a_2(\tau)\, d\tau\right). \end{aligned} \tag{6.161}$$

Let us consider a numerical example. We take the Reynolds number[2] $R = 50$, which corresponds to the supercritical flow at which vortex shedding occurs in the uncontrolled case. For this Reynolds number the numerical coefficients of (6.77), (6.78) derived from the coefficients given in [103] are $a_R = 0.156/x_d^2$, $a_I = 0$, and $b_R(x)$ and $b_I(x)$ are plotted in Figure 6.4 for $x_d = -1$, $x_d = -4$, and $x_d = -7$. Setting $f_R(x) = -0.2$ and $f_I(x) = 0$, exponential stability is assured by Theorem 6.6, and the control gain (6.161) can be calculated numerically via (6.109)–(6.110), (6.120)–(6.122), and (6.131). Figure 6.5 shows the feedback gain kernel (6.161) for $x_d = -1$, $x_d = -4$, and $x_d = -7$. It is clear that the control gains rapidly grow as $|x_d|$ increases, which is an undesirable feature, since we want to make $|x_d|$ large in order to minimize the effect of truncating the downstream subsystem. The increase can be seen in connection with Figure 6.4, which shows that the absolute value of the differences $b_R(x) - f_R(x)$ and $b_I(x) - f_I(x)$ increases as $|x_d|$ increases. In other words, the control effort needed to change the dynamics of system (6.77)–(6.80), (6.85), (6.86) into that of (6.90)–(6.93) increases with the difference between the two systems. Therefore, the functions $f_R(x)$ and $f_I(x)$ must be chosen more intelligently than by simply setting them constant.

Theorem 6.6 allows some flexibility in choosing $f_R(x)$ and $f_I(x)$, within the constraints of (6.133). In order to postpone choosing x_d, we study f_R, f_I, b_R, and

[2]The Reynolds number for the flow past a circular cylinder is usually defined as $R = \rho U_\infty D/\mu$, where U_∞ is the free stream velocity, D is the cylinder diameter, and ρ and μ are the density and viscosity of the fluid, respectively. Vortex shedding occurs when $R > 47$.

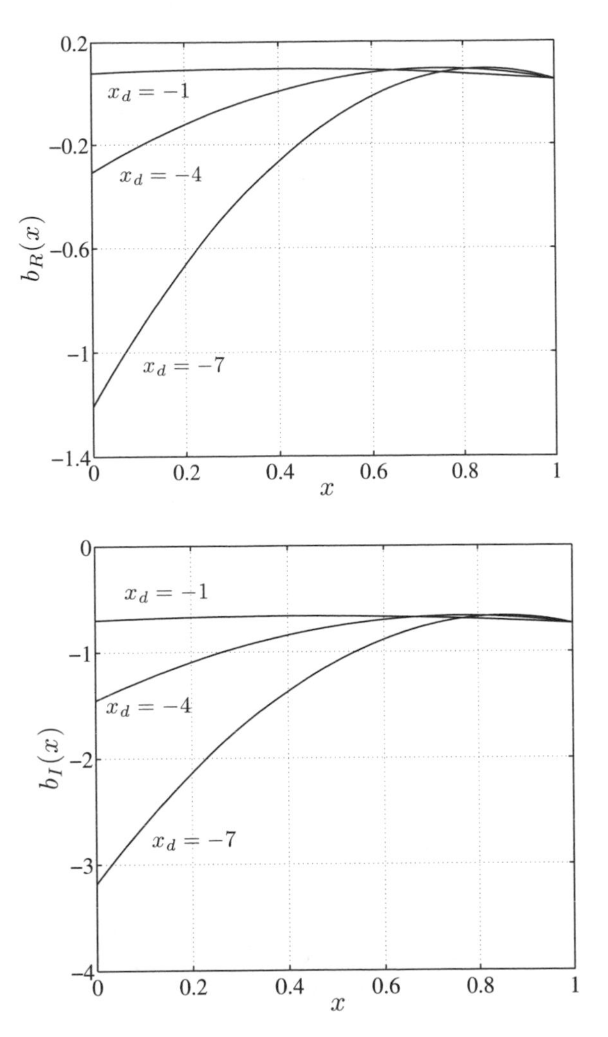

Figure 6.4 Coefficients $b_R(x)$ and $b_I(x)$ for $x_d = -1$, $x_d = -4$, and $x_d = -7$. The Reynolds number is $R = 50$.

b_I as functions of $\breve{x}$ rather than x in the following. This is convenient, since f_R, f_I, b_R, and b_I are invariant of x_d when treated as functions of $\breve{x}$. Recall that when x_d is chosen, the two domains are related by $x = (\breve{x} - x_d)/(1 - x_d)$. We propose to choose $f_R(\breve{x})$ and $f_I(\breve{x})$ as close to $b_R(\breve{x})$ and $b_I(\breve{x})$ as possible, without violating the conditions of Theorem 6.6, which we now write as

$$\sup_{\breve{x}\in[x_d,1]} \left(f_R(\breve{x}) + \frac{1}{2}\left|f_I'(\breve{x})\right| \right) \leq -\frac{1}{2}c. \tag{6.162}$$

Toward that end, we first set $f_R(\breve{x}) = b_R(\breve{x})$ and $f_I(\breve{x}) = b_I(\breve{x})$, and plot (6.162) along with $-\frac{1}{2}c = -0.2$. The result is shown in Figure 6.6, for $\breve{x} \in [-20, 1]$. The figure shows that the conditions for stability are already satisfied, without control,

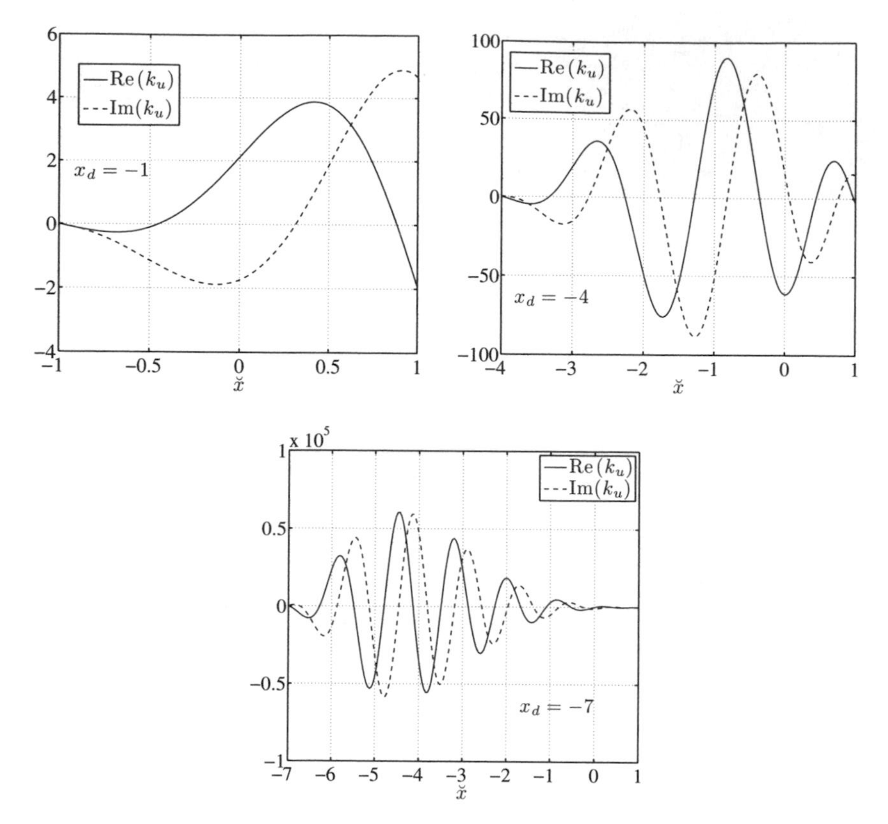

Figure 6.5 Feedback kernel (6.161) for $f_R(x) = -0.2$, $f_I(x) = 0$, and various values of x_d: $x_d = -1$ (top left), $x_d = -4$ (top right), and $x_d = -7$ (bottom).

for $\breve{x} \in [-20, x_s]$ (in fact, the stability conditions are satisfied for $\breve{x} \in (-\infty, x_s]$), which means that it suffices to alter $f_R(\breve{x})$ and $f_I(\breve{x})$ in $(x_s, 1]$ in order to satisfy (6.162). Thus, we set[3]

$$f_R(\breve{x}) = \begin{cases} -\frac{1}{2}c - \frac{1}{2}\left|b_I'(\breve{x})\right|, & \text{for } x_s < x \le 1, \\ b_R(\breve{x}), & \text{for } \breve{x} \le x_s, \end{cases} \tag{6.163}$$

$$f_I(\breve{x}) = b_I(\breve{x}), \qquad \text{for all } \breve{x}. \tag{6.164}$$

With these choices of $f_R(\breve{x})$ and $f_I(\breve{x})$, we calculate numerically the stabilizing feedback gain kernel (6.161) for $x_d = -10$, $x_d = -15$, and $x_d = -20$. Figure 6.7 shows the result. As expected, the control gains look similar, and in particular, they appear to be zero for $\breve{x}$ less than approximately -7. In fact, they are identical and have compact support, as stated in the next theorem.

[3]By the choice of x_s, $f_R(\breve{x})$ is continuous. In this example, we ignore the fact that our choice of $f_R(\breve{x})$ may not be C^1, although this can easily be achieved by smoothing $f_R(\breve{x})$ in a small neighborhood of x_s.

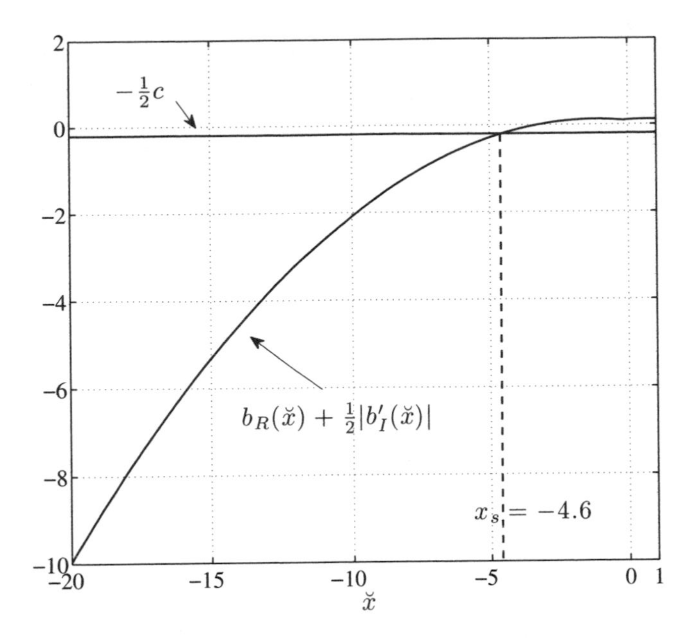

Figure 6.6 The stability criterion (6.162) when $f_R(\breve{x}) = b_R(\breve{x})$ and $f_I(\breve{x}) = b_I(\breve{x})$ is satisfied for $\breve{x} \le x_s$.

THEOREM 6.10. *Given $c > 0$, suppose there exists $x_s \in (-\infty, 1)$ such that*

$$b_R(\breve{x}) + \frac{1}{2}\left|b_I'(\breve{x})\right| \le -\frac{1}{2}c, \quad \textit{for } \breve{x} \le x_s. \tag{6.165}$$

Then $f_R(\breve{x})$ and $f_I(\breve{x})$, satisfying (6.162)*, can be chosen such that $f_R(\breve{x}) = b_R(\breve{x})$ and $f_I(\breve{x}) = b_I(\breve{x})$ for $\breve{x} \in [x_d, x_s]$. The resulting stabilizing control gain* (6.161) *has compact support contained in* $[2x_s - 1, 1]$*. Moreover, all choices of $x_d \le 2x_s - 1$ will produce the same control gain* (6.161) *in* $[2x_s - 1, 1]$.

Proof. The existence of $f_R(\breve{x})$ and $f_I(\breve{x})$ satisfying the criterion for stability (6.162) follows trivially from (6.165). To prove that the kernel has support contained in $[2x_s - 1, 1]$, we show that it is identically zero outside this interval. We have that

$$b(\xi, 0) = \beta(x, x) = 0, \tag{6.166}$$

$$b_c(\xi, 0) = \beta_c(x, x) = 0, \tag{6.167}$$

for

$$\xi \in \left[0, 2\frac{x_s - x_d}{1 - x_d}\right). \tag{6.168}$$

It follows that

$$\Delta G^0(\xi, \eta) = \Delta G_c^n(\xi, \eta) = 0, \quad \text{for } (\xi, \eta) \in \mathcal{T}_1, \xi \le 2\frac{x_s - x_d}{1 - x_d}. \tag{6.169}$$

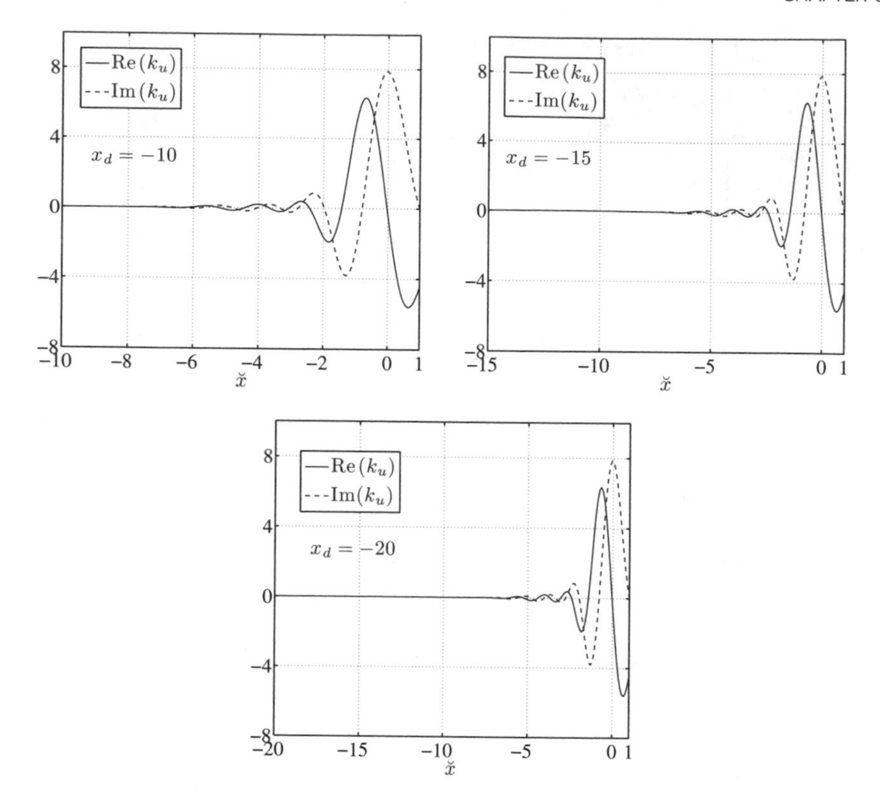

Figure 6.7 Feedback kernels (6.161) for $f_R(\breve{x})$ and $f_I(\breve{x})$ as defined in (6.163)–(6.164), and for various values of x_d: $x_d = -10$ (top left), $x_d = -15$ (top right), and $x_d = -20$ (bottom).

Now, suppose that

$$\Delta G^n(\xi,\eta) = \Delta G_c^n(\xi,\eta) = 0, \qquad \text{for } (\xi,\eta) \in \mathcal{T}_1, \xi \le 2\frac{x_s - x_d}{1 - x_d}. \tag{6.170}$$

From (6.121) and (6.122), we get

$$\Delta G^{n+1}(\xi,\eta) = \Delta G_c^{n+1}(\xi,\eta) = 0, \qquad \text{for } (\xi,\eta) \in \mathcal{T}_1, \xi \le 2\frac{x_s - x_d}{1 - x_d}. \tag{6.171}$$

Thus, (6.170) is proved by induction, and

$$\left.\begin{aligned} k(1,y) &= G(1+y, 1-y) = 0 \\ k_c(1,y) &= G_c(1+y, 1-y) = 0 \end{aligned}\right\} \qquad \text{for } 0 \le y \le 2\frac{x_s - x_d}{1 - x_d} - 1. \tag{6.172}$$

Therefore, $k_u(\breve{x}) = 0$ for

$$0 \le \frac{\breve{x} - x_d}{1 - x_d} \le 2\frac{x_s - x_d}{1 - x_d} - 1, \tag{6.173}$$

which yields

$$x_d \le \breve{x} \le 2x_s - 1. \tag{6.174}$$

In order to prove the last part of the theorem, we need to show that for any $x_{d,1}, x_{d,2} \in (-\infty, 2x_s - 1]$,

$$\left.\begin{aligned} \frac{1}{x_{d,1}} k_{x_{d,1}}\left(1, \frac{x_{d,1}-\breve{x}}{x_{d,1}}\right) &= \frac{1}{x_{d,2}} k_{x_{d,2}}\left(1, \frac{x_{d,2}-\breve{x}}{x_{d,2}}\right) \\ \frac{1}{x_{d,1}} k_{c,x_{d,1}}\left(1, \frac{x_{d,1}-\breve{x}}{x_{d,1}}\right) &= \frac{1}{x_{d,2}} k_{c,x_{d,2}}\left(1, \frac{x_{d,2}-\breve{x}}{x_{d,2}}\right) \end{aligned}\right\} \text{ for } \breve{x} \in [2x_s - 1, 1], \tag{6.175}$$

where the additional subscripts identify the domain of the problem from which they stem. We have

$$k_{x_d}\left(1, \frac{\breve{x}-x_d}{1-x_d}\right) = G_{x_d}\left(\frac{1+\breve{x}-2x_d}{1-x_d}, \frac{1-\breve{x}}{1-x_d}\right). \tag{6.176}$$

From (6.166)–(6.168) it follows that

$$(1-x_d)\Delta G^0_{x_d}\left(\frac{1+\breve{x}-2x_d}{1-x_d}, \frac{1-\breve{x}}{1-x_d}\right) = -\frac{1}{4}\int_{\breve{x}}^{2x_s-1} b_{x_d}\left(\frac{\tau-x_d}{1-x_d}+1, 0\right) d\tau, \tag{6.177}$$

for $x_d \leq 2x_s - 1$. From the definition of b, we have

$$(1-x_{d,2})^2 b_{x_{d,1}}\left(\frac{\tau-x_{d,1}}{1-x_{d,1}}+1, 0\right) = (1-x_{d,1})^2 b_{x_{d,2}}\left(\frac{\tau-x_{d,2}}{1-x_{d,2}}+1, 0\right), \tag{6.178}$$

for $\tau \in [2x_s - 1, 1]$. It follows from (6.177)–(6.178) that

$$\Delta G^0_{x_{d,1}}\left(\frac{1+\breve{x}-2x_{d,1}}{1-x_{d,1}}, \frac{1-\breve{x}}{1-x_{d,1}}\right) = \frac{1-x_{d,1}}{1-x_{d,2}}\Delta G^0_{x_{d,2}}\left(\frac{1+\breve{x}-2x_{d,2}}{1-x_{d,2}}, \frac{1-\breve{x}}{1-x_{d,2}}\right).$$

Similar arguments for ΔG^0_c, ΔG^n, and ΔG^n_c yield

$$\Delta G_{x_{d,1}}\left(\frac{1+\breve{x}-2x_{d,1}}{1-x_{d,1}}, \frac{1-\breve{x}}{1-x_{d,1}}\right) = \frac{1-x_{d,1}}{1-x_{d,2}}\Delta G_{x_{d,2}}\left(\frac{1+\breve{x}-2x_{d,2}}{1-x_{d,2}}, \frac{1-\breve{x}}{1-x_{d,2}}\right),$$

which in turn gives (6.175). □

The significance of Theorem 6.10 is that it guarantees stabilization of the system evolving on an infinite domain by solving the stabilization problem on a finite domain. The procedure for verifying the conditions of the theorem was demonstrated above, but for clarity we repeat it in the following remark.

Remark 6.1. The key to applying Theorem 6.10 is being able to find x_s that satisfies (6.165). This is most easily done by inspecting a graph, like the one shown in Figure 6.6. Once x_s is found, a possible choice of $f_R(\breve{x})$ and $f_I(\breve{x})$ is given in (6.163)–(6.164). Other choices are possible, and in particular, care should be taken to ensure the necessary smoothness properties of $f_R(\breve{x})$. Also, note that the estimate for the support of (6.161) is not tight, as suggested by Figures 6.6 and 6.7, which indicate that $k_u(\breve{x})$ is supported on approximately $[-7, 1]$, while $2x_s - 1 = -10.2$. Theorem 6.10 states that $[-7, 1] \subseteq [2x_s - 1, 1]$, which is true.

We have shown that a particular choice of f_R and f_I results in state-feedback kernel functions that are invariant of x_d and vanish in $[x_d, x_s]$, where x_s is a constant that can be deduced from the coefficients of (6.70). This implies that if the domain is truncated at some $x_d \leq x_s$ for the purpose of computing the feedback kernel functions, the resulting state feedback will stabilize the plant evolving on the semi-infinite domain $(-\infty, 1)$. This is achieved by avoiding the complete cancelation of the terms involving b_R and b_I in (6.77), (6.78) by using a target system (6.90)–(6.93) that contains the natural damping that exists in the plant downstream of x_s. It ensures that only cancelation/domination of the source of instability is performed in the design, and is similar to common practice in the design of finite-dimensional backstepping controllers, where one seeks to leave unaltered terms that add to the stability while canceling terms that don't. The result is less complexity and better robustness properties.

6.6 OBSERVER DESIGN FOR THE GINZBURG-LANDAU EQUATION

The control law (6.160) requires distributed measurements that typically are not available. Therefore, we want to design an observer that estimates the state based on the measurements at one of the boundaries. We restrict the measurements to be taken at the location of the cylinder only (via pressure sensors), collocated with actuation. This setup is more realistic than the anti-collocated case (with measurements taken at one point downstream of the cylinder).

Let us denote the measurements by $Y_R(t) = \rho(1, t)$ and $Y_I(t) = \iota(1, t)$, which leaves $\rho_x(1, t)$ and $\iota_x(1, t)$ for control input. Consider the following Luenberger-type observer:

$$\hat{\rho}_t = a_R\hat{\rho}_{xx} + b_R(x)\hat{\rho} - a_I\hat{\iota}_{xx} - b_I(x)\hat{\iota} + p_1(x)(Y_R - \hat{Y}_R) + p_{c,1}(x)(Y_I - \hat{Y}_I), \tag{6.179}$$

$$\hat{\iota}_t = a_I\hat{\rho}_{xx} + b_I(x)\hat{\rho} + a_R\hat{\iota}_{xx} + b_R(x)\hat{\iota} - p_{c,1}(x)(Y_R - \hat{Y}_R) + p_1(x)(Y_I - \hat{Y}_I), \tag{6.180}$$

for $x \in (0, 1)$, with boundary conditions

$$\hat{\rho}(0, t) = 0, \tag{6.181}$$

$$\hat{\iota}(0, t) = 0, \tag{6.182}$$

$$\hat{\rho}_x(1, t) = p_0(Y_R(t) - \hat{Y}_R(t)) + p_{c,0}(Y_I(t) - \hat{Y}_I(t)) + U_R(t), \tag{6.183}$$

$$\hat{\iota}_x(1, t) = -p_{c,0}(Y_R(t) - \hat{Y}_R(t)) + p_0(Y_I(t) - \hat{Y}_I(t)) + U_I(t). \tag{6.184}$$

In (6.179)–(6.184), $p_1(x)$, $p_{c,1}(x)$, p_0, and $p_{c,0}$ are output injection gains to be designed.

The dynamics for the observer error $\tilde{\rho} = \rho - \hat{\rho}$, $\tilde{\iota} = \iota - \hat{\iota}$ are given by

$$\tilde{\rho}_t = a_R\tilde{\rho}_{xx} + b_R(x)\tilde{\rho} - a_I\tilde{\iota}_{xx} - b_I(x)\tilde{\iota} - p_1(x)\tilde{\rho}(1, t) - p_{c,1}(x)\tilde{\iota}(1, t), \tag{6.185}$$

$$\tilde{\iota}_t = a_I\tilde{\rho}_{xx} + b_I(x)\tilde{\rho} + a_R\tilde{\iota}_{xx} + b_R(x)\tilde{\iota} + p_{c,1}(x)\tilde{\rho}(1, t) - p_1(x)\tilde{\iota}(1, t), \tag{6.186}$$

for $x \in (0, 1)$, with boundary conditions

$$\tilde{\rho}(0, t) = 0, \tag{6.187}$$

$$\tilde{\iota}(0, t) = 0, \tag{6.188}$$

$$\tilde{\rho}_x(1, t) = -p_0\tilde{\rho}(1, t) - p_{c,0}\tilde{\iota}(1, t), \tag{6.189}$$

$$\tilde{\iota}_x(1, t) = p_{c,0}\tilde{\rho}(1, t) - p_0\tilde{\iota}(1, t). \tag{6.190}$$

The output injection gains $p_1(x)$, $p_{c,1}(x)$, p_0, and $p_{c,0}$ should be chosen to make the system (6.185)–(6.190) exponentially stable at the origin. To that end, we look for a backstepping transformation

$$\tilde{\rho}(x, t) = \tilde{\sigma}(x, t) - \int_x^1 [p(x, y)\tilde{\sigma}(y, t) + p_c(x, y)\tilde{\kappa}(y, t)]\, dy, \tag{6.191}$$

$$\tilde{\iota}(x, t) = \tilde{\kappa}(x, t) - \int_x^1 [-p_c(x, y)\tilde{\sigma}(y, t) + p(x, y)\tilde{\kappa}(y, t)]\, dy, \tag{6.192}$$

that maps (6.185)–(6.190) into the exponentially stable system

$$\tilde{\sigma}_t = a_R\tilde{\sigma}_{xx} + f_R(x)\tilde{\sigma} - a_I\tilde{\kappa}_{xx} - f_I(x)\tilde{\kappa}, \tag{6.193}$$

$$\tilde{\kappa}_t = a_I\tilde{\sigma}_{xx} + f_I(x)\tilde{\sigma} + a_R\tilde{\kappa}_{xx} + f_R(x)\tilde{\kappa}, \tag{6.194}$$

for $x \in (0, 1)$, with boundary conditions

$$\tilde{\sigma}(0, t) = \tilde{\kappa}(0, t) = 0, \tag{6.195}$$

$$\tilde{\sigma}_x(1, t) = \tilde{\kappa}_x(1, t) = 0. \tag{6.196}$$

Subtracting (6.185)–(6.190) from (6.193)–(6.196) and using (6.191)–(6.192), one can show that the output injection gains have to be selected as

$$p_1(x) = -a_R p_y(x, 1) - a_I p_{c,y}(x, 1), \tag{6.197}$$

$$p_0 = -p(1, 1), \tag{6.198}$$

$$p_{c,1}(x) = a_I p_y(x, 1) - a_R p_{c,y}(x, 1), \tag{6.199}$$

$$p_{c,0} = -p_c(1, 1), \tag{6.200}$$

where the kernels $p(x, y)$ and $p_c(x, y)$ satisfy the following PDEs:

$$p_{xx} = p_{yy} - \bar{\beta}(x, y)p - \bar{\beta}_c(x, y)p_c, \tag{6.201}$$

$$p_{c,xx} = p_{c,yy} + \bar{\beta}_c(x, y)p - \bar{\beta}(x, y)p_c, \tag{6.202}$$

with boundary conditions

$$p(0, y) = p_c(0, y) = 0, \tag{6.203}$$

$$p(x, x) = -\frac{1}{2}\int_0^x \bar{\beta}(\gamma, \gamma)\, d\gamma, \tag{6.204}$$

$$p_c(x, x) = \frac{1}{2}\int_0^x \bar{\beta}_c(\gamma, \gamma)\, d\gamma, \tag{6.205}$$

where

$$\bar{\beta}(x,y) = \frac{a_R(b_R(x) - f_R(y)) + a_I(b_I(x) - f_I(y))}{a_R^2 + a_I^2}, \tag{6.206}$$

$$\bar{\beta}_c(x,y) = \frac{a_R(b_I(x) - f_I(y)) - a_I(b_R(x) - f_R(y))}{a_R^2 + a_I^2}. \tag{6.207}$$

With a change of variables $\breve{x} = y$, $\breve{y} = x$, $\breve{p}(\breve{x},\breve{y}) = p(x,y)$, $\breve{p}_c(\breve{x},\breve{y}) = p_c(x,y)$, we obtain

$$\breve{p}_{\breve{x}\breve{x}} = \breve{p}_{\breve{y}\breve{y}} + \beta(\breve{x},\breve{y})\breve{p} + \beta_c(\breve{x},\breve{y})\breve{p}_c, \tag{6.208}$$

$$\breve{p}_{c,\breve{x}\breve{x}} = \breve{p}_{c,\breve{y}\breve{y}} - \beta_c(\breve{x},\breve{y})\breve{p} + \beta(\breve{x},\breve{y})\breve{p}_c, \tag{6.209}$$

with boundary conditions

$$\breve{p}(\breve{x},0) = \breve{p}_c(\breve{x},0) = 0, \tag{6.210}$$

$$\breve{p}(\breve{x},\breve{x}) = -\frac{1}{2}\int_0^{\breve{x}} \beta(\gamma,\gamma)\,d\gamma, \tag{6.211}$$

$$\breve{p}_c(\breve{x},\breve{x}) = \frac{1}{2}\int_0^{\breve{x}} \beta_c(\gamma,\gamma)\,d\gamma. \tag{6.212}$$

Here, β and β_c are given by (6.100), (6.101). Since equations (6.208)–(6.212) are identical with (6.94)–(6.99), using Theorem 6.5 we immediately obtain the following result.

THEOREM 6.11. *The equations* (6.201)–(6.205) *have a unique* $C^2(\overline{\mathcal{T}})$ *solution satisfying*

$$|p(x,y)| \le Me^{2Mx}, \qquad |p_c(x,y)| \le Me^{2Mx}, \tag{6.213}$$

where M *is given by* (6.123).

The output injection gains can be obtained from the control gains:

$$p_1(x) = -a_R k_x(1,x) - a_I k_{c,x}(1,x), \tag{6.214}$$

$$p_0 = -k(1,1), \tag{6.215}$$

$$p_{c,1}(x) = a_I k_x(1,x) - a_R k_{c,x}(1,x), \tag{6.216}$$

$$p_{c,0} = -k_c(1,1). \tag{6.217}$$

Note that the target system (6.193)–(6.196) is exactly the same as the system (6.155)–(6.158), which is well posed and exponentially stable, as stated in Theorem 6.9. From the transformation (6.191), (6.192) and Theorems 6.11 and 6.9 we get the following result.

THEOREM 6.12. *Suppose* f_R *and* f_I *satisfy* (6.133) *and* $f_R'(1) = 0$, *and let* k, k_c *be the solutions of* (6.94)–(6.99). *Then for any initial data* $\tilde{\rho}_0, \tilde{\iota}_0 \in H^1(0,1)$ *compatible with the boundary conditions, the system* (6.185)–(6.190) *with output* *injection gains given by* (6.214)–(6.217) *has a unique classical solution* $\tilde{\rho}, \tilde{\iota} \in C^{2,1}([0,1] \times (0,\infty))$. *The solution satisfies*

$$\|(\tilde{\rho}, \tilde{\iota})\|_{H^1} \leq M \, \|(\rho_0, \iota_0)\|_{H^1} \, e^{-\frac{c}{2}t}, \tag{6.218}$$

where $M > 0$ *and* c *is an arbitrary positive constant.*

6.7 OUTPUT FEEDBACK FOR THE GINZBURG-LANDAU EQUATION

When both sensors and actuators are located on the surface of a cylinder, we can combine the control and observer designs developed in previous sections to solve the output-feedback problem. In this section we assume Neumann actuation and Dirichlet sensing. The other case (Dirichlet actuation and Neumann sensing) is treated similarly (see Section 6.8 for numerical results for that alternative architecture).

THEOREM 6.13. *Suppose* f_R *and* f_I *satisfy* (6.133) *and* $f_R'(1) = 0$, *and let* k, k_c, p_1, $p_{c,1}$, p_0, $p_{c,0}$ *be the solution of* (6.94)–(6.99), (6.214)–(6.217). *Then for any initial data* ρ_0, ι_0, $\hat{\rho}_0$, $\hat{\iota}_0 \in H^1(0,1)$ *compatible with the boundary conditions, the closed-loop system consisting of the plant* (6.77)–(6.79), (6.81) *with the controller*

$$\begin{aligned} U_R(t) = & \int_0^1 [k_x(1,y)\hat{\rho}(y,t) + k_{c,x}(1,y)\hat{\iota}(y,t)]\,dy \\ & + k(1,1)\rho(1,t) + k_c(1,1)\iota(1,t), \end{aligned} \tag{6.219}$$

$$\begin{aligned} U_I(t) = & \int_0^1 [-k_{c,x}(1,y)\hat{\rho}(y,t) + k_x(1,y)\hat{\iota}(y,t)]\,dy \\ & - k_c(1,1)\rho(1,t) + k(1,1)\iota(1,t), \end{aligned} \tag{6.220}$$

and the observer

$$\begin{aligned} \hat{\rho}_t = & a_R\hat{\rho}_{xx} + b_R(x)\hat{\rho} - a_I\hat{\iota}_{xx} - b_I(x)\hat{\iota} \\ & + p_1(x)(\rho(1,t) - \hat{\rho}(1,t)) + p_{c,1}(x)(\iota(1,t) - \hat{\iota}(1,t)), \end{aligned} \tag{6.221}$$

$$\begin{aligned} \hat{\iota}_t = & a_I\hat{\rho}_{xx} + b_I(x)\hat{\rho} + a_R\hat{\iota}_{xx} + b_R(x)\hat{\iota} \\ & - p_{c,1}(x)(\rho(1,t) - \hat{\rho}(1,t)) + p_1(x)(\iota(1,t) - \hat{\iota}(1,t)), \end{aligned} \tag{6.222}$$

$$\hat{\rho}(0,t) = 0, \tag{6.223}$$

$$\hat{\iota}(0,t) = 0, \tag{6.224}$$

$$\hat{\rho}_x(1,t) = p_0(\rho(1,t) - \hat{\rho}(1,t)) + p_{c,0}(\iota(1,t) - \hat{\iota}(1,t)) + \rho_x(1,t), \tag{6.225}$$

$$\hat{\iota}_x(1,t) = -p_{c,0}(\rho(1,t) - \hat{\rho}(1,t)) + p_0(\iota(1,t) - \hat{\iota}(1,t)) + \iota_x(1,t), \tag{6.226}$$

has a unique classical solution $\rho, \iota, \hat{\rho}, \hat{\iota} \in C^{2,1}([0,1] \times (0,\infty))$ *that satisfies*

$$\left\|(\rho, \iota, \hat{\rho}, \hat{\iota})\right\|_{H^1} \leq M \left\|(\rho_0, \iota_0, \hat{\rho}_0, \hat{\iota}_0)\right\|_{H^1} e^{-\frac{c}{2}t}, \tag{6.227}$$

where $M > 0$ *and* c *is an arbitrary positive constant.*

Proof. The coordinate transformation

$$\hat{\sigma}(x,t) = \hat{\rho}(x,t) - \int_0^x [k(x,y)\hat{\rho}(y,t) + k_c(x,y)\hat{\iota}(y,t)]\,dy, \tag{6.228}$$

$$\hat{\kappa}(x,t) = \hat{\iota}(x,t) - \int_0^x [-k_c(x,y)\hat{\rho}(y,t) + k(x,y)\hat{\iota}(y,t)]\,dy, \tag{6.229}$$

maps (6.221)–(6.226) into the system

$$\hat{\sigma}_t = a_R\hat{\sigma}_{xx} + f_R(x)\hat{\sigma} - a_I\hat{\kappa}_{xx} - f_I(x)\hat{\kappa} - \gamma(x)\tilde{\sigma}(1) - \gamma_c(x)\tilde{\kappa}(1), \tag{6.230}$$

$$\hat{\kappa}_t = a_I\hat{\sigma}_{xx} + f_I(x)\hat{\sigma} + a_R\hat{\kappa}_{xx} + f_R(x)\hat{\kappa} + \gamma_c(x)\tilde{\sigma}(1) - \gamma(x)\tilde{\kappa}(1), \tag{6.231}$$

for $x \in (0,1)$, with boundary conditions

$$\hat{\sigma}(0,t) = \hat{\kappa}(0,t) = 0, \tag{6.232}$$

$$\hat{\sigma}_x(1,t) = \hat{\kappa}_x(1,t) = 0, \tag{6.233}$$

where

$$\gamma(x) = \int_0^x [k(x,y)p_1(y) - k_c(x,y)p_{c,1}(y)]\,dy, \tag{6.234}$$

$$\gamma_c(x) = \int_0^x [k(x,y)p_{c,1}(y) + k_c(x,y)p_1(y)]\,dy. \tag{6.235}$$

Note that systems (6.193)–(6.196) and (6.230)–(6.233) form a cascade, where the $(\hat{\sigma},\hat{\kappa})$ subsystem is driven by the $(\tilde{\sigma},\tilde{\kappa})$ subsystem. Well-posedness of the $(\tilde{\sigma},\tilde{\kappa})$ subsystem is established in Theorem 6.12. From standard results for uniformly parabolic equations [41], it follows that system (6.230)–(6.233) with initial data $\hat{\sigma}_0, \hat{\kappa}_0 \in H^1(0,1)$ has a unique classical solution $\hat{\sigma}, \hat{\kappa} \in C^{2,1}([0,1]\times(0,\infty))$. The smoothness of k, k_c and of the kernels for the inverse transformation, l, l_c, as stated in Theorems 6.5 and 6.7, provide well-posedness of system (6.77)–(6.79), (6.81) in a closed loop with (6.219)–(6.226).

Next, we establish stability. Consider the Lyapunov function

$$\begin{aligned} E &= \frac{1}{2}\int_0^1 \left(\hat{\sigma}^2 + \hat{\kappa}^2 + \mu\tilde{\sigma}^2 + \mu\tilde{\kappa}^2\right)dx \\ &= \frac{1}{2}\left(\|\hat{\sigma}\|^2 + \|\hat{\kappa}\|^2 + \mu\|\tilde{\sigma}\|^2 + \mu\|\tilde{\kappa}\|^2\right), \end{aligned} \tag{6.236}$$

where μ is a strictly positive constant to be determined. Owing to (6.133), the time derivative of $E(t)$ along the solutions of systems (6.193)–(6.196) and (6.230)–(6.233) satisfies

$$\begin{aligned} \dot{E} \le & -\frac{c}{2}\left(\|\hat{\sigma}\|^2 + \|\hat{\kappa}\|^2 + \mu\|\tilde{\sigma}\|^2 + \mu\|\tilde{\kappa}\|^2\right) \\ & + \bar{\gamma}\int_0^1 \left(|\hat{\sigma}\tilde{\sigma}(1)| + |\hat{\sigma}\tilde{\kappa}(1)| + |\hat{\kappa}\tilde{\sigma}(1)| + |\hat{\kappa}\tilde{\kappa}(1)|\right)dx \\ & - a_R\|\hat{\sigma}_x\|^2 - a_R\|\hat{\kappa}_x\|^2 - \mu a_R\|\tilde{\sigma}_x\|^2 - \mu a_R\|\tilde{\kappa}_x\|^2, \end{aligned} \tag{6.237}$$

where $\bar{\gamma} \triangleq \max\{\sup_{x\in[0,1]} |\gamma(x)|, \sup_{x\in[0,1]} |\gamma_c(x)|\}$. Using the Schwartz and Poincaré inequalities and noting that $|\tilde{\sigma}(1)| \leq \|\tilde{\sigma}_x\|$, we get

$$\begin{aligned}
\dot{E} \leq & -cE + \alpha\|\hat{\sigma}\|^2 + \alpha\|\hat{\kappa}\|^2 + \frac{\bar{\gamma}^2}{\alpha}\|\tilde{\sigma}_x\|^2 + \frac{\bar{\gamma}^2}{\alpha}\|\tilde{\kappa}_x\|^2 \\
& -a_R\|\hat{\sigma}_x\|^2 - a_R\|\hat{\kappa}_x\|^2 - \mu a_R\|\tilde{\sigma}_x\|^2 - \mu a_R\|\tilde{\kappa}_x\|^2 \\
\leq & -cE + \left(\frac{4\alpha}{\pi^2} - a_R\right)\|\hat{\sigma}_x\|^2 + \left(\frac{4\alpha}{\pi^2} - a_R\right)\|\hat{\kappa}_x\|^2 \\
& + \left(\frac{\bar{\gamma}^2}{\alpha} - \mu a_R\right)\|\tilde{\sigma}_x\|^2 + \left(\frac{\bar{\gamma}^2}{\alpha} - \mu a_R\right)\|\tilde{\kappa}_x\|^2. \qquad (6.238)
\end{aligned}$$

Setting $\alpha = \pi^2 a_R/8$, $\mu = 16\bar{\gamma}^2/(\pi^2 a_R^2)$, and applying the comparison principle, we obtain

$$E(t) \leq E(0)e^{-ct}, \qquad (6.239)$$

which proves exponential stability of the $(\hat{\sigma}, \hat{\kappa}, \tilde{\sigma}, \tilde{\kappa})$-system in the $L^2(0, 1)$ norm. Next, consider

$$V = \frac{1}{2}\left(\|\hat{\sigma}_x\|^2 + \|\hat{\kappa}_x\|^2 + \mu\|\tilde{\sigma}_x\|^2 + \mu\|\tilde{\kappa}_x\|^2\right). \qquad (6.240)$$

As a result of (6.133), the derivative of $V(t)$ along the solutions of system (6.193)–(6.196) and (6.230)–(6.233) satisfies

$$\begin{aligned}
\dot{V}(t) \leq & -a_R\left(\|\hat{\sigma}_{xx}\|^2 + \|\hat{\kappa}_{xx}\|^2 + \mu\|\tilde{\sigma}_{xx}\|^2 + \mu\|\tilde{\kappa}_{xx}\|^2\right) \\
& +\bar{\gamma}(\|\hat{\sigma}_{xx}\|\|\tilde{\sigma}_x\| + \|\hat{\sigma}_{xx}\|\|\tilde{\kappa}_x\| + \|\hat{\kappa}_{xx}\|\|\tilde{\sigma}_x\| + \|\hat{\kappa}_{xx}\|\|\tilde{\kappa}_x\|) \\
& -\frac{c}{2}(\|\hat{\sigma}_x\|^2 + \|\hat{\kappa}_x\|^2 + \mu\|\tilde{\sigma}_x\|^2 + \mu\|\tilde{\kappa}_x\|^2) \\
& +\frac{1}{2\beta}\int_0^1 (f_R'(x)^2 + f_I'(x)^2)(\hat{\sigma}^2 + \hat{\kappa}^2 + \mu\tilde{\sigma}^2 + \mu\tilde{\kappa}^2)\,dx \\
& +\beta(\|\hat{\sigma}_x\|^2 + \|\hat{\kappa}_x\|^2 + \mu\|\tilde{\sigma}_x\|^2 + \mu\|\tilde{\kappa}_x\|^2). \qquad (6.241)
\end{aligned}$$

Defining $\beta = \pi^2 a_R/8$ and $c_2 = (1/\beta)\sup_{x\in[0,1]}\{f_R'(x)^2 + f_I'(x)^2\}$, we have

$$\begin{aligned}
\dot{V} \leq & -cV + c_2E \\
& -\left(a_R - \frac{4\alpha}{\pi^2} - \frac{4\beta}{\pi^2}\right)\left(\|\hat{\sigma}_{xx}\|^2 + \|\hat{\kappa}_{xx}\|^2\right) \\
& -\left(\mu a_R - \frac{\bar{\gamma}^2}{\alpha} - \frac{4\mu\beta}{\pi^2}\right)\left(\|\tilde{\sigma}_{xx}\|^2 + \|\tilde{\kappa}_{xx}\|^2\right) \\
\leq & -cV + c_2E. \qquad (6.242)
\end{aligned}$$

Applying the comparison principle and the Poincaré inequality, we obtain

$$V(t) \leq \left(1 + \frac{c_2}{c}\right)V(0)e^{-ct}, \qquad (6.243)$$

which proves exponential stability of the $\left(\hat{\sigma}, \hat{\kappa}, \tilde{\sigma}, \tilde{\kappa}\right)$-system in $H^1(0, 1)$. From the transformation (6.228)–(6.229) and its inverse we get equivalence of $L^2(0, 1)$ and

$H^1(0,1)$ norms of $(\hat{\rho}, \hat{\iota}, \tilde{\rho}, \tilde{\iota})$ and $(\hat{\sigma}, \hat{\kappa}, \tilde{\sigma}, \tilde{\kappa})$. Therefore, the properties proven for system (6.230)–(6.233) also hold for the systems (6.221)–(6.226) and (6.77)–(6.79), (6.81), (6.219)–(6.220), which gives (6.227). □

6.8 SIMULATIONS WITH THE NONLINEAR GINZBURG-LANDAU EQUATION

In this section we present the results of numerical simulations of the full nonlinear Ginzburg-Landau model with backstepping controllers and observers.

The plant model (6.77), (6.78) with additional nonlinear terms from (6.70) becomes

$$\rho_t = a_R\rho_{xx} + (b_R(x) + c_R(x)(\rho^2+\iota^2))\rho - a_I\iota_{xx} - (b_I(x) + c_I(x)(\rho^2+\iota^2))\iota, \quad (6.244)$$

$$\iota_t = a_I\rho_{xx} + (b_I(x) + c_I(x)(\rho^2+\iota^2))\rho + a_R\iota_{xx} + (b_R(x) + c_R(x)(\rho^2+\iota^2))\iota, \quad (6.245)$$

for $x \in (0,1)$, where

$$c_R(x) = \mathrm{Re}\{a_4\}\exp(-r(x)), \qquad c_I(x) = \mathrm{Im}\{a_4\}\exp(-r(x)), \quad (6.246)$$

$$r(x) = \mathrm{Re}\left\{\frac{1}{a_1}\int_{x_d}^{(1-x_d)x+x_d} a_2(\tau)\,d\tau\right\}. \quad (6.247)$$

We take the following values of the parameters (taken from [103]):

$$R_c = 47, \qquad x^t = 1.183 - 0.031j,$$

$$\omega^t_{kk} = -0.292j, \qquad \omega^t_{xx} = 0.108 - 0.057j,$$

$$\omega^t_0 = 0.690 + 0.080j + (-0.00159 + 0.00447j)(R - R_c),$$

$$k^t_0 = 1.452 - 0.844j + (0.00341 + 0.011j)(R - R_c),$$

$$k^t_x = 0.164 - 0.006j, \qquad k_0(\breve{x}) = k^t_0 + k^t_x(\breve{x} - x^t),$$

$$\omega_0(\breve{x}) = \omega^t_0 + \frac{1}{2}\omega^t_{xx}(\breve{x} - x^t)^2,$$

$$a_1 = \frac{1}{2}j\omega^t_{kk}, \qquad a_2(\breve{x}) = \omega^t_{kk}k_0(\breve{x}),$$

$$a_3(\breve{x}) = -\left(\omega_0(\breve{x}) + \frac{1}{2}\omega^t_{kk}k_0^2(\breve{x})\right)j, \qquad a_4 = -0.0225 + 0.0671j.$$

The open-loop plant response for the nonlinear system at the Reynolds number $Re = 60$ for $x_d = -7$ is shown in Figure 6.8 (only ρ is shown; ι looks qualitatively the same). The system is linearly unstable but is kept bounded by cubic

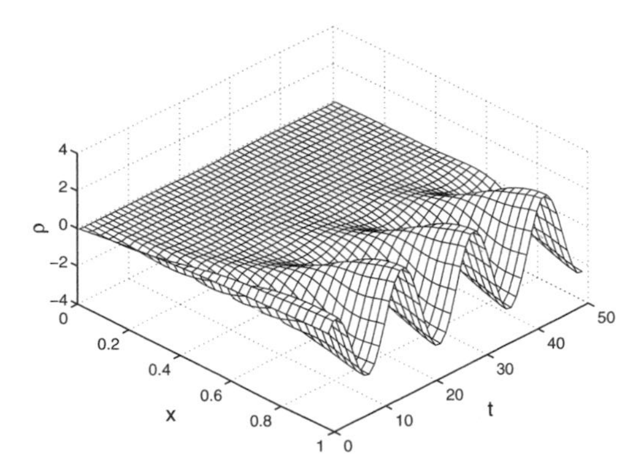

Figure 6.8 Open-loop response of the nonlinear Ginzburg-Landau equation.

nonlinearities and goes into a quasi-steady/limit-cycling motion reminiscent of vortex shedding.

Observer. If we are using an observer for state estimation only, without a controller that suppresses vortex shedding, the observer must incorporate the *nonlinearities* in the system (in the same manner as an extended Kalman filter), in addition to linear output injection designed by the backstepping method.

Therefore, we propose an observer that consists of a copy of (6.244), (6.245) plus linear output injection as in (6.179), (6.180). Figure 6.9 (top) shows the convergence of that observer, despite the plant undergoing unsteady motion. In Figure 6.9 (bottom), output injection gains are shown.

Output-feedback controller. The top graph in Figure 6.10 shows the control gains $k_x(1, y)$ and $k_{c,x}(1, y)$ used in (6.219)–(6.220). When a stabilizing controller is present, simulations show that one can use either a linear or a nonlinear observer. In Figure 6.10 (bottom), the closed-loop response with a nonlinear observer is shown. Although our controller (6.219)–(6.220) is linear, it is easy to understand why it is stabilizing for large initial conditions (the initial conditions of the uncontrolled vortex shedding). This is due to the nonlinearities being cubic *damping* terms, which have a stabilizing effect on large states. The ability of our linear controller to stabilize vortex shedding is in agreement with recent results by Lauga and Bewley [78], where linear $\mathcal{H}_2/\mathcal{H}_\infty$ optimal control methods were used for a spatially discretized Ginzburg-Landau model, and stabilization was achieved up to $Re = 97$. Their controller is structurally similar to ours—a linear state-feedback controller plus an observer consisting of a copy of the nonlinear system and linear output injection. The difference is twofold: our design is for the continuum model, and it places both the sensor(s) (in addition to the actuator(s)) on the bluff body. It was pointed out in [78] that stabilization becomes increasingly difficult when the Reynolds number and the number of open-loop unstable modes are increased, as these unstable modes become nearly uncontrollable and unobservable. When designed for the exact Reynolds number of the plant, our controller with nonlinear observer stabilizes the nonlinear plant (6.244)–(6.245) up to $Re = 127$. A controller designed for

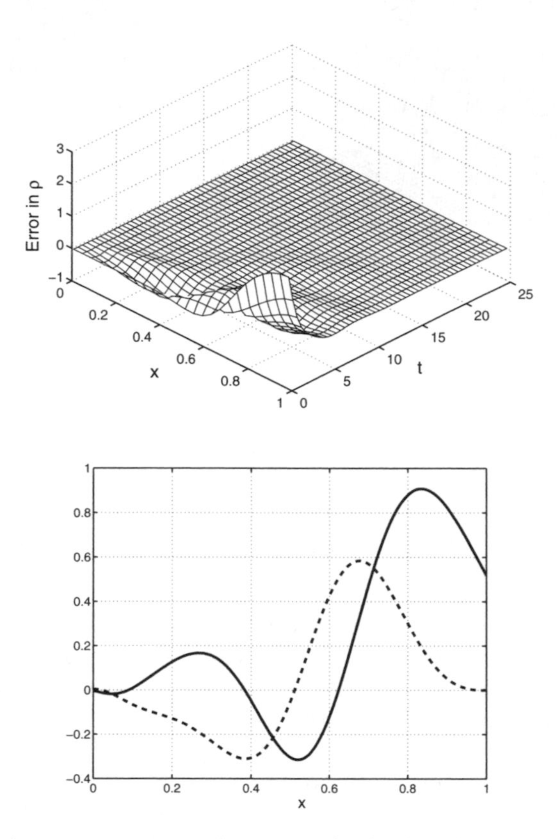

Figure 6.9 Top: observer error converging to zero for the linearly unstable nonlinear plant. Bottom: output injection gains $p_1(x)$ (solid line) and $p_{c,1}(x)$ (dashed line).

$Re = 60$ stabilizes the nonlinear plant up to $Re = 78$, while a controller designed for $Re = 80$ stabilizes the nonlinear plant up to $Re = 88$. This indicates some degree of robustness.

A fully linear compensator. As mentioned above, simulations show that either a nonlinear or a linear observer suffices in the presence of a stabilizing controller. When the observer is linear one can take a Laplace transform of the observer and get a transfer function of the linear compensator. The compensator is two-input-two-output; however, owing to the symmetry in the plant, only two of the four transfer functions are different. Figure 6.11 shows the Bode plots of the transfer function of the linear compensator, as well as the stable closed-loop response of the nonlinear plant under the linear compensator. The linear compensator can be approximated very accurately with a 10th-order reduced model, which is stable and minimum phase.

An alternative actuator/sensor architecture. In Figure 6.11 we used an opportunity to show that actuation/sensing can also be done in a Dirichlet/Neumann configuration (in addition to the Neumann/Dirichlet configuration used in Fig. 6.10).

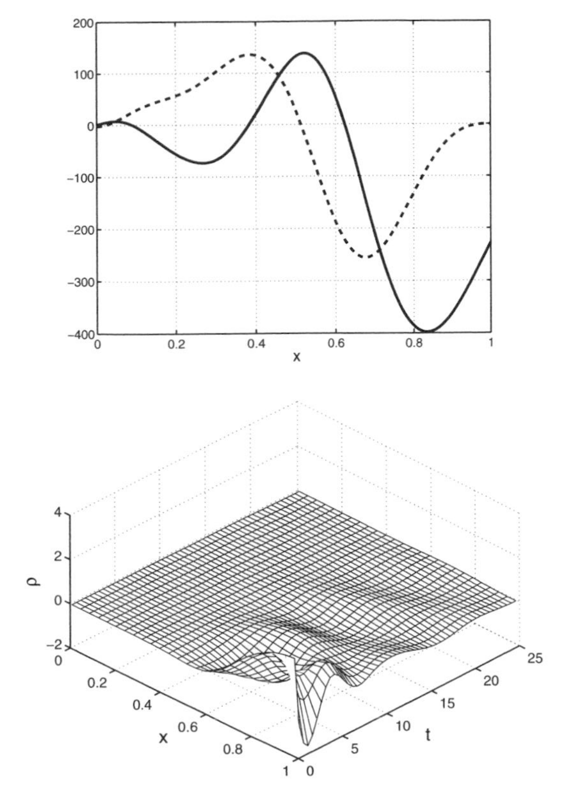

Figure 6.10 Top: state-feedback gain kernels $k_x(1,x)$ (solid line) and $k_{c,x}(1,x)$ (dashed line). Bottom: closed-loop response for the nonlinear Ginzburg-Landau equation with linear output-feedback controller and nonlinear observer.

In this case we used the measurements of $y_R(t) = \rho_x(1,t)$ and $y_I(t) = \iota_x(1,t)$, the linear observer (6.179)–(6.180) with output injection gains

$$p_1(x) = a_R k_1(x) + a_I k_{c,1}(x), \tag{6.248}$$

$$p_{c,1}(x) = -a_I k_1(x) + a_R k_{c,1}(x), \tag{6.249}$$

and actuation via $\rho(1,t) = \hat{\rho}(1,t) = U_R(t)$ and $\iota(1,t) = \hat{\iota}(1,t) = U_I(t)$. Physically, this corresponds, for instance, to micro- or synthetic jet actuators and pressure sensors distributed on the cylinder surface. Figure 6.11 shows the compensator Bode plots from $\rho_x(1)$ to $\rho(1)$ (solid line) and from $\iota_x(1)$ to $\rho(1)$ (dashed line).

6.9 NOTES AND REFERENCES

The material presented in this chapter is based on [2], [3], and [67]. Besides backstepping, simple passivity-based boundary feedbacks stabilize the Schrödinger equation [45, 87]; however, they do not achieve so-called "rapid stabilization," that

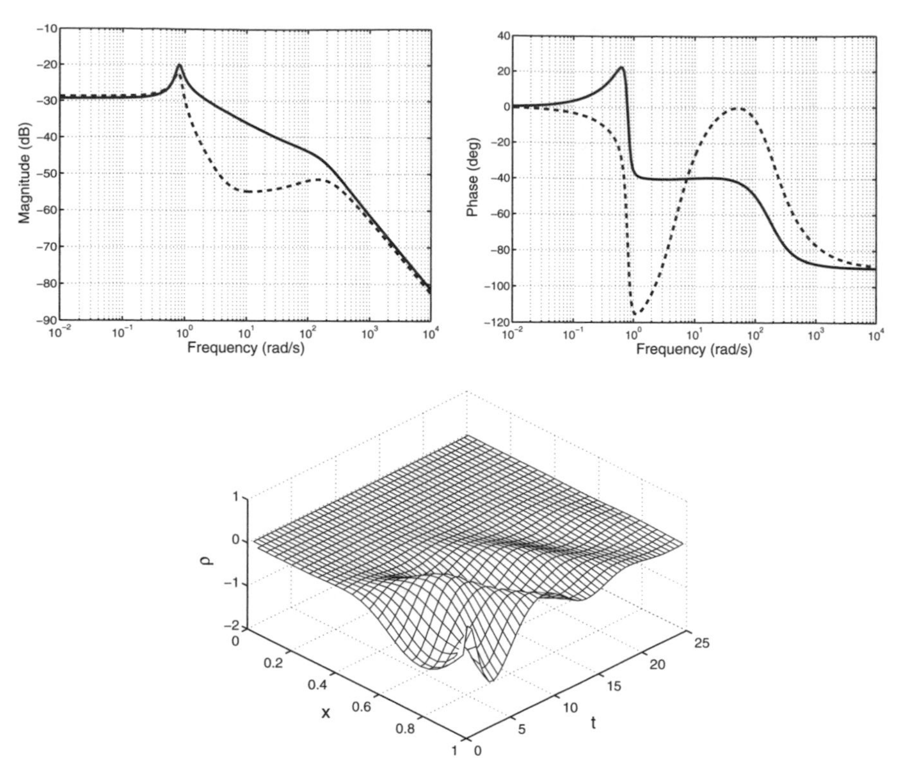

Figure 6.11 Top: compensator transfer functions from $\rho_x(1)$ to $\rho(1)$ (solid line) and $\iota_x(1)$ to $\rho(1)$ (dashed line). Bottom: closed-loop plant response.

is, the closed-loop system cannot have an arbitrary decay rate. For exact controllability and observability results for the Schrödinger equation, see [74, 86, 97].

Controllers for the Ginzburg-Landau model have previously been designed for finite-dimensional approximations of equation (6.70) in [76, 77, 78] for the linearized model and in [1] for the nonlinear model. Numerical investigations based on the Navier-Stokes equation are numerous; see, for instance, [44, 47, 96]. In [95, 103], it was shown numerically that the Ginzburg-Landau model for Reynolds numbers close to the critical Reynolds number for the onset of vortex shedding can be stabilized using proportional feedback from a single measurement downstream of the cylinder, to local forcing at the location of the cylinder. Lauga and Bewley [78] provide an excellent review of modeling aspects using (6.70), as well as an overview of previous work on stabilization of bluff body flows. We refer the reader to that reference for further details. In [43], an optimal solution to a boundary control problem formulated for a stationary Ginzburg-Landau model of superconductivity defined on a bounded domain was shown to exist, and the optimality system of equations was solved by employing the finite element method.

PART II
Adaptive Schemes

Chapter Seven

Systematization of Approaches to Adaptive Boundary Stabilization of PDEs

Having presented our underlying nonadaptive control designs in Chapters 2–6, we are now ready to broach the main subject of the book—control of PDEs with parametric uncertainties.

This tutorial chapter is an informal but rather comprehensive catalog of design tools, meant to serve as an entry point for a reader with little background in adaptive control. The chapter presents an introduction to the key ideas employed in the synthesis of adaptive boundary controllers for PDEs. Its purpose is also to help the reader of the subsequent technical Chapters 8–13, which contain the detailed proofs for most of the designs presented here, to understand the broader context of the individual approaches and the trade-offs and interrelations between the approaches. For this reason, our presentation proceeds through a series of benchmark examples.

7.1 CATEGORIZATION OF ADAPTIVE CONTROLLERS AND IDENTIFIERS

Stability is the central issue in adaptive control because one often starts with an unstable plant and no knowledge of its parameters. Approaches to adaptive control can be divided on the basis of how closed-loop stability is achieved into two groups:

- Lyapunov approach,
- certainty equivalence approach.

The *Lyapunov approach* directly addresses the issue of closed-loop stability and results in controllers and identifiers designed jointly, with all the states of the closed-loop system (plant, parameter estimator, state estimator) incorporated into a single Lyapunov function. Lyapunov adaptive controllers possess the best transient performance properties; however, they are often more complex, and this approach is not as broadly applicable as the certainty equivalence approach.

The *certainty equivalence* approach refers to a broad group of methods where the controller and the identifier are designed separately. The controller is designed in a form parametrized by the unknown parameters[1] as if they were known. The

[1]In this book we consider only indirect approaches, as the direct approaches do not naturally extend from ODEs to PDEs (one may have an infinite-dimensional parametrization of the controller, even if the plant has only one scalar unknown parameter).

parameter identifier is designed separately, without taking closed-loop stability into account but only with the objective that the parameter estimation error be bounded and the output estimation error and the derivative of the parameter estimate be square integrable in time. That the controller-identifier pair would guarantee closed-loop stability is highly nonobvious and typically difficult to prove, resulting in transient performance inferior to the Lyapunov design. However, certainty equivalence approaches have an advantage in implementation because they combine easier-to-design controller and identifier modules.

Parameter identifiers for use in the certainty equivalence approach can be split into two classes:

- passivity-based identifiers,
- swapping identifiers.

The *passivity-based* method uses a copy of the plant, with the unknown parameter replaced by its estimate, to generate a parametric model that is passive from the parameter estimation error to the error between the plant state and the state of its copy. Sometimes this method is referred to as the observer-based method because it uses a copy of the plant. We avoid this name because it is misleading—the observer does not serve the purpose of state estimation—and in fact, this method is seldom used in output-feedback adaptive control problems.

The *swapping* method is perhaps the most common method of parameter estimation in adaptive control. Filters of the regressor and of the measured part of the plant are implemented to convert a dynamic parametrization of the problem (given by the plant's dynamic model) into a static parametrization where standard gradient and least squares estimation techniques can be used. Because of the prevalence of this method, it is often (incorrectly) referred to in the literature as simply the gradient or least squares method, even though such terms describe only the form of the update law and not the approach used to eliminate the dynamics from the parametrization of the problem. The swapping method uses the highest order of dynamics of all identifier approaches. Lyapunov is the lowest in this respect as it incorporates only the dynamics of the parameter update, and the passivity-based approach is better than swapping because it uses only one filter, as opposed to one filter per unknown parameter in the case of the swapping approach. Despite its high dynamic order, the swapping approach is popular because it is the most transparent (its stability proof is the simplest, owing to the static parametrization) and it is the only method that allows least squares estimation.

Both the passivity-based approach and the swapping approach can be applied to plant models that are linear (affine, to be precise) in the unknown parameter (we refer to them as u-models, since the state of a plant is denoted by $u(x, t)$). However, these two methods are also applicable to models that arise in stability analysis of the controlled PDE systems ("target systems," or w-models). Such models are linear in parameter estimation errors. While more complicated than the basic plant models, and thus leading to somewhat more complicated identifiers, they result in easier closed-loop analysis because the error systems corresponding to the control problem and those corresponding to the identification problem are the same.

We will therefore be developing four categories of identifiers for the certainty equivalence approach: u-passive, w-passive, u-swapping, and w-swapping. In summary, taking into account also the Lyapunov approach, adaptive controllers will be developed in the following categories:

Lyapunov	Certainty Equivalence			
	passive		swapping	
	u-passive	w-passive	u-swapping	w-swapping

This categorization is consistent with the structure of backstepping tools for adaptive control of nonlinear ODEs, summarized in Appendix A.

7.2 BENCHMARK SYSTEMS

We illustrate different approaches to adaptive control design on three benchmark plants with parametric uncertainties appearing in various ways in the domain and in the boundary condition:

λ-system

$$u_t = u_{xx} + \lambda u, \tag{7.1}$$

$$u(0,t) = 0, \tag{7.2}$$

g-system

$$u_t = u_{xx} + gu(0,t), \tag{7.3}$$

$$u_x(0,t) = 0, \tag{7.4}$$

q-system

$$u_t = u_{xx}, \tag{7.5}$$

$$u_x(0,t) = -qu(0,t). \tag{7.6}$$

The parameters λ, g, and q in (7.1), (7.3), and (7.6), respectively, are assumed to be unknown. In this chapter the arguments of the function $u(x,t)$ will be suppressed whenever possible, to reduce notation. Symbols $u(0)$ and $u_x(0)$ refer to the boundary conditions at $x = 0$, where the dependence on time is suppressed in the notation.

All three systems are controlled at the boundary $x = 1$:

$$u(1,t) = U(t). \tag{7.7}$$

In the absence of control ($U = 0$), the corresponding systems are unstable for $\lambda > \pi^2$, $g > 2$, $q > 1$.

7.3 CONTROLLERS

For λ, g, q known, explicit controllers for the corresponding systems were derived in Chapter 3. With λ, g, q replaced by their estimates $\hat{\lambda}$, $\hat{g}$, $\hat{q}$, the controllers become:

λ-controller

$$U(t) = -\hat{\lambda} \int_0^1 x \frac{I_1\left(\sqrt{\hat{\lambda}(1-x^2)}\right)}{\sqrt{\hat{\lambda}(1-x^2)}} u(x,t)\, dx, \tag{7.8}$$

g-controller

$$U(t) = -\int_0^1 \sqrt{\hat{g}} \sinh\left(\sqrt{\hat{g}}(1-x)\right) u(x,t)\, dx, \tag{7.9}$$

q-controller

$$U(t) = -\int_0^1 \hat{q} e^{\hat{q}(1-x)} u(x,t)\, dx. \tag{7.10}$$

The respective transformations and their inverses are given by:

λ-transformation

$$w(x,t) = u(x,t) + \int_0^x \hat{\lambda}\xi \frac{I_1\left(\sqrt{\hat{\lambda}(x^2-\xi^2)}\right)}{\sqrt{\hat{\lambda}(x^2-\xi^2)}} u(\xi,t)\, d\xi, \tag{7.11}$$

$$u(x,t) = w(x,t) - \int_0^x \hat{\lambda}\xi \frac{J_1\left(\sqrt{\hat{\lambda}(x^2-\xi^2)}\right)}{\sqrt{\hat{\lambda}(x^2-\xi^2)}} w(\xi,t)\, d\xi, \tag{7.12}$$

g-transformation

$$w(x,t) = u(x,t) + \int_0^x \sqrt{\hat{g}} \sinh\left(\sqrt{\hat{g}}(x-\xi)\right) u(\xi,t)\, d\xi, \tag{7.13}$$

$$u(x,t) = w(x,t) + \hat{g} \int_0^x (x-\xi) w(\xi,t)\, d\xi, \tag{7.14}$$

q-transformation

$$w(x,t) = u(x,t) + \int_0^x \hat{q} e^{\hat{q}(x-\xi)} u(\xi,t)\, d\xi, \tag{7.15}$$

$$u(x,t) = w(x,t) + \hat{q} \int_0^x w(\xi,t)\, d\xi. \tag{7.16}$$

While for $\hat{\lambda} = \lambda$, $\hat{g} = g$, $\hat{q} = q$ the transformed variables are governed by the heat equation $w_t = w_{xx}$, when the estimates are imperfect and, moreover, time varying, the transformed systems are much more complicated:

λ-target system

$$w_t = w_{xx} + \dot{\hat{\lambda}} \int_0^x \frac{\xi}{2} w(\xi, t)\, d\xi + \tilde{\lambda} w, \tag{7.17}$$

$$w(0, t) = 0, \tag{7.18}$$

$$w(1, t) = 0, \tag{7.19}$$

g-target system

$$w_t = w_{xx} + \dot{\hat{g}} \int_0^x \frac{\sinh\left(\sqrt{\hat{g}}(x-\xi)\right)}{\sqrt{\hat{g}}} w(\xi, t)\, d\xi + \tilde{g} w(0, t) \cosh\left(\sqrt{\hat{g}} x\right), \tag{7.20}$$

$$w_x(0, t) = 0, \tag{7.21}$$

$$w(1, t) = 0, \tag{7.22}$$

q-target system

$$w_t = w_{xx} + \dot{\hat{q}} \int_0^x e^{\hat{q}(x-\xi)} w(\xi, t)\, d\xi, \tag{7.23}$$

$$w_x(0, t) = -\tilde{q} w(0, t), \tag{7.24}$$

$$w(1, t) = 0, \tag{7.25}$$

where $\tilde{\lambda} = \lambda - \hat{\lambda}$, $\tilde{g} = g - \hat{g}$, $\tilde{q} = q - \hat{q}$ are the parameter estimation errors and $\dot{\hat{\lambda}}$, $\dot{\hat{g}}$, $\dot{\hat{q}}$ are the derivatives of the parameter estimates, which will be defined by the parameter update laws, yet to be designed.

7.4 LYAPUNOV DESIGN

Even for linear finite-dimensional systems, quadratic Lyapunov functions work for adaptive stabilization only for a very restrictive class of systems of relative degree one. For systems of higher relative degree, Lyapunov-based adaptive controllers become nonlinear even when the plants are linear, and the corresponding Lyapunov functions are highly nonlinear [68].

We use the following Lyapunov function:[2]

$$V = \frac{1}{2} \log\left(1 + \|w\|^2\right) + \frac{1}{2\gamma} \tilde{\theta}^2, \tag{7.26}$$

where γ is a positive adaptation gain, $\|w\|$ denotes the spatial L^2-norm of $w(x, t)$, and $\tilde{\theta}$ denotes a generic parameter estimation error, that is, $\tilde{\theta} = \tilde{\lambda}$, $\tilde{g}$, or $\tilde{q}$. The

[2]The unusual form of this Lyapunov function was inspired by Praly's idea for finite-dimensional systems with growth conditions [100], which is discussed in more detail in Chapter 8.

logarithm in (7.26) is crucial. Because of this term one can tolerate the potentially destabilizing effect of the derivatives $\dot{\hat{\lambda}}$, $\dot{\hat{g}}$, and $\dot{\hat{q}}$ in (7.17), (7.20), and (7.23).

As will be shown in Chapter 8, the update laws designed with the Lyapunov function (7.26) are:

λ-update

$$\dot{\hat{\lambda}} = \gamma \frac{\|w\|^2}{1 + \|w\|^2}, \qquad 0 < \gamma < 1, \tag{7.27}$$

g-update

$$\dot{\hat{g}} = \frac{\gamma}{1 + \|w\|^2} \mathrm{Proj}_{[\underline{g}, \bar{g}]} \left\{ w(0,t) \int_0^1 w(x,t) \cosh\left(\sqrt{\hat{g}} x\right) dx \right\}, \quad \gamma < \frac{1}{2} e^{-2\sqrt{\bar{g}}}, \tag{7.28}$$

q-update

$$\dot{\hat{q}} = \frac{\gamma}{1 + \|w\|^2} \mathrm{Proj}_{[\underline{q}, \bar{q}]} \left\{ w(0,t)^2 \right\}, \qquad \gamma < \frac{\sqrt{2}}{2} e^{-\bar{q}}. \tag{7.29}$$

Note that the update laws (7.28) and (7.29) employ parameter projection defined as

$$\mathrm{Proj}_{[\underline{\theta}, \bar{\theta}]}\{a\} = \begin{cases} 0, & \hat{\theta} = \underline{\theta} \text{ and } a < 0, \\ 0, & \hat{\theta} = \bar{\theta} \text{ and } a > 0, \\ a, & \text{else}. \end{cases} \tag{7.30}$$

It is assumed that bounds $\bar{g} > \underline{g} \geq 0$ and $\bar{q} > \underline{q} \geq 0$ are a priori known for g and q. In addition, the update laws require restrictions on the size of the adaptation gain γ. The upper bounds on γ that guarantee stability are known to the designer.

In addition to the limit on the adaptation gain, the Lyapunov update laws incorporate normalization by $1 + \|w\|^2$. This normalization slows down the adaptation to prevent the harmful effect of the derivatives $\dot{\hat{\lambda}}$, $\dot{\hat{g}}$, $\dot{\hat{q}}$ in (7.17), (7.20), and (7.23). While normalization is common in adaptive laws of the swapping type, it is seldom possible to incorporate it into Lyapunov schemes, where much more commonly the effect of fast adaptation is compensated for by additional terms in the control law.

Closed-loop adaptive systems are nonlinear even when the plants are linear. For example, in the simplest case (the λ-plant), the (transformed) closed-loop system is given by

$$w_t = w_{xx} + \frac{\gamma}{2} \frac{\|w\|^2}{1 + \|w\|^2} \int_0^x \xi w(\xi, t)\, d\xi + \tilde{\lambda} w, \tag{7.31}$$

$$w(0,t) = 0, \tag{7.32}$$

$$w(1,t) = 0, \tag{7.33}$$

$$\dot{\tilde{\lambda}} = -\gamma \frac{\|w\|^2}{1 + \|w\|^2}. \tag{7.34}$$

Besides the quadratic nonlinearities in the update law, the system has a product nonlinearity $\tilde{\lambda} w$ and the nonlinearity that has arisen from $\dot{\hat{\lambda}}$ on the right-hand side of w_t. Despite these nonlinearities, boundedness and regulation to zero are achieved globally (for arbitrarily large initial conditions of the plant $u(x,t)$). By boundedness and regulation we mean not just the properties that hold for the spatial L^2-norm $\sqrt{\int_0^1 u(x,t)^2 dx}$ but also pointwise in x. This property requires H^1 stability analysis, which goes beyond the Lyapunov function (7.26). It is shown in Chapter 8.

7.5 CERTAINTY EQUIVALENCE DESIGNS

7.5.1 u-Passive Identifier

The u-passive identifiers are designed on the basis of the parametric models (7.1)–(7.6). While the Lyapunov identifiers are finite dimensional for finite-dimensional unknown parameters (in fact, they are one-dimensional in our examples), the u-passive identifiers each employ a copy of the PDE plant, which makes them infinite dimensional even when the unknown parameter is a scalar. The observers and the update laws are stated next:

λ-system

$$\hat{u}_t = \hat{u}_{xx} + \hat{\lambda} u + \gamma \|u\|^2 (u - \hat{u}), \tag{7.35}$$

$$\hat{u}(0,t) = 0, \tag{7.36}$$

$$\hat{u}(1,t) = u(1,t), \tag{7.37}$$

$$\dot{\hat{\lambda}} = \gamma \int_0^1 (u(x,t) - \hat{u}(x,t)) u(x,t)\, dx, \tag{7.38}$$

g-system

$$\hat{u}_t = \hat{u}_{xx} + \hat{g} u(0,t) + \gamma u(0,t)^2 (u - \hat{u}), \tag{7.39}$$

$$\hat{u}_x(0,t) = 0, \tag{7.40}$$

$$\hat{u}(1,t) = u(1,t), \tag{7.41}$$

$$\dot{\hat{g}} = \gamma u(0,t) \int_0^1 (u(x,t) - \hat{u}(x,t))\, dx, \tag{7.42}$$

q-system

$$\hat{u}_t = \hat{u}_{xx}, \tag{7.43}$$

$$\hat{u}_x(0,t) = -\hat{q} u(0,t) - \gamma^2 u(0,t)^2 (u(0,t) - \hat{u}(0,t)), \tag{7.44}$$

$$\hat{u}(1,t) = u(1,t), \tag{7.45}$$

$$\dot{\hat{q}} = \gamma u(0,t)(u(0,t) - \hat{u}(0,t)). \tag{7.46}$$

The properties of these identifiers are established with the Lyapunov function

$$V = \frac{1}{2}\|u - \hat{u}\|^2 + \frac{1}{2\gamma}\tilde{\theta}^2, \tag{7.47}$$

where $\tilde{\theta}$ denotes $\tilde{\lambda}$, $\tilde{g}$, or $\tilde{q}$. It can be shown that

$$\dot{V} \leq -\left\|(u - \hat{u})_x\right\|^2 - \dot{\hat{\theta}}^2. \tag{7.48}$$

This establishes that $\|u - \hat{u}\|$ and $\tilde{\theta}$ are bounded and $\|(u - \hat{u})_x\|$ and $\dot{\hat{\theta}}$ are square integrable over infinite time. These properties are essential for proving boundedness and regulation of $u(x, t)$.

The term "passive identifier" comes from the fact that the nonlinear operator from, say, $\tilde{\lambda}$ to $\int_0^1 (u(x) - \hat{u}(x))u(x)dx$ is strictly passive. This property is achieved by adding the observer $\hat{u}$.

The terms like $+\gamma\|u\|^2(u - \hat{u})$ in (7.35) act as nonlinear damping terms whose task is to ensure square integrability of $\dot{\hat{\lambda}}$. They slow down the adaptation and act as an alternative to update law normalization.

7.5.2 w-Passive Identifier

Consider the target systems (7.17)–(7.25). These systems incorporate the unknown parameters through the parameter errors $\tilde{\lambda}$, $\tilde{g}$, and $\tilde{q}$ and thus are valid parametric models. It can be noted that, for example, (7.17) would be strictly passive from $\tilde{\lambda}$ to $\|w\|^2$ if it weren't for the perturbation $\dot{\hat{\lambda}} \int_0^x \frac{\xi}{2} w(\xi)\, d\xi$. The observers in w-passive identifiers serve the purpose of eliminating those perturbations. The identifiers are defined as follows:

λ-system

$$\widehat{w}_t = \widehat{w}_{xx} + \dot{\hat{\lambda}} \int_0^x \frac{\xi}{2} w(\xi, t)\, d\xi + \gamma^2 \|w\|^2 (w - \widehat{w}), \tag{7.49}$$

$$\widehat{w}(0, t) = 0, \tag{7.50}$$

$$\widehat{w}(1, t) = 0, \tag{7.51}$$

$$\dot{\hat{\lambda}} = \gamma \int_0^1 (w(x, t) - \widehat{w}(x, t))\, w(x, t)\, dx, \tag{7.52}$$

g-system

$$\widehat{w}_t = \widehat{w}_{xx} + \dot{\hat{g}} \int_0^x \frac{\sinh\left(\sqrt{\hat{g}}(x - \xi)\right)}{\sqrt{\hat{g}}} w(\xi, t)\, d\xi + \gamma^2 w(0, t)^2 \left(w - \widehat{w}\right), \tag{7.53}$$

$$\widehat{w}_x(0, t) = 0, \tag{7.54}$$

$$\widehat{w}(1, t) = 0, \tag{7.55}$$

$$\dot{\hat{g}} = \gamma w(0, t) \int_0^1 (w(x, t) - \widehat{w}(x, t)) \cosh\left(\sqrt{\hat{g}} x\right)\, dx, \tag{7.56}$$

q-system

$$\widehat{w}_t = \widehat{w}_{xx} + \dot{\hat{q}} \int_0^x e^{\hat{q}(x - \xi)} w(\xi, t)\, d\xi, \tag{7.57}$$

$$\widehat{w}_x(0, t) = -\gamma^2 w(0, t)^2 (w(0, t) - \widehat{w}(0, t)), \tag{7.58}$$

$$\widehat{w}(1, t) = 0, \tag{7.59}$$

$$\dot{\hat{q}} = \gamma w(0, t)\, (w(0, t) - \widehat{w}(0, t)). \tag{7.60}$$

The properties of these identifiers are established with the Lyapunov function

$$V = \frac{1}{2}\|w - \widehat{w}\|^2 + \frac{1}{2\gamma}\tilde{\theta}^2, \tag{7.61}$$

where $\tilde{\theta}$ denotes $\tilde{\lambda}$, $\tilde{g}$, or $\tilde{q}$. It can be shown that

$$\dot{V} \leq -\|(w - \widehat{w})_x\|^2 - \dot{\hat{\theta}}^2, \tag{7.62}$$

which implies boundedness of $\|w - \widehat{w}\|$ and $\tilde{\theta}$ and square integrability of $\|(w - \widehat{w})_x\|$ and $\dot{\hat{\theta}}$.

It is evident in the target system observers (7.49), (7.53), and (7.57) that the explicit form of the backstepping transformations is the key to designing the w-passive identifiers.

7.5.3 *u*-Swapping Identifier

This class of identifiers would be the most readily recognizable for a reader with lay knowledge of identification. Filters are employed that convert the dynamic models (7.1)–(7.6) into static parametric models. For all three problems the update law

is chosen as the normalized gradient scheme

$$\dot{\hat{\theta}} = \gamma \frac{\int_0^1 (u(x,t) - \hat{\theta}v(x,t) - \eta(x,t))v(x,t)\,dx}{1 + \|v\|^2}, \tag{7.63}$$

where $\hat{\theta}$ denotes $\hat{\lambda}$, $\hat{g}$, or $\hat{q}$. Of the two filters η and v, one is common to all three systems,[3]

$$\eta_t = \eta_{xx}, \tag{7.64}$$

$$\eta_x(0,t) = 0, \tag{7.65}$$

$$\eta(1,t) = u(1,t), \tag{7.66}$$

and the other is given as:

<u>λ-system</u>

$$v_t = v_{xx} + u, \tag{7.67}$$

$$v(0,t) = 0, \tag{7.68}$$

$$v(1,t) = 0, \tag{7.69}$$

<u>g-system</u>

$$v_t = v_{xx} + u(0,t), \tag{7.70}$$

$$v_x(0,t) = 0, \tag{7.71}$$

$$v(1,t) = 0, \tag{7.72}$$

[3]For the λ-system, $\eta(0) = 0$ rather than $\eta_x(0) = 0$.

q-system

$$v_t = v_{xx}, \tag{7.73}$$
$$v_x(0,t) = -u(0,t), \tag{7.74}$$
$$v(1,t) = 0. \tag{7.75}$$

For the Lyapunov function

$$V = \frac{1}{2}\|u - \theta v - \eta\|^2 + \frac{1}{2\gamma}\tilde{\theta}^2, \tag{7.76}$$

it can be proved that

$$\dot{V} \le -\frac{1}{2}\left\|(u - \theta v - \eta)_x\right\|^2 - \frac{1}{2\gamma^2}\dot{\hat{\theta}}^2. \tag{7.77}$$

The above identifiers look extremely simple; however, they employ the highest dynamic order, and the proof of stability for the u-swapping scheme is the most complicated of all the schemes because the regressor in the output estimation error $v\tilde{\theta}$ is not closely related to the regressor in the target system. For instance, in the g-system, the former regressor is v (which is a filtered version of $u(0)$), whereas the latter regressor is $u(0)\cosh(\sqrt{\hat{g}}x)$.

7.5.4 u-Swapping Identifier

The w-swapping identifiers use the target systems (7.17)–(7.25) as the parametric models. They all employ the update law

$$\dot{\hat{\theta}} = \gamma \frac{\int_0^1 (w(x,t) - \hat{\theta}p(x,t) - \psi(x,t))p(x,t)\,dx}{1 + \|p\|^2} \tag{7.78}$$

and two separate filters, p and ψ, defined as:

λ-system

$$p_t = p_{xx} + w, \tag{7.79}$$
$$p(0,t) = 0, \tag{7.80}$$
$$p(1,t) = 0, \tag{7.81}$$
$$\psi_t = \psi_{xx} + \dot{\hat{\lambda}} \int_0^x \frac{\xi}{2} w(\xi,t)\,d\xi - w\hat{\lambda}, \tag{7.82}$$
$$\psi(0,t) = 0, \tag{7.83}$$
$$\psi(1,t) = 0, \tag{7.84}$$

g-system

$$p_t = p_{xx} + w(0,t)\cosh\left(\sqrt{\hat{g}}x\right), \tag{7.85}$$

$$p_x(0,t) = 0, \tag{7.86}$$

$$p(1,t) = 0, \tag{7.87}$$

$$\psi_t = \psi_{xx} + \dot{\hat{g}} \int_0^x \frac{\sinh\left(\sqrt{\hat{g}}(x-\xi)\right)}{\sqrt{\hat{g}}} w(\xi,t)\,d\xi - \hat{g}w(0,t)\cosh\left(\sqrt{\hat{g}}x\right), \tag{7.88}$$

$$\psi_x(0,t) = 0, \tag{7.89}$$

$$\psi(1,t) = 0, \tag{7.90}$$

q-system

$$p_t = p_{xx}, \tag{7.91}$$

$$p_x(0,t) = -w(0,t), \tag{7.92}$$

$$p(1,t) = 0, \tag{7.93}$$

$$\psi_t = \psi_{xx} + \dot{\hat{q}} \int_0^x e^{\hat{q}(x-\xi)} w(\xi,t)\,d\xi, \tag{7.94}$$

$$\psi(0,t) = \hat{q}w(0,t), \tag{7.95}$$

$$\psi(1,t) = 0. \tag{7.96}$$

For the Lyapunov function

$$V = \frac{1}{2}\|w - \theta p - \psi\|^2 + \frac{1}{2\gamma}\tilde{\theta}^2, \tag{7.97}$$

it can be proved that

$$\dot{V} \le -\frac{1}{2}\left\|(w - \theta p - \psi)_x\right\|^2 - \frac{1}{2\gamma^2}\dot{\hat{\theta}}^2. \tag{7.98}$$

7.6 TRADE-OFFS BETWEEN THE DESIGNS

With five designs per benchmark problem, the reader probably wonders why so many different designs are needed. As we explain below, each design has some advantage over the others, so it is important to be aware of all five design options.

7.6.1 Dynamic Order

The Lyapunov identifier clearly has the lowest dynamic order. It employs only one scalar differential equation, whereas the other designs in addition incorporate PDEs. Between the certainty equivalence identifiers, the passive identifiers have a lower dynamic order than the swapping identifiers because, while the former incorporate only one PDE (the observer $\hat{u}$ or $\widehat{w}$), the latter incorporate two PDEs (two filters). While one PDE versus two PDEs may not seem a substantial difference, when the unknown parameters are spatially varying (functional) rather than constant, this becomes an important issue. In that case the swapping identifiers require a whole infinite-dimensional array of filters (since in general they use one filter per unknown parameter) or, alternatively, a certain PDE has to be solved at each time step (see Chapters 11 and 13 for more details).

7.6.2 Update Law

Both Lyapunov and swapping identifiers use simple normalization in the update law, whereas the passive identifiers utilize an unusual form of nonlinear damping in the observer. The main benefit of swapping identifiers is that they are able to employ the standard gradient and least squares estimation techniques (although the latter is not considered here).

Note also that the Lyapunov identifier in general requires knowledge of a priori bounds on the unknown parameters (with the notable exception of the λ-plant) and limits the adaptation gain. Certainty equivalence identifiers do not need these bounds, except for a special case of the problems with an unknown diffusion coefficient (considered in Chapters 8–10). In that case the parameter estimate must be kept positive at all times in all three designs.

7.6.3 Transient Performance

The Lyapunov adaptive controllers possess the best transient performance properties since all the states of the closed-loop system are incorporated into a single Lyapunov function. However, this comes at the cost of increased complexity compared to the u-passive and u-swapping identifiers because the Lyapunov identifier incorporates the change of variable $u(x) \mapsto w(x)$, which is nondynamic but nevertheless high dimensional (integration in x).

Unlike the u-passive and u-swapping identifiers, the w-passive and w-swapping identifiers are as functionally complex as the Lyapunov identifier. However, their advantage lies in the fact that they are based on the w-system as the parametric model. This quality endows them with transient performance properties that are easier to quantify, as demonstrated in [68] for finite-dimensional nonlinear systems (note that the x-models in [68] correspond to u-models here and the z-models in [68] correspond to w-models here).

7.7 STABILITY

While stability of the identifiers (taken separately from the plant stability) is quite immediate for some of the schemes presented above, the *closed-loop* stability of the complete dynamics—consisting of the plant, controller, parameter update law, and the filters—is far from immediate. Even in the simplest of cases the stability analysis is quite involved, as in the case of classical adaptive control for linear ODEs [51]. In this section we outline a stability proof for one of the schemes presented earlier to give an idea of what is involved in such analysis. This proof is not contained in Chapters 8–10, where only u-schemes are presented.

Consider the w-passive identifier (7.49)–(7.52) for the λ-system given in its error form by (7.17)–(7.19). From (7.61), (7.62) we get boundedness of $\|w - \widehat{w}\|$ and $\tilde{\theta}$ and square integrability of $\|(w - \widehat{w})_x\|$ and $\dot{\hat{\lambda}}$. However, it is not the boundedness of $w-\widehat{w}$ that we need but the boundedness of both w and $\hat{w}$. Consider the Lyapunov function

$$U = \frac{1}{2}\int_0^1 \widehat{w}^2\,dx. \tag{7.99}$$

With some calculations that involve integration by parts and the Cauchy-Schwartz and triangle inequalities, we get

$$\dot{U} \le -\|\hat{w}_x\|^2 + \frac{|\dot{\hat{\lambda}}|}{2}\|\hat{w}\|(\|\hat{w}\| + \|e\|) + \gamma^2\|w\|(\|\hat{w}\| + \|e\|)\|\hat{w}\|\|e\|, \tag{7.100}$$

where $e = w - \hat{w}$. Denoting

$$l = \left(\frac{1}{2}|\dot{\hat{\lambda}}| + \gamma^2\|w\|\|e\|\right)^2 \tag{7.101}$$

and using Poincaré's and Young's inequalities, we get

$$\dot{U} \le -\frac{1}{4}U + 6l(t)U + \frac{1}{4}\|e(t)\|^2. \tag{7.102}$$

Since the functions $l(t)$ and $\|e(t)\|^2$ are integrable over infinite time,[4] using Lemma D.3, we get from (7.102) that $\|\hat{w}(t)\|$ is bounded and square integrable, which, together with the boundedness and square integrability of $\|e(t)\|$, implies the same properties for $\|w(t)\|$ and $\dot{\hat{\lambda}}$.

One of the difficulties in working with PDEs is that the boundedness of $\|w\|$ does not imply boundedness of $w(x,t)$ pointwise in x. To show the boundedness of $|w(x,t)|$ for all $x \in [0,1]$, we can show the boundedness of $\|w_x(t)\|$ and use Agmon's inequality. To this end, we first derive the bound

$$\frac{1}{2}\frac{d}{dt}\|w_x(t)\|^2 \le \left(\tilde{\lambda}^2 + \frac{|\dot{\hat{\lambda}}|^2}{4}\right)\|w(t)\|^2. \tag{7.103}$$

Since $\tilde{\lambda}$, $\dot{\hat{\lambda}}$ are bounded and $\|w(t)\|$ is integrable, integrating (7.103) with respect to time we get the boundedness of $\|w_x(t)\|$.

To prove regulation, we note from (7.17)–(7.19) that

$$\frac{1}{2}\left|\frac{d}{dt}\|w\|^2\right| \le \|w_x\|^2 + |\dot{\hat{\lambda}}|\|w\|^2 + |\tilde{\lambda}|\|w\|^2 < \infty. \tag{7.104}$$

So, $\|w(t)\|^2$, $\frac{d}{dt}\|w(t)\|^2$ are bounded and $\|w(t)\|^2$ is integrable. By Barbalat's lemma (Lemma D.1), $\|w(t)\| \to 0$ as $t \to \infty$. To show the convergence to zero

[4] It can be shown independently of (7.62) that $\|w\|^2\|e\|^2$ in l is integrable.

for all $x \in [0, 1]$, we use Agmon's inequality and the fact that $\|w_x(t)\|$ is bounded:

$$\lim_{t\to\infty} \max_{x\in[0,1]} |w(x,t)| \le \lim_{t\to\infty} (2\|w(t)\|\|w_x(t)\|)^{1/2} = 0. \tag{7.105}$$

The remaining step is to show that the properties established for w also hold for u. This is done by proving from (7.12) that $\|u\| \le C\|w\|$ and $\max_{x\in[0,1]} |u(x,t)| \le C \max_{x\in[0,1]} |w(x,t)|$ for some finite C, which yields boundedness and regulation of $u(x,t)$ pointwise in $x \in [0, 1]$.

7.8 NOTES AND REFERENCES

Our systematization of the adaptive schemes for PDEs, first presented in [63], follows standard classification in finite-dimensional nonlinear adaptive control; see [68] (or Appendix A for a summary of finite-dimensional adaptive backstepping designs).

We should point out that the u-passive designs for g- and q-plants, as well as all w-passive and w-swapping designs, are presented only in this chapter and are not pursued in the rest of the book.

Chapter Eight

Lyapunov-Based Designs

In this section we present the first of our three adaptive design approaches, the Lyapunov design.

The Lyapunov design that we employ for linear PDEs is inspired by an idea Praly [100] developed for adaptive nonlinear control under growth conditions. Since PDE problems in this book are linear, we significantly simplify Praly's approach; however, we retain its main feature—a logarithm weight on the plant state in the Lyapunov function. This results in a normalization of the update law by a norm of the plant state, which is uncommon for Lyapunov designs employing backstepping controllers.

Except for some special examples, projection is needed in our approach in this chapter to keep the parameters within an a priori set. Projection is not used as a robustification tool but to prevent adaptation transients that would require overly conservative restrictions on the size of the adaptation gain. The projection set may be taken conservatively (large); however, in order for stability to be guaranteed, the adaptation gain needs to be taken inversely proportional to the size of the parameter set. The bounds on the gain can be derived explicitly and are a priori verifiable.

We start the chapter with a design for a benchmark system in Sections 8.1–8.2. In Section 8.3 we use this benchmark plant to show how to prove well-posedness of the closed-loop systems given a priori bounds guaranteed by the designs (well-posedness is assumed for the rest of Part II of this book) . The robustness properties of the adaptation laws are discussed in Section 8.4. In Section 8.5 we present an alternative approach to the Lyapunov design. Extensions to other systems are pursued in Sections 8.6–8.7. In Section 8.8 we numerically demonstrate that our design is successful in the presence of additive disturbances in the plant, measurement noise, and non-smooth initial conditions.

8.1 PLANT WITH UNKNOWN REACTION COEFFICIENT

Consider the plant

$$u_t(x,t) = u_{xx}(x,t) + \lambda u(x,t), \tag{8.1}$$

$$u(0,t) = 0, \tag{8.2}$$

$$u_x(1,t) = U(t), \tag{8.3}$$

where λ is an unknown constant parameter. We use a Neumann boundary controller designed in Chapter 2 in the form

$$U(t) = -\frac{\hat{\lambda}}{2}u(1,t) - \hat{\lambda}\int_0^1 \xi \frac{I_2\left(\sqrt{\hat{\lambda}(1-\xi^2)}\right)}{1-\xi^2} u(\xi,t)\,d\xi, \tag{8.4}$$

which employs the measurements of $u(x)$ for $x \in [0,1]$ and an estimate $\hat{\lambda}$ of λ. Consider an invertible change of variable:

$$w(x,t) = u(x,t) - \int_0^x k(x,\xi,\hat{\lambda})u(\xi,t)\,d\xi, \tag{8.5}$$

$$k(x,\xi,\hat{\lambda}) = -\hat{\lambda}\xi \frac{I_1\left(\sqrt{\hat{\lambda}(x^2-\xi^2)}\right)}{\sqrt{\hat{\lambda}(x^2-\xi^2)}}. \tag{8.6}$$

Lemma 8.1. *The transformation* (8.5)–(8.6) *maps the system* (8.1)–(8.4) *into*

$$w_t = w_{xx} + \dot{\hat{\lambda}}\int_0^x \frac{\xi}{2} w(\xi)\,d\xi + \tilde{\lambda}w, \tag{8.7}$$

$$w(0,t) = 0, \tag{8.8}$$

$$w_x(1,t) = 0, \tag{8.9}$$

where $\tilde{\lambda} = \lambda - \hat{\lambda}$ *is the parameter estimation error.*

Proof. Boundary conditions (8.8) and (8.9) are obviously satisfied. Substituting (8.5) into (8.1), we get

$$w_t = w_{xx} - \dot{\hat{\lambda}}\int_0^x k_{\hat{\lambda}}(x,\xi,\hat{\lambda})u(\xi,t)\,d\xi + \tilde{\lambda}w, \tag{8.10}$$

To replace u with w we use an inverse transformation (8.32) with a kernel (8.33). We have

$$\begin{aligned}
&\int_0^x k_{\hat{\lambda}}(x,\xi,\hat{\lambda})u(\xi,t)\,d\xi \\
&\quad = \int_0^x k_{\hat{\lambda}}(x,\xi,\hat{\lambda})\left(w(\xi,t) + \int_0^\xi l(\xi,\eta,\hat{\lambda})w(\eta,t)\,d\eta\right)d\xi \\
&\quad = \int_0^x \left(k_{\hat{\lambda}}(x,\xi,\hat{\lambda}) + \int_\xi^x k_{\hat{\lambda}}(x,\eta,\hat{\lambda})l(\eta,\xi,\hat{\lambda})\,d\eta\right)w(\xi,t)\,d\xi.
\end{aligned} \tag{8.11}$$

The inner integral is computed as follows:

$$\begin{aligned}
&\int_{\xi}^{x} k_{\hat{\lambda}}(x,\eta,\hat{\lambda})l(\eta,\xi,\hat{\lambda})\,d\eta \\
&= \int_{\xi}^{x} \frac{\hat{\lambda}\eta\xi}{2} I_0\left(\sqrt{\hat{\lambda}(x^2-\eta^2)}\right) \frac{J_1\left(\sqrt{\hat{\lambda}(\eta^2-\xi^2)}\right)}{\sqrt{\hat{\lambda}(\eta^2-\xi^2)}}\,d\eta \\
&= \frac{\xi}{2}\int_0^{\sqrt{\hat{\lambda}(x^2-\xi^2)}} I_0\left(\sqrt{\hat{\lambda}(x^2-\xi^2)-s^2}\right) J_1(s)\,ds \\
&= \frac{\xi}{2}\left(I_0\left(\sqrt{\hat{\lambda}(x^2-\xi^2)}\right)-1\right).
\end{aligned} \tag{8.12}$$

Here the last integral was computed with the help of [101]. Finally, substituting (8.12) into (8.11), we get

$$\int_0^x k_{\hat{\lambda}}(x,\xi,\hat{\lambda})u(\xi,t)\,d\xi = -\int_0^x \frac{\xi}{2} w(\xi,t)\,d\xi, \tag{8.13}$$

which, combined with (8.10), gives (8.7). □

We will show that the update law,

$$\dot{\hat{\lambda}} = \gamma\frac{\|w\|^2}{1+\|w\|^2}, \qquad 0<\gamma<1, \tag{8.14}$$

achieves regulation of $u(x,t)$ to zero for all $x \in [0,1]$, for arbitrarily large initial data $u(x,0)$ and for an arbitrarily poor initial estimate $\hat{\lambda}(0)$.

Theorem 8.1. *For any initial condition $u_0 \in H^2(0,1)$ compatible with boundary conditions, and any $\hat{\lambda}(0) \in \mathbb{R}$, the classical solution $(u,\hat{\lambda})$ of the closed-loop system* (8.1)–(8.4), (8.14) *is bounded for all $x \in [0,1]$, $t \geq 0$ and*

$$\lim_{t\to\infty}\max_{x\in[0,1]}|u(x,t)| = 0. \tag{8.15}$$

Moreover, the following performance bounds hold:

$$\begin{aligned}
u(x,t)^2 \leq\, &32\left(1+3\lambda^2+\tilde{\lambda}(0)^2+\gamma\log\left(1+\|w_0\|^2\right)\right) \\
&\times\left[\|w_0'\|^2+3\sqrt{\gamma}\left(1+\|w_0\|^2\right)e^{\frac{1}{\gamma}\tilde{\lambda}(0)^2}\right. \\
&\left.\times\left(\log\left(1+\|w_0\|^2\right)+\frac{1}{\gamma}\tilde{\lambda}(0)^2\right)^{3/2}\right]
\end{aligned} \tag{8.16}$$

for all $x \in [0,1]$, $t \geq 0$, *and*

$$\int_0^\infty u(x,t)^2 dt \leq 48\left(1+3\lambda^2+\tilde{\lambda}(0)^2+\gamma\log\left(1+\|w_0\|^2\right)\right)$$

$$\times\left(1+\|w_0\|^2\right)e^{\frac{1}{\gamma}\tilde{\lambda}(0)^2}\left(\log\left(1+\|w_0\|^2\right)+\frac{1}{\gamma}\tilde{\lambda}(0)^2\right) \quad (8.17)$$

for all $x \in [0,1]$, *where* $w_0(x) = w(x,0)$.

Well-posedness of the closed-loop system, assumed in Theorem 8.1, will be shown later in Section 8.3, which can be skipped if the reader is interested only in the design procedure.

While the bound (8.16) obviously quantifies the "peak transient" performance, the bound (8.17) quantifies the rate of convergence to zero.

The nonnegative form of the adaptive law (8.14) is coincidental for this particular benchmark plant and is discussed further in Section 8.4.

It is important to note that the update law (8.14) contains normalization. Normalization is uncommon in Lyapunov designs and is the result of including the logarithm in the Lyapunov function. Normalization is necessary because the control law (8.4) is of certainty equivalence type—unlike the Lyapunov adaptive controllers in [68] which employ nonnormalized adaptation and strengthened nonlinear controllers that compensate for time-varying effects of adaptation. An additional measure to prevent overly fast adaptation in (8.14) is the restriction on the adaptation gain ($\gamma < 1$).

8.2 PROOF OF THEOREM 8.1

Consider a Lyapunov function candidate

$$V = \frac{1}{2}\log\left(1+\|w\|^2\right)+\frac{1}{2\gamma}\tilde{\lambda}^2. \quad (8.18)$$

The time derivative along the solutions of (8.7)–(8.14) can be shown to be

$$\dot{V} = -\frac{\|w_x\|^2}{1+\|w\|^2}+\frac{\dot{\hat{\lambda}}}{2}\frac{\int_0^1 w(x)\left(\int_0^x \xi w(\xi)\,d\xi\right)dx}{1+\|w\|^2} \quad (8.19)$$

(the calculation involves one step of integration by parts). Since

$$
\begin{aligned}
\left|\int_0^1 w(x)\left(\int_0^x \xi w(\xi)\,d\xi\right)dx\right| &\le \int_0^1 |w(x)|\left(\int_0^x \xi|w(\xi)|\,d\xi\right)dx \\
&\le \int_0^1 |w(x)|\left(\int_0^x \xi^2\,d\xi\right)^{1/2}\left(\int_0^x w(\xi)^2\,d\xi\right)^{1/2} dx \\
&\le \|w\| \int_0^1 |w(x)|\frac{1}{\sqrt{3}}x^{3/2}\,dx \\
&\le \frac{\|w\|}{\sqrt{3}}\|w\|\left(\int_0^1 x^3\,dx\right)^{1/2} \\
&\le \frac{1}{2\sqrt{3}}\|w\|^2 \le \frac{2}{\sqrt{3}}\|w_x\|^2,
\end{aligned}
\tag{8.20}
$$

using the fact that $|\dot{\hat{\lambda}}| < \gamma$ and substituting (8.14) into (8.19), we get

$$
\dot{V} \le -\left(1 - \frac{\gamma}{\sqrt{3}}\right)\frac{\|w_x\|^2}{1 + \|w\|^2}\,. \tag{8.21}
$$

This implies that $V(t)$ remains bounded for all time whenever $0 < \gamma \le \sqrt{3}$. From the definition of V it follows that $\|w\|$ and $\hat{\lambda}$ remain bounded for all time. However, we need to show that $w(x,t)$ is bounded for all time and for all x. To do this, consider

$$
\begin{aligned}
\frac{1}{2}\frac{d}{dt}\|w_x\|^2 &= \int_0^1 w_x w_{xt}\,dx \\
&= w_x(1,t)w_t(1,t) - w_x(0,t)w_t(0,t) - \int_0^1 w_{xx}w_t\,dx \\
&= -\int_0^1 w_{xx}^2\,dx - \tilde{\lambda}\int_0^1 w_{xx}w\,dx - \frac{\dot{\hat{\lambda}}}{2}\int_0^1 w_{xx}\int_0^x \xi w(\xi)\,d\xi \\
&= -\|w_{xx}\|^2 + \tilde{\lambda}\int_0^1 w_x^2\,dx + \frac{\dot{\hat{\lambda}}}{2}\int_0^1 x w w_x\,dx \\
&= -\|w_{xx}\|^2 + \tilde{\lambda}\|w_x\|^2 + \frac{\dot{\hat{\lambda}}}{4}\left(w(1)^2 - \|w\|^2\right).
\end{aligned}
\tag{8.22}
$$

Integration by parts was used several times to obtain the above equalities. Using Agmon's inequality (noting that $w(0) = 0$), then Young's inequality, and finally the Poincaré inequality (noting that $w_x(1) = 0$), one gets

$$
w(1)^2 - \|w\|^2 \le \|w_x\|^2 \le 4\|w_{xx}\|^2\,. \tag{8.23}
$$

Substituting (8.23) into (8.22), it follows that

$$
\frac{1}{2}\frac{d}{dt}\|w_x\|^2 \le -(1-\gamma)\|w_{xx}\|^2 + \tilde{\lambda}\|w_x\|^2 \le \tilde{\lambda}\|w_x\|^2\,. \tag{8.24}
$$

Integrating the last inequality, we obtain

$$\|w_x(t)\|^2 \leq \|w_0'\|^2 + 2 \sup_{0\leq\tau\leq t} |\tilde{\lambda}(\tau)| \int_0^t \|w_x(\tau)\|^2 \, d\tau. \tag{8.25}$$

To obtain this bound, on the one hand we have from (8.18) and (8.21) that

$$\tilde{\lambda}(t)^2 \leq \tilde{\lambda}(0)^2 + \gamma \log\left(1 + \|w_0\|^2\right). \tag{8.26}$$

On the other hand,

$$\int_0^t \|w_x(\tau)\|^2 \, d\tau \leq \sup_{0\leq\tau\leq t} \left(1 + \|w(\tau)\|^2\right) \int_0^t \frac{\|w_x(\tau)\|^2}{1 + \|w(\tau)\|^2} \, d\tau. \tag{8.27}$$

From (8.21) we have $V(\tau) \leq V(0)$, and therefore it follows from (8.18) that

$$1 + \|w(\tau)\|^2 \leq \left(1 + \|w_0\|^2\right) e^{\frac{1}{\gamma}\tilde{\lambda}(0)^2}. \tag{8.28}$$

Integrating (8.21) we get

$$\int_0^t \frac{\|w_x(\tau)\|^2}{1 + \|w(\tau)\|^2} \, d\tau \leq \frac{\log\left(1 + \|w_0\|^2\right) + \frac{1}{\gamma}\tilde{\lambda}(0)^2}{2\left(1 - \frac{\gamma}{\sqrt{3}}\right)}. \tag{8.29}$$

Substituting (8.28) and (8.29) into (8.27), and then, along with (8.26), into (8.25), we get

$$\begin{aligned} \|w_x(t)\|^2 \leq \|w_0'\|^2 &+ \frac{\sqrt{\gamma}}{1 - \frac{\gamma}{\sqrt{3}}} \left(1 + \|w_0\|^2\right) e^{\frac{1}{\gamma}\tilde{\lambda}(0)^2} \\ &\times \left(\log\left(1 + \|w_0\|^2\right) + \frac{\tilde{\lambda}(0)^2}{\gamma}\right)^{3/2}. \end{aligned} \tag{8.30}$$

By combining Agmon's and Poincaré's inequalities (and using the fact that $w(0) = 0$), we get $\max_{x\in[0,1]} |w(x)|^2 \leq 4\|w_x\|^2$; thus, $w(x,t)$ is uniformly bounded.

Next, we prove the regulation of $w(x,t)$ to zero. Using (8.7)–(8.9) and (8.20) we obtain

$$\frac{1}{2}\left|\frac{d}{dt}\|w\|^2\right| \leq \|w_x\|^2 + \left(|\tilde{\lambda}| + \frac{\gamma}{4\sqrt{3}}\right) \|w\|^2. \tag{8.31}$$

Since $\|w\|$ and $\|w_x\|$ have been proven bounded, it follows that $\frac{d}{dt}\|w\|^2$ is bounded, and thus $\|w(t)\|$ is uniformly continuous. By combining (8.27)–(8.29) with Poincaré's inequality, we also get that $\|w(t)\|^2$ is integrable in time over the infinite time interval. By Barbalat's lemma (Lemma D.1), it follows that $\|w(t)\| \to 0$ as $t \to \infty$.

To show regulation also in the maximum norm, we note that, from Agmon's inequality, $|w(x,t)|^2 \leq 2\|w(t)\|\|w_x(t)\|$. Since $\|w_x\|$ is bounded and $\|w(t)\|$ has been shown convergent to zero, the regulation in maximum norm follows.

Having proved the boundedness and regulation of w, we now set out to establish the same for u. We start by noting that

$$u(x) = w(x) + \int_0^x l(x, \xi, \hat{\lambda}) w(\xi)\, d\xi \tag{8.32}$$

$$l(x, \xi, \hat{\lambda}) = -\hat{\lambda}\xi \frac{J_1\left(\sqrt{\hat{\lambda}(x^2 - \xi^2)}\right)}{\sqrt{\hat{\lambda}(x^2 - \xi^2)}}. \tag{8.33}$$

It is straightforward to show that

$$\|u_x\|^2 \leq 2\left(1 + \hat{\lambda}^2 + 4M\right) \|w_x\|^2, \tag{8.34}$$

where

$$M = \int_0^1 \left(\int_0^1 |l_x(x, \xi, \hat{\lambda})|\, d\xi\right)^2 dx \tag{8.35}$$

$$l_x(x, \xi, \hat{\lambda}) = \hat{\lambda} x \xi \frac{J_2\left(\sqrt{\hat{\lambda}(x^2 - \xi^2)}\right)}{x^2 - \xi^2}. \tag{8.36}$$

One can show that

$$\int_0^1 |l_x(x, \xi, \hat{\lambda})|\, d\xi \leq |\hat{\lambda}| x + 1, \tag{8.37}$$

which implies

$$M \leq \int_0^1 (|\hat{\lambda}| x + 1)^2\, dx = \frac{1}{3}\hat{\lambda}^2 + |\hat{\lambda}| + 1 \leq \frac{\hat{\lambda}^2 + 3}{2}. \tag{8.38}$$

Thus, it follows that

$$\|u_x\|^2 \leq 2(4 + 3\hat{\lambda}^2)\|w_x\|^2 \leq 8(1 + 3\lambda^2 + \tilde{\lambda}^2)\|w_x\|^2. \tag{8.39}$$

Noting that $u(x, t)^2 \leq 4\|u_x\|^2$ for all $(x, t) \in [0, 1] \times [0, \infty)$, by combining (8.39), (8.26), and (8.30) and using the fact that $\frac{1}{1 - \frac{\gamma}{\sqrt{3}}} < 3$ for $\gamma < 1$, we get (8.16), which proves the uniform boundedness of u.

To prove the regulation of $u(x, t)$ to zero for all $x \in [0, 1]$, we start by noting that

$$\|u\|^2 \leq 2(1 + L)\|w\|^2, \tag{8.40}$$

where $L = \max_{0 \leq \xi \leq x \leq 1} l(x, \xi, \hat{\lambda})^2$ is finite whenever $\hat{\lambda}$ is finite (which we have proved using Lyapunov analysis). Since $\|w\|$ is regulated to zero, so is $\|u\|$. By Agmon's inequality, $u(x, t)^2 \leq 2\|u\|\|u_x\|$, where $\|u_x\|$ is bounded by (8.39), (8.26), and (8.30). This completes the proof of regulation of u.

The bound (8.17) is obtained in a similar manner to (8.16), by combining (8.39) with (8.26)–(8.29).

8.3 WELL-POSEDNESS OF THE CLOSED-LOOP SYSTEM

Owing to the nonlinear character of the adaptive closed-loop systems, the proof of boundedness and regulation is the main challenge and our primary focus in the book. The well-posedness of the closed-loop systems, although important, is of less interest because the systems are parabolic, and provided the initial data are sufficiently smooth and global a priori bounds are established, they have classical solutions. For the sake of completeness, in this section we show how well-posedness is proved for the plant considered in the previous section.

Consider the closed-loop system $(w, \tilde{\lambda})$:

$$w_t = w_{xx} + \frac{\gamma}{2}\frac{\|w\|^2}{1+\|w\|^2}\int_0^x \xi w(\xi)\,d\xi + \tilde{\lambda} w, \tag{8.41}$$

$$w(0) = w_x(1) = 0, \tag{8.42}$$

$$\dot{\tilde{\lambda}} = -\gamma \frac{\|w\|^2}{1+\|w\|^2}. \tag{8.43}$$

Let us denote $z = (w, \tilde{\lambda})^T$ and introduce the operator

$$A = \begin{pmatrix} -\dfrac{\partial^2}{\partial x^2} & 0 \\ 0 & 0 \end{pmatrix} \tag{8.44}$$

on $H = L^2(0,1) \times \mathbb{R}$, with $D(A) = \{f, g : g \in \mathbb{R}, f \in H^2(0,1), f(0) = 0, f_x(1) = 0\}$ and with the norm $\|z\|_H = \|w\|^2 + \tilde{\lambda}^2$. Let us also denote

$$F(z) \equiv \begin{pmatrix} \dfrac{\gamma}{2}\dfrac{\|w\|^2}{1+\|w\|^2}\displaystyle\int_0^x \xi w(\xi)\,d\xi + \tilde{\lambda} w \\ -\gamma \dfrac{\|w\|^2}{1+\|w\|^2} \end{pmatrix}. \tag{8.45}$$

Then (8.41)–(8.43) can be written in abstract form as

$$z_t = -Az + F(z), \tag{8.46}$$

$$z(0) = z_0. \tag{8.47}$$

We will use the following result to establish the well-posedness of (8.46)–(8.47).

Theorem 8.2 (Theorem 2.5.6 from [121]). *Consider the problem* (8.46)–(8.47). *Suppose that A is a maximal accretive operator from a dense subset $D(A)$ in a Banach space H into H. Suppose also that F is a nonlinear operator from $D(A)$ to $D(A)$ and satisfies the local Lipschitz condition. Then for any $z_0 \in D(A)$, the problem* (8.46)–(8.47) *admits a unique classical solution z such that*

$$z \in C^1([0, T_{max}), H) \cap C([0, T_{max}), D(A)). \tag{8.48}$$

Moreover, there is an alternative: either

(i) $T_{max} = +\infty$, *i.e., there is a unique global classical solution, or*
(ii) $T_{max} < +\infty$ *and* $\lim_{t \to T_{max}-0} \|z(t)\| = +\infty$.

We first note that A is a maximal accretive operator (see, e.g., [121, p. 71] for the proof of this fact). It is straightforward to show that for any $z_1, z_2 \in H$,

$$\|F(z_1) - F(z_2)\|_H \le C\|z_1 - z_2\|_H \max\{\|z_1\|_H, \|z_2\|_H\}, \tag{8.49}$$

where the constant C is independent of z_1, z_2. Therefore, F is locally Lipschitz on H. By Theorem 8.2 we get the existence and uniqueness of the classical solution.

To show that $T_{max} = +\infty$, that is, that there is no blowup, we make a priori estimates in H^2-norm (such estimates in L^2- and H^1-norms were obtained in the previous section). We first observe from the first line of (8.24) that $\|w_{xx}\|$ is square integrable over infinite time. The same property holds for $\|w_t\|$. It is then shown that

$$\begin{aligned} \frac{1}{2}\frac{d}{dt}\|w_t\|^2 + \|w_{tx}\|^2 = \tilde{\lambda}\|w_t\|^2 + \frac{\ddot{\hat{\lambda}}}{2}\int_0^1 w_t(x)\int_0^x \xi w(\xi)\,d\xi\,dx \\ + \dot{\hat{\lambda}}\int_0^1 w_t(x)\left(\int_0^x \frac{\xi}{2} w_t(\xi)\,d\xi - w(x)\right)dx \end{aligned} \tag{8.50}$$

and

$$\begin{aligned} \frac{1}{2}\frac{d}{dt}\|w_{tx}\|^2 + \|w_{txx}\|^2 = \tilde{\lambda}\|w_{tx}\|^2 + \frac{\ddot{\hat{\lambda}}}{2}\int_0^1 x w_{tx}(x) w(x)\,dx \\ + \dot{\hat{\lambda}}\int_0^1 w_{tx}(x)\left(\frac{x}{2} w_t(x) - w_x(x)\right)dx, \end{aligned} \tag{8.51}$$

where

$$\ddot{\hat{\lambda}} = \frac{\gamma}{\left(1 + \|w\|^2\right)^2}\frac{d}{dt}\|w\|^2 \tag{8.52}$$

is bounded because of (8.31). From the boundedness of $\|w\|$, $\|w_x\|$, $\dot{\hat{\lambda}}$, $\ddot{\hat{\lambda}}$ and from the square integrability in time of $\|w\|$, $\|w_t\|$, by integrating (8.50) it follows that $\|w_t\|$ is bounded and $\|w_{tx}\|$ is square integrable. Then, by integrating (8.51) and using the square integrability of $\|w_x\|$ and the other functions mentioned above, it follows that $\|w_{tx}\|$ is bounded and $\|w_{txx}\|$ is square integrable. By Agmon's inequality, we get that $w_t(x, t)$ is uniformly bounded for all values of its arguments, and the same holds for $w_{xx}(x, t)$. Therefore, we have proved that $\|z\|_H$ is bounded, so that $T_{max} = +\infty$ and the solution (8.52) is actually global.

The existence and uniqueness of the classical solution of the closed-loop system (8.1)–(8.4) now follow from the invertibility of the transformation (8.5)–(8.6), which can be used to show that $u_t(x, t)$, $u_{xx}(x, t)$ have the same properties as $w_t(x, t)$, $w_{xx}(x, t)$.

For the rest of the book we shall simply assume well-posedness of the nonlinear adaptive schemes. Using our a priori estimates, it is a matter of writing these systems in abstract form and applying Theorem 8.2.

8.4 PARAMETRIC ROBUSTNESS

Let us suppose that the adaptation is turned off, that is, $\gamma = 0$, $\dot{\hat{\lambda}} \equiv 0$. Then the closed-loop system is

$$w_t = w_{xx} + (\lambda - \hat{\lambda})w, \tag{8.53}$$

$$w(0) = 0, \tag{8.54}$$

$$w_x(1) = 0, \tag{8.55}$$

where $\hat{\lambda}$ is a constant parameter estimate. By studying the eigenvalue problem of this system, it can be shown that parameter estimates $\hat{\lambda}$ that are greater than $\lambda - \frac{\pi^2}{4}$ are exponentially stabilizing, whereas those smaller than $\lambda - \frac{\pi^2}{4}$ are destabilizing. This means that, if an upper bound on λ is known—let us denote this bound by $\bar{\lambda}$—then (8.4) is a stabilizing linear controller whenever $\hat{\lambda}$ is replaced by $\bar{\lambda}$ (or any constant value higher than $\bar{\lambda}$).

This robustness property explains why $\dot{\hat{\lambda}}$ in the adaptation law (8.14) is nonnegative: overestimating $\hat{\lambda}$ cannot be harmful within the controller structure (8.4).[1] A caveat, however, is that in the presence of measurement noise, the parameter estimate will drift. In the update law (8.14) the estimate has nowhere to drift but up[2] (which is consistent with the structure of the control law but still undesirable). In practical implementation one would add leakage, deadzone, or projection [51] to reduce or completely stop the drift.

The linear/frozen-parameter robustness is an unusual feature of the control formula (8.4). It is different from the "infinite gain margin" property of inverse optimal controllers (considered in Chapter 14), which allows an arbitrary increase of a scalar gain in front of the optimal control law. The infinite gain margin allows only an unplanned increase in the "control authority" and does not guarantee robustness to changes in the physical parameters of the system. The robustness exhibited with (8.4) is with respect to the physical parameter λ.

Owing to the ability of the controller (8.4) to remain stabilizing when λ is overestimated, it might be tempting to view the backstepping design as "high gain." This would not be appropriate, because (8.4) resorts to high gain only when λ generates a high number of unstable eigenvalues in the plant.

The form of high gain that controller (8.4) is capable of employing should not be confused with the adaptive high-gain controllers surveyed in [84], where a multiplicative gain is tuned for a controller of the form

$$u_x(1) = G\{Cu\}, \tag{8.56}$$

---

[1] While the update law (8.14) can take the estimate $\hat{\lambda}$ only "up," the growth of the estimate stops as $\|w(t)\|$ goes to zero. Since $V(t)$ is nonincreasing and bounded from below (by zero), it has a limit. Hence $\tilde{\lambda}(t)^2$ has a limit. So does $\hat{\lambda}(t)$, and it is higher than $\lambda - \frac{\pi^2}{4}$. The size of $\hat{\lambda}(\infty)$ depends on the size of the initial condition u_0.

[2] This issue is no less critical with update laws that are sign-indefinite; however, with (8.14) it is obvious.

where G is the gain and C is an output operator such that $u_x(1) \mapsto Cu$ is of relative degree one. For the present system, an operator C independent of the unknown λ cannot be found; therefore, tuning of a multiplicative gain G would not be successful.

8.5 AN ALTERNATIVE APPROACH

The use of a logarithm in the Lyapunov function (8.18) was inspired by Praly's Lyapunov adaptation designs in [100]. We do not follow that method exactly in this chapter because our PDE plants are linear. It is, however, of interest to see what an exact application of that method results in, as it has potential beyond the class of problems considered in this chapter.

Let us start by denoting

$$A = \frac{\int_0^1 w(x)\left(\int_0^x \xi w(\xi)\,d\xi\right)dx}{1+\|w\|^2}, \qquad B = 2\frac{A}{1+\|w\|^2}, \tag{8.57}$$

$$H = -A^2 + \frac{1}{1+\|w\|^2}\int_0^1 \left(\left(\int_0^x \xi w(\xi)\,d\xi\right)^2 + w(x)\left(\int_0^x \xi\left(\int_0^\xi w(\eta)d\eta\right)d\xi\right)\right)dx. \tag{8.58}$$

We employ two estimates working in tandem, $\hat{\lambda}$ and $\hat{\theta}$. A long Lyapunov-based derivation, briefly justified after the statement of the theorem below, yields

$$\dot{\hat{\lambda}} = \gamma\frac{\beta\gamma}{\beta\gamma(1-\gamma H)-1}\left[\frac{\frac{3}{2}\|w\|^2 + 2A\|w_x\|^2}{1+\|w\|^2} - \left(1+\frac{1}{\gamma^2}-\frac{1}{\beta\gamma^2}\right)\beta\gamma B(\hat{\lambda}-\hat{\theta}-\gamma A) - \sigma(\hat{\lambda}-\hat{\theta}-\gamma A)\right], \tag{8.59}$$

$$\dot{\hat{\theta}} = \gamma\left(\frac{2\|w\|^2}{1+\|w\|^2} - \beta\gamma B(\hat{\lambda}-\hat{\theta}-\gamma A)\right). \tag{8.60}$$

We have written the two update laws in a way to highlight as much as possible the parts that are similar about them. Three gains are employed that need to satisfy the conditions $\gamma < 3$, $\beta > 3/[\gamma(3-\gamma)]$, $\sigma > 0$. The last two conditions are related to the fact that $|H| \le 1/3$. These conditions ensure that the denominator in the first line of (8.59) remains positive.

Besides its complexity, a disadvantage of the update law (8.59) is that it employs $\|w_x\|$, that is, it requires the measurement of the spatial derivative $u_x(x,t)$.

THEOREM 8.3. *Suppose that the system* (8.1)–(8.4), (8.59), (8.60) *has a unique classical solution for all* $t \ge 0$. *Then, for any initial condition* $u_0 \in H^2(0,1)$ *compatible with boundary conditions, and any* $\hat{\lambda}(0), \hat{\theta}(0) \in \mathbb{R}$, *the spatial* L^2 *norm*

$\|u(t)\|$ remains bounded and the spatial H^1 norm $\|u_x(t)\|$ is square integrable over an infinite time interval. Moreover, the estimates $\hat{\lambda}(t), \hat{\theta}(t)$ are uniformly bounded.

The proof of this result employs a Lyapunov function:

$$V = \frac{\beta\gamma^2}{2}\frac{\beta\gamma+1}{\beta\gamma-1} + \log\left(1+\|w\|^2\right) + \frac{\beta}{2}(\hat{\lambda}-\hat{\theta}-\gamma A)^2$$
$$-\frac{1}{2\gamma}(\hat{\lambda}-\hat{\theta})^2 + \frac{1}{2\gamma}(\lambda-\hat{\theta})^2. \tag{8.61}$$

It is possible to prove that

$$V \geq \log\left(1+\|w\|^2\right) + \frac{1}{2\gamma}\left((\hat{\lambda}-\hat{\theta})^2 + \frac{\beta\gamma-1}{2}(\lambda-\hat{\theta})^2\right), \tag{8.62}$$

that is, V is positive definite around the equilibrium $w(x) \equiv 0, \hat{\lambda} = \hat{\theta} = \lambda$. Then a very long calculation yields

$$\dot{V} = -2\frac{\|w_x\|^2}{1+\|w\|^2}. \tag{8.63}$$

The properties stated in Theorem 8.3 readily follow from this equation.

8.6 OTHER BENCHMARK PROBLEMS

In this section we extend the Lyapunov adaptive design beyond the basic reaction-diffusion class of parabolic PDEs. We consider two benchmark problems, one with a boundary value appearing on the right-hand side of the PDE model and another with a parametric uncertainty in an uncontrolled boundary condition. Both benchmark problems are unstable in the absence of feedback.

These benchmarks will expose one limitation of the log-Lyapunov paradigm: in general, it requires not only a restriction on the value of the adaptation gain γ but also the use of parameter projection. A small γ is the main tool for preventing destabilizing transients. Projection is only used to make the restriction on γ a priori verifiable.

The projection operator that would be used in implementation is defined as

$$\mathrm{Proj}_{[\underline{\theta},\bar{\theta}]}\{\tau\} = \begin{cases} 0, & \hat{\theta} = \underline{\theta} \text{ and } \tau < 0 \\ 0, & \hat{\theta} = \bar{\theta} \text{ and } \tau > 0 \\ \tau, & \text{else,} \end{cases} \tag{8.64}$$

where $\hat{\theta}$ is the parameter estimate (θ is used as a generic symbol for an unknown parameter that in subsequent presentation will be replaced by specific parameters labeled by $g, q, \varepsilon, b, \lambda$), the interval $\left[\underline{\theta}, \bar{\theta}\right]$ is the interval within which $\hat{\theta}$ is being kept by projection, and τ denotes the nominal update law.

Unfortunately, the projection operator (8.64) is discontinuous. This presents a problem at two levels: (1) in the analysis, it is not possible to obtain classical solutions, only Filippov solutions; (2) in implementation the presence of noise may induce frequent switching of the update law. This issue is not as serious as controller switching in sliding mode control because the projection operator does not drive an actuator. Since the projection drives only the update law $\dot{\hat{\theta}}$, there would be no discontinuities in $\hat{\theta}(t)$, and therefore no jumps in the control action. However, obtaining classical solutions and not having to deal with Filippov solutions is a good enough reason to consider a continuous version of the projection operator where, instead of a hard switch, a boundary layer of width $\delta > 0$ is introduced:

$$\mathrm{Proj}^{\delta}_{[\underline{\theta},\bar{\theta}]}\{\tau\} = \begin{cases} \dfrac{\hat{\theta} - \underline{\theta} + \delta}{\delta}, & \underline{\theta} - \delta \leq \hat{\theta} < \underline{\theta} \text{ and } \tau < 0 \\ \dfrac{\bar{\theta} + \delta - \hat{\theta}}{\delta}, & \bar{\theta} < \hat{\theta} \leq \bar{\theta} + \delta \text{ and } \tau > 0 \\ \tau, & \text{else} \end{cases} \tag{8.65}$$

where the update law τ is scaled linearly with θ in the boundary layer. With the help of [68, Lemma E.1] we get:

LEMMA 8.2. *The following properties of the projection operator* (8.65) *are guaranteed:*

(i) The operator is a locally Lipschitz function of $\hat{\theta}, \tau$ *on* $[\underline{\theta} - \delta, \bar{\theta} + \delta] \times \mathbb{R}$.

(ii) $\left(\mathrm{Proj}^{\delta}_{[\underline{\theta},\bar{\theta}]}\{\tau\}\right)^2 \leq \tau^2$.

(iii) For $\hat{\theta}(0) \in [\underline{\theta} - \delta, \bar{\theta} + \delta]$, *the solution of* $\dot{\hat{\theta}} = \mathrm{Proj}^{\delta}_{[\underline{\theta},\bar{\theta}]}\{\tau\}$ *remains in* $[\underline{\theta} - \delta, \bar{\theta} + \delta]$.

(iv) $-\tilde{\theta}\mathrm{Proj}^{\delta}_{[\underline{\theta},\bar{\theta}]}\{\tau\} \leq -\tilde{\theta}\tau$ *for all* $\hat{\theta} \in [\underline{\theta} - \delta, \bar{\theta} + \delta]$, $\theta \in [\underline{\theta}, \bar{\theta}]$.

All of the properties in Lemma 8.2 except Lipschitzness also hold for the discontinuous projection (8.64), with $\delta = 0$. The discontinuous projection would be preferable in applications for its simplicity, which does not come at the expense of control switching, and because it is a standard feature in the integrator block in Simulink. For these reasons, and to avoid clutter in our further presentation, we employ (8.64) where projection is needed.

Now we return to our presentation of the benchmark problems.

8.6.1 Plant with Uncertain Parameter in the Domain

Consider the plant

$$u_t = u_{xx} + gu(0,t), \tag{8.66}$$

$$u_x(0,t) = 0, \tag{8.67}$$

$$u(1,t) = U(t), \tag{8.68}$$

where g is a constant, unknown parameter and $u(0,t)$ is the boundary value of $u(x,t)$ at $x = 0$. This system is inspired by a model of thermal instability in solid propellant rockets [13]. In the absence of control, the system is unstable if and only if $g > 2$. We assume that this is indeed the case, $g > 2$. Let us further assume that an upper bound $\bar{g}$ on g is known to us. (It is important to note that such an assumption was not made on λ in Section 8.1.) In this section we design an adaptive controller whose update law incorporates the standard projection operator [68] to keep the parameter estimate $\hat{g}$ in the interval $[2, \bar{g}]$ while driving $u(x, t)$ to zero.

The following controller was designed in Chapter 3:[3]

$$U = -\int_0^1 \sqrt{\hat{g}} \sinh\left(\sqrt{\hat{g}}(1-\xi)\right) u(\xi)\, d\xi. \tag{8.69}$$

Consider the variable change

$$w(x) = u(x) + \int_0^x \sqrt{\hat{g}} \sinh\left(\sqrt{\hat{g}}(x-\xi)\right) u(\xi)\, d\xi. \tag{8.70}$$

It can be shown that

$$w_t = w_{xx} + \dot{\hat{g}} \int_0^x w(\xi) \frac{\sinh\left(\sqrt{\hat{g}}(x-\xi)\right)}{\sqrt{\hat{g}}}\, d\xi + \tilde{g} w(0) \cosh\left(\sqrt{\hat{g}}x\right), \tag{8.71}$$

$$w_x(0) = 0, \tag{8.72}$$

$$w(1) = 0, \tag{8.73}$$

where $\tilde{g} = g - \hat{g}$. Consider the Lyapunov function candidate

$$V = \frac{1}{2} \log\left(1 + \|w\|^2\right) + \frac{1}{2\gamma} \tilde{g}^2. \tag{8.74}$$

Taking its time derivative, we arrive at the update law

$$\dot{\hat{g}} = \frac{\gamma}{1 + \|w\|^2} \text{Proj}_{[2,\bar{g}]} \left\{ w(0) \int_0^1 w(x) \cosh\left(\sqrt{\hat{g}}x\right) dx \right\}. \tag{8.75}$$

The derivative of the Lyapunov function is

$$\dot{V} = -\frac{\|w_x\|^2}{1 + \|w\|^2} + \dot{\hat{g}} \frac{\int_0^1 w(x) \int_0^x w(\xi) \frac{\sinh\left(\sqrt{\hat{g}}(x-\xi)\right)}{\sqrt{\hat{g}}}\, d\xi\, dx}{1 + \|w\|^2}. \tag{8.76}$$

It can be shown that

$$\dot{V} \le -\left(1 - 2\gamma e^{2\sqrt{\bar{g}}}\right) \frac{\|w_x\|^2}{1 + \|w\|^2}. \tag{8.77}$$

Stability is thus achieved whenever

$$\gamma < \frac{1}{2} e^{-2\sqrt{\bar{g}}}. \tag{8.78}$$

[3]This formula is written with the implicit understanding that $\sinh(jx) = -j \sin(x)$, so that when $\hat{g}$ becomes negative, $\sqrt{\hat{g}} \sinh(\sqrt{\hat{g}}(1-\xi)) = \sqrt{|\hat{g}|} \sin(\sqrt{|\hat{g}|}(1-\xi))$.

This condition highlights the key differences between the design for the PDE in Section 8.1 and for the PDE (8.66):

1. The adaptation gain, which was limited by 1 in Section 8.1, needs to decrease as g increases in (8.66).
2. Knowledge of the parameter's upper bound is needed for the plant (8.66). Projection is used to keep the parameter within the a priori bound, such that the condition (8.78) is sufficient to achieve stability. It should also be noted that stability can be achieved without projection, by selecting γ to satisfy

$$\gamma < \frac{1}{2} e^{-2\left(\sqrt{2\bar{g}+\hat{g}(0)}+\left(\gamma \log\left(1+\|w_0\|^2\right)\right)^{1/4}\right)}, \tag{8.79}$$

where $w_0(x)$ is determined using the initial state $u_0(x)$ and the initial parameter estimate $\hat{g}(0)$. While it may be unusual to choose the adaptation gain based on the initial state u_0, it is acceptable as a theoretical result and consistent with the Lyapunov function (8.74), yielding estimates on $\|u(t)\|$ and $\tilde{g}(t)$ that depend on $\|u_0\|$ and $\tilde{g}(0)$. However, in application one would prefer projection, owing to its added assurance against parameter drift.

Other than the use of projection, the rest of the results of this section are qualitatively the same as those in Section 8.1. One can prove boundedness in the maximum norm in a similar manner as in Section 8.2. A lengthy calculation yields

$$\begin{aligned}
\frac{1}{2}\frac{d}{dt}\|w_x\|^2 = & -\|w_{xx}\|^2 - w_x(1)\tilde{g}\cosh\left(\sqrt{\hat{g}}\right) w(0) \\
& - w_x(1)\dot{\hat{g}} \int_0^1 w(x) \frac{\sinh\left(\sqrt{\hat{g}}(1-x)\right)}{\sqrt{\hat{g}}} dx \\
& - \dot{\hat{g}}\sqrt{\hat{g}} \int_0^1 w(x) \int_0^x \sinh\left(\sqrt{\hat{g}}(x-\xi)\right) w(\xi)\, d\xi\, dx \\
& - \tilde{g}\hat{g} w(0) \int_0^1 w(x)\cosh\left(\sqrt{\hat{g}x}\right) dx,
\end{aligned} \tag{8.80}$$

which can be majorized by

$$\frac{1}{2}\frac{d}{dt}\|w_x\|^2 \le 8\left(\gamma^2 + \tilde{g}^2\right) e^{4\sqrt{\hat{g}}}\|w_x\|^2. \tag{8.81}$$

Integrating (8.77) and (8.81), one gets the boundedness of $\|w_x\|$. Regulation is shown similarly as in Section 8.2. The results in the $u(x,t)$ variable follow from the inverse transformation

$$u(x) = w(x) + \hat{g} \int_0^x (x-\xi) w(\xi)\, d\xi. \tag{8.82}$$

Theorem 8.4. *For any initial condition $u_0 \in H^2(0,1)$ compatible with boundary conditions, and any $\hat{g}(0) \in [2, \bar{g}]$, the classical solution $(u, \hat{g})$ of the system* (8.66)–(8.69), (8.75) *is bounded for all $x \in [0,1]$, $t \ge 0$ and*

$$\lim_{t\to\infty} \max_{x\in[0,1]} |u(x,t)| = 0. \tag{8.83}$$

8.6.2 Plant with Uncertain Parameter at the Uncontrolled Boundary

Consider the plant

$$u_t = u_{xx} \tag{8.84}$$

$$u_x(0) = -qu(0,t), \tag{8.85}$$

where q is a constant, unknown parameter. This system is also inspired by the solid propellant rocket instability [13]. We will control this system via Dirichlet actuation, $u(1,t)$. In the absence of control, $u(1,t) \equiv 0$, the system is unstable if and only if $q > 1$. We assume that $q > 1$ and also that an upper bound $\bar{q}$ on q is known to us. We will design an adaptive controller with projection [68] to keep the parameter estimate $\hat{q}$ in the interval $[1, \bar{q}]$, while achieving stability.

A stabilizing control formula for this system was obtained in Chapter 3 as

$$U = -\int_0^1 \hat{q} e^{\hat{q}(1-\xi)} u(\xi)\, d\xi. \tag{8.86}$$

Consider the variable change

$$w(x) = u(x) + \int_0^x \hat{q} e^{\hat{q}(x-\xi)} u(\xi)\, d\xi. \tag{8.87}$$

It can be shown that

$$w_t = w_{xx} + \dot{\hat{q}} \int_0^x w(\xi) e^{\hat{q}(x-\xi)}\, d\xi, \tag{8.88}$$

$$w_x(0) = -\tilde{q} w(0), \tag{8.89}$$

$$w(1) = 0, \tag{8.90}$$

where $\tilde{q} = q - \hat{q}$. Consider the Lyapunov function candidate

$$V = \frac{1}{2} \log\left(1 + \|w\|^2\right) + \frac{1}{2\gamma} \tilde{q}^2. \tag{8.91}$$

Taking its time derivative, we arrive at the update law

$$\dot{\hat{q}} = \frac{\gamma}{1 + \|w\|^2} \mathrm{Proj}_{[1,\bar{q}]} \left\{ w(0)^2 \right\}. \tag{8.92}$$

The derivative of the Lyapunov function is

$$\dot{V} = -\frac{\|w_x\|^2}{1 + \|w\|^2} + \dot{\hat{q}} \frac{\int_0^1 w(x) \int_0^x w(\xi) e^{\hat{q}(x-\xi)}\, d\xi\, dx}{1 + \|w\|^2}. \tag{8.93}$$

With a lengthy, careful calculation, applying twice the Cauchy-Schwartz inequality, one can show that

$$\left| \int_0^1 w(x) \int_0^x w(\xi) e^{\hat{q}(x-\xi)}\, d\xi\, dx \right| \le \frac{e^{\hat{q}}}{\sqrt{2}} \|w\|^2. \tag{8.94}$$

Using projection and Agmon's inequality, it then follows that

$$\dot{V} \le -\left(1 - \sqrt{2}\gamma e^{\bar{q}}\right) \frac{\|w_x\|^2}{1 + \|w\|^2}. \tag{8.95}$$

Stability is thus achieved whenever

$$\gamma < \frac{\sqrt{2}}{2} e^{-\bar{q}}. \tag{8.96}$$

Again, projection and slow adaptation are needed to mitigate the effect of $\dot{\hat{q}}$ in the Lyapunov analysis.

We have thus proved L^2 stability in the w-variable. The square integrability of $\|w_x(t)\|$ in time also readily follows from (8.95). To show boundedness of $w(x,t)$ for all $x \in [0, 1]$, let us introduce the transformation

$$\overline{w}(x) = w(x) + \tilde{q}(1-x)\int_0^x w(y)\,dy, \tag{8.97}$$

with the inverse

$$w(x) = \overline{w}(x) - \tilde{q}(1-x)e^{\frac{1}{2}\tilde{q}(1-x)^2}\int_0^x \overline{w}(y)\,dy. \tag{8.98}$$

The purpose of this transformation is to move the uncertain term from the boundary condition (8.89) into the domain, so that a standard H^1-norm $\|\overline{w}_x\|$ succeeds as a Lyapunov function. The new variable $\overline{w}$ satisfies the following PDE:

$$\overline{w}_t = \overline{w}_{xx} + \tilde{q}^2(1-x)\overline{w}(0) + 2\tilde{q}w$$
$$+\dot{\hat{q}}\int_0^x w(\xi)\left(e^{\hat{q}(x-\xi)} + (x-1)(\tilde{q}e^{\hat{q}(x-\xi)} - \tilde{q} - 1)\right)d\xi, \tag{8.99}$$

$$\overline{w}_x(0) = 0, \tag{8.100}$$

$$\overline{w}(1) = 0. \tag{8.101}$$

We get the following estimate:

$$\frac{d}{dt}\|\overline{w}_x\|^2 \le -\|\overline{w}_{xx}\|^2 + (|q|+\bar{q})^2|\overline{w}(0)|\|\overline{w}_{xx}\| + 2(|q|+\bar{q})\|w\|\|\overline{w}_{xx}\|^2$$
$$+ |\dot{\hat{q}}|\|w\|\|\overline{w}_{xx}\|(1+|q|+\bar{q})(e^{\bar{q}}+1). \tag{8.102}$$

Note that $\overline{w}(0) = \int_0^1 \overline{w}(x)\,dx \le \|\overline{w}_x\|$. From (8.97) we get the square integrability of $\|\overline{w}_x\|$, because $\|w_x\|$ is square integrable over time. Using Young's inequality in (8.102), we obtain

$$\frac{d}{dt}\|\overline{w}_x\|^2 \le -\|\overline{w}_{xx}\|^2 + \frac{1}{4}\|\overline{w}_{xx}\|^2 + (|q|+\bar{q})^4\|\overline{w}_x\|^2 + \frac{1}{4}\|\overline{w}_{xx}\|^2$$
$$+ 4(|q|+\bar{q})^2\|w\|^2 + \frac{1}{4}\|\overline{w}_{xx}\|^2 + (1+|q|+\bar{q})^2(e^{\bar{q}}+1)^2\|\overline{w}_x\|^2\|\overline{w}_x\|^2$$
$$\le -\frac{1}{4}\|\overline{w}_x\|^2 + l_1(t)\|\overline{w}_x\|^2 + l_2(t), \tag{8.103}$$

where $l_1(t) = (|q|+\bar{q})^4\|\overline{w}_x\|^2 + 4(|q|+\bar{q})^2\|w\|^2$ and $l_2(t) = (1+|q|+\bar{q})^2(e^{\bar{q}} + 1)^2\|\overline{w}_x\|^2$ are integrable over infinite time. By Lemma D.3 we get that $\|\overline{w}_x\|^2$ is bounded and square integrable. From the transformation (8.98) we get that $\|w_x\|^2$

is bounded and square integrable. By Agmon's inequality, $w(x,t)$ is bounded uniformly in $x \in [0,1]$.

To show regulation to zero, we compute

$$\frac{d}{dt}\|w\|^2 = -\frac{\|w_x\|^2}{1+\|w\|^2} + \dot{\hat{q}}\frac{\int_0^1 w(x)\int_0^x w(\xi)e^{\hat{q}(x-\xi)}\,d\xi\,dx}{1+\|w\|^2} + \frac{\tilde{q}\text{Proj}_{[1,\bar{q}]}\left\{w(0)^2\right\}}{1+\|w\|^2}. \tag{8.104}$$

Since $\|w_x\|$ is bounded (and hence $\dot{\hat{q}}$ is bounded), it is clear from (8.104) that $|(d/dt)\|w\|^2|$ is bounded. By Barbalat's lemma (Lemma D.1), we get $\|w\| \to 0$ as $t \to \infty$. Using Agmon's inequality and the boundedness of $\|w_x\|$, we get that $w(x,t) \to 0$ uniformly in $x \in [0,1]$. All of the above boundedness and regulation properties for the w-variable are also valid for the original u-variable, owing to the inverse transformation

$$u(x) = w(x) + \hat{q}\int_0^x w(\xi)\,d\xi. \tag{8.105}$$

Theorem 8.5. *For any initial condition $u_0 \in H^2(0,1)$ compatible with boundary conditions, and any $\hat{q}(0) \in [1,\bar{q}]$, the classical solution $(u,\hat{q})$ of the system* (8.84)–(8.86), (8.92) *is bounded for all $x \in [0,1]$, $t \geq 0$ and*

$$\lim_{t\to\infty}\max_{x\in[0,1]}|u(x,t)| = 0. \tag{8.106}$$

Let us now consider the "frozen adaptation" version of (8.88)–(8.90), with $\dot{\hat{q}} = 0$ and a constant parameter error $\tilde{q}$. This system is exponentially stable if and only if the estimate is $\hat{q} > q - 1$. The same parametric robustness observations as those made in Section 8.4 hold for the plant–controller pair (8.84)–(8.86). Likewise, those observations justify the use of the estimator of the type (8.92), where the product of the estimation error and regressor is always nonnegative.

8.7 SYSTEMS WITH UNKNOWN DIFFUSION AND ADVECTION COEFFICIENTS

In this section we show how to incorporate adaptation for unknown diffusion and advection coefficients. Consider the system

$$u_t = \varepsilon u_{xx} + bu_x + \lambda u, \tag{8.107}$$

$$u(0) = 0, \tag{8.108}$$

$$u(1) = U(t), \tag{8.109}$$

where ε, b, and λ are unknown constants.

The control law for this system is (see Chapter 3)

$$U(t) = -\int_0^1 \frac{\hat{\lambda}+c}{\hat{\varepsilon}} \xi e^{-\frac{\hat{b}}{2\hat{\varepsilon}}(1-\xi)} \frac{I_1\left(\sqrt{\frac{\hat{\lambda}+c}{\hat{\varepsilon}}(1-\xi^2)}\right)}{\sqrt{\frac{\hat{\lambda}+c}{\hat{\varepsilon}}(1-\xi^2)}} u(\xi)\, d\xi, \tag{8.110}$$

where $\hat{\varepsilon}$, $\hat{b}$, and $\hat{\lambda}$ are the estimates of ε, b, and λ and $c \geq 0$ is a design gain. Using the transformation

$$w(x) = u(x) - \int_0^x k(x,\xi) u(\xi)\, d\xi, \tag{8.111}$$

$$k(x,\xi) = -\frac{\hat{\lambda}+c}{\hat{\varepsilon}} \xi e^{-\frac{\hat{b}}{2\hat{\varepsilon}}(x-\xi)} \frac{I_1\left(\sqrt{\frac{\hat{\lambda}+c}{\hat{\varepsilon}}(x^2-\xi^2)}\right)}{\sqrt{\frac{\hat{\lambda}+c}{\hat{\varepsilon}}(x^2-\xi^2)}}, \tag{8.112}$$

and its inverse

$$u(x) = w(x) + \int_0^x l(x,\xi) w(\xi)\, d\xi, \tag{8.113}$$

$$l(x,\xi) = -\frac{\hat{\lambda}+c}{\hat{\varepsilon}} \xi e^{-\frac{\hat{b}}{2\hat{\varepsilon}}(x-\xi)} \frac{J_1\left(\sqrt{\frac{\hat{\lambda}+c}{\hat{\varepsilon}}(x^2-\xi^2)}\right)}{\sqrt{\frac{\hat{\lambda}+c}{\hat{\varepsilon}}(x^2-\xi^2)}}, \tag{8.114}$$

we get

$$\begin{aligned} w_t = {} & \varepsilon w_{xx} + b w_x - c w + \dot{\hat{\varepsilon}} \int_0^x \varphi_0(x,\xi) w(\xi)\, d\xi \\ & + \dot{\hat{b}} \int_0^x \varphi_1(x,\xi) w(\xi)\, d\xi + \dot{\hat{\lambda}} \int_0^x \varphi_2(x,\xi) w(\xi)\, d\xi \\ & - \tilde{\varepsilon} \left(\frac{\hat{\lambda}+c}{\hat{\varepsilon}} w + \frac{\hat{b}}{\hat{\varepsilon}} \int_0^x \varphi_3(x,\xi) w(\xi)\, d\xi \right) \\ & + \tilde{b} \int_0^x \varphi_3(x,\xi) w(\xi)\, d\xi + \tilde{\lambda} w, \end{aligned} \tag{8.115}$$

$$w(0) = 0, \tag{8.116}$$

$$w(1) = 0, \tag{8.117}$$

where

$$\begin{aligned} \varphi_0(x,\xi) &= -\frac{\hat{\lambda}+c}{\hat{\varepsilon}} \varphi_2(x,\xi) - \frac{\hat{b}}{\hat{\varepsilon}} \varphi_1(x,\xi), \\ \varphi_1(x,\xi) &= \frac{x-\xi}{2\hat{\varepsilon}} k(x,\xi) + \frac{1}{2\hat{\varepsilon}} \int_\xi^x (x-\sigma) k(x,\sigma) l(\sigma,\xi)\, d\sigma, \\ \varphi_2(x,\xi) &= \frac{\xi}{2\hat{\varepsilon}} e^{-\frac{\hat{b}}{2\hat{\varepsilon}}(x-\xi)}, \\ \varphi_3(x,\xi) &= \operatorname{div} k(x,\xi) + \int_\xi^x (\operatorname{div} k(x,\sigma)) l(\sigma,\xi)\, d\sigma, \end{aligned} \tag{8.118}$$

and

$$\operatorname{div} k(x,\xi) = \frac{1}{\xi}k(x,\xi) + \frac{\hat{\lambda}+c}{\hat{\varepsilon}} e^{-\frac{\hat{b}}{2\hat{\varepsilon}}(x-\xi)} \frac{\xi}{x+\xi} I_2\left(\sqrt{\frac{\hat{\lambda}+c}{\hat{\varepsilon}}(x^2-\xi^2)}\right). \tag{8.119}$$

Based on (8.115) and the Lyapunov function

$$V = \frac{1}{2}\left(\log(1+\|w\|^2) + \frac{\tilde{\varepsilon}^2 + \tilde{b}^2 + \tilde{\lambda}^2}{\gamma}\right) \tag{8.120}$$

we choose the update laws

$$\dot{\hat{\lambda}} = \gamma \frac{\|w\|^2}{1+\|w\|^2}, \tag{8.121}$$

$$\dot{\hat{b}} = \gamma \frac{\int_0^1 w(x) \int_0^x \varphi_3(x,\xi) w(\xi)\, d\xi dx}{1+\|w\|^2}, \tag{8.122}$$

$$\dot{\hat{\varepsilon}} = -\gamma \frac{(\hat{\lambda}+c)\|w\|^2 + \hat{b}\int_0^1 w(x) \int_0^x \varphi_3(x,\xi) w(\xi)\, d\xi dx}{\hat{\varepsilon}(1+\|w\|^2)}, \tag{8.123}$$

where projection is used (though we don't explicitly include it in the definition of the update laws) to keep the parameter estimates within a priori bounds $[\underline{\lambda}, \bar{\lambda}]$, $[\underline{b}, \bar{b}]$, and $[\underline{\varepsilon}, \bar{\varepsilon}]$, where $\underline{\varepsilon} > 0$. As in the previous problems, γ is limited by an upper bound that can be a priori computed.

Theorem 8.6. *There exists $\gamma^* > 0$ such that for all $\gamma \in (0, \gamma^*)$, for any initial condition $u_0 \in H^2(0,1)$ compatible with boundary conditions, and any $\hat{\lambda}(0) \in [\underline{\lambda}, \bar{\lambda}]$, $\hat{b}(0) \in [\underline{b}, \bar{b}]$, and $\hat{\varepsilon}(0) \in [\underline{\varepsilon}, \bar{\varepsilon}]$, the classical solution $(u, \hat{\lambda}, \hat{b}, \hat{\varepsilon})$ of the system* (8.107)–(8.110), (8.121)–(8.123) *is uniformly bounded, and*

$$\lim_{t\to\infty} \max_{x\in[0,1]} |u(x,t)| = 0. \tag{8.124}$$

Proof. It can be shown that

$$\dot{V} = \frac{1}{1+\|w\|^2}\left(-\varepsilon\|w_x\|^2 - c\|w\|^2 + \dot{\hat{\varepsilon}} F_0 + \dot{\hat{b}} F_1 + \dot{\hat{\lambda}} F_2\right), \tag{8.125}$$

where

$$F_i(x) = \int_0^1 w(x) \int_0^x \varphi_i(x,\xi) w(\xi)\, d\xi\, dx \tag{8.126}$$

for $i = 0, 1, 2, 3$. By applying the Cauchy-Schwartz inequality twice to (8.126), we get

$$|F_i| \le \|w\|^2 \left(\int_0^1 \int_0^x \varphi_i(x,\xi)^2\, d\xi\, dx\right)^{1/2}. \tag{8.127}$$

Because the functions $\varphi_i(x,\xi)$ are continuous in $x, \xi, \hat{\varepsilon}, \hat{b}, \hat{\lambda}$ over the domain of their definition given by $\mathcal{T} \times [\underline{\varepsilon}, \bar{\varepsilon}] \times [\underline{b}, \bar{b}] \times [\underline{\lambda}, \bar{\lambda}]$, where $\mathcal{T} = \{x, \xi \in \mathbb{R} | 0 \leq \xi \leq x \leq 1\}$ and $\underline{\varepsilon} > 0$, it can be shown that there exist continuous, nonnegative-valued, nondecreasing functions M_i: $\mathbb{R}_+^5 \to \mathbb{R}_+$ such that

$$\int_0^1 w(x) \int_0^x \varphi_i(x,\xi) w(\xi)\, d\xi dx \leq M_i \left(\frac{1}{\underline{\varepsilon}}, |\underline{b}|, |\bar{b}|, |\underline{\lambda}|, |\bar{\lambda}| \right). \tag{8.128}$$

The simplest one among these functions is

$$M_2 = \frac{1}{4\sqrt{3}\underline{\varepsilon}} e^{\frac{1}{\underline{\varepsilon}} \max\{|\underline{b}|, |\bar{b}|\}}. \tag{8.129}$$

From (8.125)–(8.128), it follows that

$$\begin{aligned} \dot{V} \leq & \frac{1}{1+\|w\|^2} \Bigg[-\varepsilon \|w_x\|^2 - c\|w\|^2 + \frac{\gamma \|w\|^4}{1+\|w\|^2} \\ & \times \left(\frac{|\bar{\lambda}| + c}{\underline{\varepsilon}} M_0 + \frac{|\bar{b}|}{\underline{\varepsilon}} M_3 M_0 + M_3 M_1 + M_2 \right) \Bigg], \end{aligned} \tag{8.130}$$

where we emphasize the emergence of the fourth power of $\|w\|$ in the last term of the first line of (8.130). By applying Poincaré's inequality we obtain

$$\dot{V} \leq -\frac{\underline{\varepsilon}\,(1 - \gamma/\gamma^*)\, \|w_x\|^2 + c\|w\|^2}{1+\|w\|^2}, \tag{8.131}$$

where

$$\gamma^* = \frac{\varepsilon}{4} \left(\frac{|\bar{\lambda}| + c}{\underline{\varepsilon}} M_0 + \frac{|\bar{b}|}{\underline{\varepsilon}} M_3 M_0 + M_3 M_1 + M_2 \right)^{-1}. \tag{8.132}$$

This establishes the boundedness of $\|w\|$ for $\gamma < \gamma^*$.

To prove the boundedness of $\|w_x\|^2$, we show that

$$\frac{1}{2}\frac{d}{dt} \|w_x\|^2 = -\varepsilon \|w_{xx}\|^2 - \frac{\varepsilon c + \varepsilon \tilde{\lambda} + \lambda \tilde{\varepsilon}}{\hat{\varepsilon}} \|w_x\|^2 - \int_0^1 w_{xx}(x) G(x)\, dx, \tag{8.133}$$

where

$$\begin{aligned} G(x) = & \, b w_x + \dot{\hat{\varepsilon}} \int_0^x \varphi_0(x,\xi) w(\xi)\, d\xi \\ & + \dot{\hat{b}} \int_0^x \varphi_1(x,\xi) w(\xi)\, d\xi + \dot{\hat{\lambda}} \int_0^x \varphi_2(x,\xi) w(\xi)\, d\xi \\ & + \frac{\varepsilon \hat{b} - \hat{\varepsilon} b}{\hat{\varepsilon}} \int_0^x \varphi_3(x,\xi) w(\xi)\, d\xi. \end{aligned} \tag{8.134}$$

Next we note that

$$|\dot{\hat{\lambda}}| \leq \gamma, \quad |\dot{\hat{b}}| \leq \gamma M_3, \quad |\dot{\hat{\varepsilon}}| \leq \gamma M_4, \quad \left| \frac{\varepsilon \hat{b} - \hat{\varepsilon} b}{\hat{\varepsilon}} \right| \leq M_5, \tag{8.135}$$

where

$$M_4 = \frac{\max\{|\underline{\lambda}|, |\bar{\lambda}|\} + c + M_3 \max\{|\underline{b}|, |\bar{b}|\}}{\underline{\varepsilon}} \tag{8.136}$$

$$M_5 = 2\frac{\bar{\varepsilon}}{\underline{\varepsilon}} \max\{|\underline{b}|, |\bar{b}|\}. \tag{8.137}$$

With Young's inequality we get

$$-\int_0^1 w_{xx}(x)G(x)\,dx \le \varepsilon \|w_{xx}\|^2 + \frac{1}{4\varepsilon}\|G\|^2. \tag{8.138}$$

Let us denote

$$H_i(x) = \int_0^x \varphi_i(x,\xi)w(\xi)\,d\xi \tag{8.139}$$

for $i = 0, 1, 2, 3$, for which, with the Cauchy-Schwartz inequality, we get

$$\|H_i\| \le M_i \|w\|. \tag{8.140}$$

Then, from (8.134)–(8.140), with the triangle inequality and Poincaré's inequality we obtain

$$\|G\|^2 \le 8\big[b + \gamma(M_4 M_0^2 + M_3 M_1^2 + M_2^2) + M_5 M_3^2\big]\|w_x\|^2.$$

Substituting the last inequality into (8.138) and then into (8.133), we get

$$\frac{1}{2}\frac{d}{dt}\|w_x\|^2 \le N\|w_x\|^2, \tag{8.141}$$

where

$$N(t) = \frac{2}{\varepsilon}\big[b + \gamma(M_4 M_0^2 + M_3 M_1^2 + M_2^2) + M_5 M_3^2\big] - \frac{\varepsilon c + \varepsilon\tilde{\lambda} + \lambda\tilde{\varepsilon}}{\hat{\varepsilon}} \tag{8.142}$$

is bounded. With $\|w\|$ bounded, from (8.131) we get that $\|w_x\|^2$ is integrable over infinite time. By integrating (8.141), it follows that $\|w_x\|$ is bounded. By Agmon's inequality, $w(x,t)$ is also bounded for all $t \ge 0$ and for all $x \in [0, 1]$.

To show regulation, we calculate

$$\begin{aligned}\frac{1}{2}\frac{d}{dt}\|w\|^2 &= -\varepsilon\|w_x\|^2 - c\|w\|^2 + \dot{\hat{\varepsilon}}F_0 + \dot{\hat{b}}F_1 + \dot{\hat{\lambda}}F_2 \\ &\quad - \tilde{\varepsilon}\left(\frac{\hat{\lambda}+c}{\hat{\varepsilon}}\|w\|^2 + \frac{\hat{b}}{\hat{\varepsilon}}F_3\right) + \tilde{b}F_3 + \tilde{\lambda}\|w\|^2. \end{aligned} \tag{8.143}$$

All of the terms on the right-hand side of this inequality have been proved to be bounded. Therefore, $\frac{d}{dt}\|w\|^2$ is bounded from above. Since $\|w\|^2$ is also integrable over infinite time, by Lemma D.2 $\|w(t)\| \to 0$ as $t \to \infty$. Regulation in maximum norm follows from Agmon's inequality and the boundedness of $\|w_x\|$.

To infer the results for the original variable $u(x,t)$ from those for $w(x,t)$, we recall the inverse transformation (8.113)–(8.114), which is a bounded operator in both L^2 and H^1.

□

While the Lyapunov design requires the use of projection and a low adaptation gain, one of its remarkable properties is that, even though the plant has parametric uncertainties multiplying u_x and u_{xx}, the adaptive scheme does not require the measurement of either u_x or u_{xx}. The update laws (8.121)–(8.123) employ only the measurement of u.

Note that the integral in (8.118) would be calculated numerically, just like the other integrals appearing in the update laws and depending on the measured state $u(x,t)$.

Remark 8.1. It should be pointed out that in the Lyapunov approach, the diffusion coefficient ε need not be estimated directly. This is analogous to the finite-dimensional adaptive control [68], where the "high-frequency gain" need not be estimated directly in the Lyapunov approach, whereas in the passive or swapping approaches it needs to be estimated. The estimation of ε is avoided by denoting the unknown parameters $\alpha = (\lambda + c)/\varepsilon$ and $\beta = b/\varepsilon$ and by replacing the adaptive controller (8.110) by

$$U = -\int_0^1 \hat{\alpha}\xi e^{-\frac{\hat{\beta}}{2}(1-\xi)} \frac{I_1\left(\sqrt{\hat{\alpha}(1-\xi^2)}\right)}{\sqrt{\hat{\alpha}(1-\xi^2)}} u(\xi)\, d\xi, \tag{8.144}$$

by replacing the update laws (8.121)–(8.123) by

$$\dot{\hat{\alpha}} = \gamma \frac{\|w\|^2}{1+\|w\|^2}, \tag{8.145}$$

$$\dot{\hat{\beta}} = \gamma \frac{\int_0^1 w(x) \int_0^x \varphi_3(x,\xi) w(\xi)\, d\xi dx}{1+\|w\|^2} \tag{8.146}$$

(equipped with appropriate projection), and by using w and φ_3 as defined in (8.111) and (8.118), respectively, with $k(x,\xi)$, $l(x,\xi)$, and $\operatorname{div} k(x,\xi)$, redefined as

$$k(x,\xi) = -\hat{\alpha}\xi e^{-\frac{\hat{\beta}}{2}(x-\xi)} \frac{I_1\left(\sqrt{\hat{\alpha}(x^2-\xi^2)}\right)}{\sqrt{\hat{\alpha}(x^2-\xi^2)}}, \tag{8.147}$$

$$l(x,\xi) = -\hat{\alpha}\xi e^{-\frac{\hat{\beta}}{2}(x-\xi)} \frac{J_1\left(\sqrt{\hat{\alpha}(x^2-\xi^2)}\right)}{\sqrt{\hat{\alpha}(x^2-\xi^2)}}, \tag{8.148}$$

$$\operatorname{div} k(x,\xi) = \frac{k(x,\xi)}{\xi} + \hat{\alpha} e^{-\frac{\hat{\beta}}{2}(x-\xi)} \frac{\xi}{x+\xi} I_2\left(\sqrt{\hat{\alpha}(x^2-\xi^2)}\right). \tag{8.149}$$

8.8 SIMULATION RESULTS

To illustrate the designs, we present the simulation results for the plant (8.1)–(8.2) with the unknown parameter $\lambda = 20$. The initial estimate is set to $\hat{\lambda}(0) = 5$. The initial condition of the plant is chosen to be non-smooth to show that theorems' restriction for initial conditions to be in H^2 is not crucial. Even though the bound on the adaptation gain used in the proof of the closed-loop stability is $\gamma \le \sqrt{3}$, in

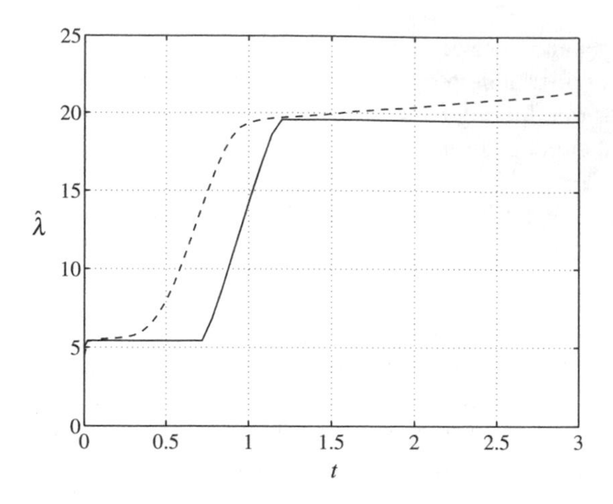

Figure 8.1 The evolution of the parameter estimate (dashed line, without deadzone; solid line, with deadzone).

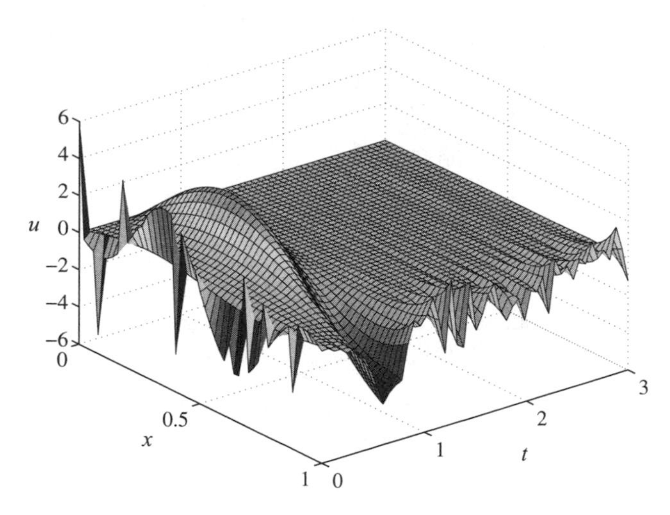

Figure 8.2 The closed-loop response. The initial condition is non-smooth. The control input $u(1, t)$ fluctuates because of measurement noise and additive noise on the right-hand side of the PDE.

simulation we took $\gamma = 30$ so that the adaptation would not be too slow. To model the realistic situation, we assume there is an additive disturbance in the plant: zero-mean white noise (both in time and space) of intensity 3. We also assume that measurements are corrupted by noise. In Figure 8.1 the evolution of the parameter estimate is shown. We can see that the estimate reaches the stabilizing value and starts to drift due to the noise. To prevent the drift we use the standard robustification tool—deadzone. As can be seen from Figure 8.1 (solid line), introducing the deadzone prevents the parameter from drifting. In Figure 8.2 the closed-loop response is shown.

8.9 NOTES AND REFERENCES

The need for projection and a bound on the adaptation gain are the key limitations of the Lyapunov approach. In the next two chapters we present methods that do not require projection and that work without limits on the adaptation gain. However, the Lyapunov schemes are unique in their ability to stabilize the reaction-advection-diffusion plants without using spatial derivatives of the state.

While, for the sake of clarity, we chose to present our Lyapunov design tools through benchmark problems, it is possible to develop a Lyapunov adaptive controller for the class of systems with five unknown parameters

$$u_t = \varepsilon u_{xx} + b u_x + \lambda u + g u(0),$$
$$u_x(0) = -q u(0),$$

with either Dirichlet or Neumann actuation.

Chapter Nine

Certainty Equivalence Design with Passive Identifiers

In this chapter we present the first of the estimation-based adaptive design approaches, the design with passive identifiers.

We start our presentation with a benchmark 1D plant with only one uncertain parameter to illustrate the main ideas of the passivity-based approach in a tutorial way without the extensive notation that is needed in higher dimensions, such as 2D and 3D, and with more than one physical parameter.

After the 1D benchmark plant we present the adaptive design for a 3D reaction-advection-diffusion plant. We prove that passive identifiers are stable with all three (reaction, advection, and diffusion) coefficients unknown.

A fundamental obstacle arises with the estimation-based designs when the *diffusion coefficient* (the coefficient multiplying the second spatial derivatives) is unknown. In that case, closed-loop stability is very hard to prove. For closed-loop stability with unknown diffusion, one seems to need a Sobolev bound on the estimation error that is one order higher than what stability analysis for the identifiers provides. Thus, we state closed-loop stability for known diffusion, though we illustrate it in simulations for unknown diffusion.

9.1 BENCHMARK PLANT

Consider a reaction-diffusion equation

$$u_t(x,t) = u_{xx}(x,t) + \lambda u(x,t), \tag{9.1}$$

$$u(0,t) = 0, \tag{9.2}$$

$$u(1,t) = U(t), \tag{9.3}$$

where λ is an unknown parameter.

Let us introduce the auxiliary system

$$\hat{u}_t(x,t) = \hat{u}_{xx}(x,t) + \hat{\lambda} u(x,t) + \gamma^2[u(x,t) - \hat{u}(x,t)] \int_0^1 u^2(x,t)\,dx, \tag{9.4}$$

$$\hat{u}(0,t) = 0, \tag{9.5}$$

$$\hat{u}(1,t) = U(t). \tag{9.6}$$

Here $\hat{\lambda} = \hat{\lambda}(t)$ is the parameter estimate and $\gamma > 0$ is a constant. Such systems are often called "observers" because they incorporate a copy of the plant, though they are not used for state estimation. This identifier employs a copy of the PDE plant and an additional nonlinear term. The term *passive identifier* comes from the fact that an operator from the parameter estimation error $\tilde{\lambda} = \lambda - \hat{\lambda}$ to the inner product of u with $u - \hat{u}$ is strictly passive. The additional nonlinear term in (9.4) acts as nonlinear damping whose task is to ensure square integrability of $\dot{\hat{\lambda}}$ over an infinite time interval (i.e., in our notation, $\dot{\hat{\lambda}} \in \mathcal{L}_2$). This slows down the adaptation and serves as an alternative to the update law normalization needed to achieve certainty equivalence.

The error signal $e = u - \hat{u}$ satisfies the PDE

$$e_t(x,t) = e_{xx}(x,t) + \tilde{\lambda} u(x,t) - \gamma^2 e(x,t)\|u\|^2, \tag{9.7}$$

$$e(0,t) = 0, \tag{9.8}$$

$$e(1,t) = 0. \tag{9.9}$$

With the Lyapunov function

$$V(t) = \frac{1}{2}\int_0^1 e^2(x,t)\,dx + \frac{\tilde{\lambda}(t)^2}{2\gamma}, \tag{9.10}$$

we get

$$\dot{V} = -\|e_x\|^2 - \gamma^2\|e\|^2\|u\|^2 + \tilde{\lambda}\int_0^1 e(x,t)u(x,t)\,dx - \frac{\tilde{\lambda}\dot{\hat{\lambda}}}{\gamma}. \tag{9.11}$$

Choosing the update law

$$\dot{\hat{\lambda}} = \gamma \int_0^1 (u(x,t) - \hat{u}(x,t))u(x,t)\,dx, \tag{9.12}$$

we obtain

$$\dot{V} = -\|e_x\|^2 - \gamma^2\|e\|^2\|u\|^2, \tag{9.13}$$

which implies $V(t) \le V(0)$, and from the definition of V we get that $\tilde{\lambda}$ and $\|e\|$ are bounded. Integrating (9.13) with respect to time from zero to infinity we get the properties $\|e_x\|$, $\|e\|\|u\| \in \mathcal{L}_2$. From the update law (9.12) we get $|\dot{\hat{\lambda}}| \le \gamma\|e\|\,\|u\|$, and so $\dot{\hat{\lambda}} \in \mathcal{L}_2$.

For the case of known λ, the controller and the transformation were obtained in Chapter 3. Using the certainty equivalence principle, we modify the transformation and the controller as follows:

$$\hat{w}(x,t) = \hat{u}(x,t) - \int_0^x \hat{k}(x,\xi)\hat{u}(\xi,t)\,d\xi, \tag{9.14}$$

$$\hat{k}(x,\xi) = -\hat{\lambda}\xi \frac{I_1\left(\sqrt{\hat{\lambda}(x^2-\xi^2)}\right)}{\sqrt{\hat{\lambda}(x^2-\xi^2)}}, \tag{9.15}$$

and

$$U(t) = \int_0^1 \hat{k}(1,\xi)\hat{u}(\xi,t)\,d\xi. \tag{9.16}$$

The above transformation maps (9.4)–(9.6) into the following target system (see Lemma 9.2 from Section 9.3):

$$\hat{w}_t(x,t) = \hat{w}_{xx}(x,t) + \dot{\hat{\lambda}} \int_0^x \frac{\xi}{2}\hat{w}(\xi,t)\,d\xi + (\hat{\lambda} + \gamma^2\|u\|^2)e_1(x,t), \tag{9.17}$$

$$\hat{w}(0,t) = 0, \tag{9.18}$$

$$\hat{w}(1,t) = 0, \tag{9.19}$$

where

$$e_1(x,t) = e(x,t) - \int_0^x \hat{k}(x,\xi)e(\xi,t)\,d\xi. \tag{9.20}$$

We observe that, in comparison to the nominal target system (heat equation), two additional terms appear in (9.17)–(9.19), both going to zero in some sense, since the identifier guarantees $\|e\|, \dot{\hat{\lambda}} \in \mathcal{L}_2$.

Next we prove boundedness of all the signals based on the joint analysis of e- and $\hat{w}$-systems. Let us denote a bound on $\hat{\lambda}$ by λ_0. The function $\hat{k}(x,\xi)$ is bounded and twice continuously differentiable with respect to x and ξ; therefore there exist constants M_1, M_2, M_3 such that

$$\|e_1\| \le M_1\|e\|, \tag{9.21}$$

$$\|u\| \le \|\hat{u}\| + \|e\| \le M_2\|\hat{w}\| + \|e\|, \tag{9.22}$$

$$\|u_x\| \le \|\hat{u}_x\| + \|e_x\| \le M_3\|\hat{w}_x\| + \|e_x\|. \tag{9.23}$$

To prove the boundedness of all the signals, we first estimate the time derivative of $\|\hat{w}\|^2$ using Agmon's, Poincaré's, and Young's inequalities:

$$\begin{aligned}
\frac{1}{2}\frac{d}{dt}\|\hat{w}\|^2 =& -\int_0^1 \hat{w}_x^2(x,t)\,dx + \dot{\hat{\lambda}}\int_0^1 \hat{w}(x,t)\int_0^x \frac{\xi}{2}\hat{w}(\xi,t)\,d\xi\,dx \\
&+ (\hat{\lambda}+\gamma^2\|u\|^2)\int_0^1 e_1(x,t)\hat{w}(x,t)\,dx \\
\le& -\|\hat{w}_x\|^2 + \frac{|\dot{\hat{\lambda}}|}{2}\|\hat{w}\|^2 + M_1\lambda_0\|\hat{w}\|\|e\| \\
&+ \gamma^2 M_1\|u\|(M_2\|\hat{w}\| + \|e\|)\|\hat{w}\|\|e\| \\
\le& -\|\hat{w}\|^2 + \frac{1}{4}\|\hat{w}\|^2 + \frac{1}{4}|\dot{\hat{\lambda}}|^2\|\hat{w}\|^2 + \frac{1}{4}\|\hat{w}\|^2 + M_1^2\lambda_0^2\|e\|^2 + \frac{1}{4}\|\hat{w}\|^2 \\
&+ 2\gamma^4 M_1^2 M_2^2\|u\|^2\|e\|^2\|\hat{w}\|^2 + \frac{\|e\|^2}{4M_2^2} \\
\le& -\frac{1}{4}\|\hat{w}\|^2 + \left(M_1^2\lambda_0^2 + \frac{1}{4M_2^2}\right)\|e\|^2 \\
&+ \left(\frac{1}{4}|\dot{\hat{\lambda}}|^2 + 2\gamma^4 M_1^2 M_2^2\|u\|^2\|e\|^2\right)\|\hat{w}\|^2 \\
\le& -\frac{1}{4}\|\hat{w}\|^2 + l_1\|\hat{w}\|^2 + l_1,
\end{aligned} \tag{9.24}$$

where l_1 denotes a generic function in $\mathcal{L}_1$. The last inequality follows from the properties $\dot{\hat{\lambda}}$, $\|u\|\|e\|$, $\|e\| \in \mathcal{L}_2$. Using Lemma D.3, we get $\|\hat{w}\| \in \mathcal{L}_\infty \cap \mathcal{L}_2$. From (9.22) we get $\|u\|$, $\|\hat{u}\| \in \mathcal{L}_\infty \cap \mathcal{L}_2$, and (9.12) implies that $\dot{\hat{\lambda}}$ is bounded.

In order to get pointwise in x boundedness, we show that $\|\hat{w}_x\|$ and $\|e_x\|$ are bounded:

$$\begin{aligned}
\frac{1}{2}\frac{d}{dt}\int_0^1 \hat{w}_x^2(x,t)\,dx =& \int_0^1 \hat{w}_x(x,t)\hat{w}_{xt}(x,t)\,dx = -\int_0^1 \hat{w}_{xx}(x,t)\hat{w}_t(x,t)\,dx \\
=& -\int_0^1 \hat{w}_{xx}^2(x,t)\,dx - \frac{\dot{\hat{\lambda}}}{2}\int_0^1 \hat{w}_{xx}(x,t)\int_0^x \xi w(\xi,t)\,d\xi\,dx \\
&- (\hat{\lambda}+\gamma^2\|u\|^2)\int_0^1 e_1(x,t)\hat{w}_{xx}(x,t)\,dx \\
\le& -\frac{1}{2}\|\hat{w}_x\|^2 + \frac{|\dot{\hat{\lambda}}|^2\|\hat{w}\|^2}{4} + (\lambda_0+\gamma^2\|u\|^2)^2 M_1^2\|e\|^2
\end{aligned} \tag{9.25}$$

and

$$\begin{aligned}
\frac{1}{2}\frac{d}{dt}\int_0^1 e_x^2(x,t)\,dx =& -\int_0^1 e_{xx}(x,t)e_t(x,t)\,dx \\
\le& -\|e_{xx}\|^2 + |\tilde{\lambda}|\|e_{xx}\|\|u\| - \gamma^2\|e_x\|^2\|u\|^2 \\
\le& -\frac{1}{8}\|e_x\|^2 + \frac{1}{2}|\tilde{\lambda}|^2\|u\|^2.
\end{aligned} \tag{9.26}$$

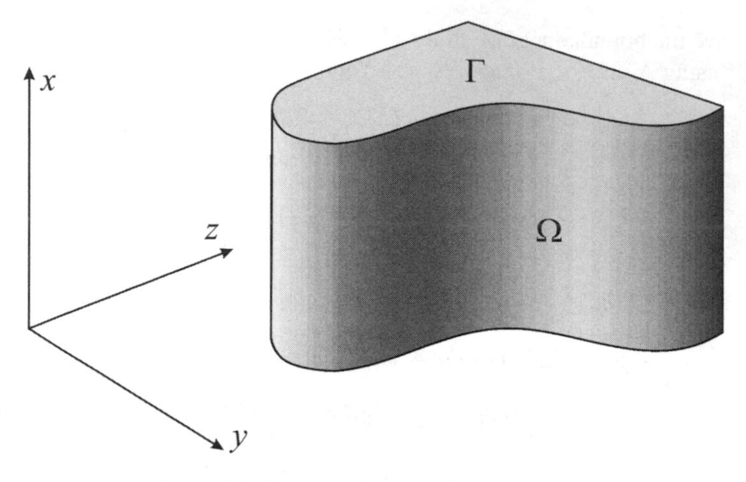

Figure 9.1 The domain Ω for the plant (9.30).

Since the right-hand sides of (9.25) and (9.26) are square integrable, using Lemma D.3 we get $\|\hat{w}_x\|, \|e_x\| \in \mathcal{L}_\infty \cap \mathcal{L}_2$. Using (9.23) we get $\|u_x\|, \|\hat{u}_x\| \in \mathcal{L}_\infty \cap \mathcal{L}_2$. Using Agmon's inequality we get the boundedness of u and $\hat{u}$ for all $x \in [0, 1]$.

To show regulation of the state to zero, we note that

$$\frac{1}{2}\frac{d}{dt}\|e\|^2 < \infty, \qquad \frac{1}{2}\frac{d}{dt}\|\hat{w}\|^2 < \infty, \tag{9.27}$$

and using Lemma D.2 we get $\|\hat{w}\| \to 0$, $\|e\| \to 0$ as $t \to \infty$. From (9.22) it follows that $\|\hat{u}\| \to 0$ and $\|u\| \to 0$. Using Agmon's inequality and the fact that $\|u_x\|$ is bounded, we get the regulation of u to zero for all $x \in [0, 1]$:

$$\lim_{t\to\infty} \max_{x\in[0,1]} |u(x, t)| \le \lim_{t\to\infty} (2\|u\|\|u_x\|)^{1/2} = 0. \tag{9.28}$$

The result can be summarized in the following theorem.

Theorem 9.1. *For any initial data $u_0, \hat{u}_0 \in H^2(0, 1)$ compatible with boundary conditions, the classical solution $(\hat{\lambda}, u, \hat{u})$ of the closed-loop system* (9.1)–(9.6), (9.12), (9.16) *is bounded for all $x \in [0, 1]$, $t \ge 0$ and*

$$\lim_{t\to\infty} \max_{x\in[0,1]} |u(x, t)| = 0. \tag{9.29}$$

9.2 3D REACTION-ADVECTION-DIFFUSION PLANT

We now present a passivity-based design for a plant in a three-dimensional setting:

$$u_t = \varepsilon(u_{xx} + u_{yy} + u_{zz}) + b_1 u_x + b_2 u_y + b_3 u_z + \lambda u \tag{9.30}$$

for $(x, y, z) \in \Omega$, where the domain Ω is a cylinder with top and bottom of arbitrary shape Γ (Fig. 9.1). This configuration of the domain Ω is essential because it allows us to view the problem as many 1D problems with $0 \le x \le 1$ and fixed y, z. We assume Dirichlet boundary conditions on the boundary $\partial\Omega$,

$$u = 0, \quad (x, y, z) \in \partial\Omega \backslash \{x = 1\}, \tag{9.31}$$

except at the top of the cylinder $x = 1$ where the actuation is applied,

$$u(1, y, z) = U(y, z, t), \quad (y, z) \in \Gamma. \tag{9.32}$$

The parameters $\varepsilon > 0, b_1, b_2, b_3, \lambda$ are unknown.

For notational convenience, let us use the following notation in this section:

$$\begin{aligned}
\Delta u &= u_{xx} + u_{yy} + u_{zz}, \\
\nabla u &= (u_x, u_y, u_z)^T, \\
\mathbf{b} &= (b_1, b_2, b_3)^T, \\
\|u\|^2 &= \int_\Gamma \int \int_0^1 u^2(x, y, z)\, dx dy dz = \int_\Omega u^2 \, d\Omega, \\
\|\nabla u\|^2 &= \int_\Omega \nabla u \cdot \nabla u \, d\Omega.
\end{aligned} \tag{9.33}$$

We employ the following observer:

$$\hat{u}_t = \hat{\varepsilon}\Delta\hat{u} + \hat{\mathbf{b}} \cdot \nabla\hat{u} + \hat{\lambda}u + \gamma^2(u - \hat{u})\|\nabla u\|^2, \qquad (x, y, z) \in \Omega, \tag{9.34}$$

$$\hat{u} = 0, \quad (x, y, z) \in \partial\Omega \backslash \{x = 1\}, \tag{9.35}$$

$$\hat{u} = u, \quad x = 1, \quad (y, z) \in \Gamma. \tag{9.36}$$

There are two main differences compared to the 1D case with one parameter considered in Section 9.1. First, since the diffusion coefficient ε is unknown, we must use projection to ensure $\hat{\varepsilon} > \underline{\varepsilon} > 0$. Note that $\hat{\varepsilon}$ does not require projection from above, and all other parameters do not require projection at all.

Second, we can see in (9.34) that while the diffusion and advection coefficients multiply the operators of $\hat{u}$, the reaction coefficient multiplies u in the observer. This is necessary in order to eliminate any λ-dependence in the error system so that it is stable.

The error signal $e = u - \hat{u}$ satisfies the following PDE:

$$e_t = \hat{\varepsilon}\Delta e + \hat{\mathbf{b}} \cdot \nabla e + \tilde{\varepsilon}\Delta u + \tilde{\mathbf{b}} \cdot \nabla u + \tilde{\lambda}u - \gamma^2 e\|\nabla u\|^2, \quad (x, y, z) \in \Omega, \tag{9.37}$$

$$e = 0, \quad (x, y, z) \in \partial\Omega. \tag{9.38}$$

Using the Lyapunov function

$$V = \frac{1}{2}\int_\Omega e^2 \, d\Omega + \frac{\tilde{\varepsilon}^2}{2\gamma_1} + \frac{|\hat{\mathbf{b}}|^2}{2\gamma_2} + \frac{\tilde{\lambda}^2}{2\gamma_3}, \tag{9.39}$$

we get

$$\begin{aligned}
\dot{V} = &-\hat{\varepsilon}\|\nabla e\|^2 - \gamma^2\|e\|^2\|\nabla u\|^2 + \tilde{\varepsilon}\int_\Omega e\Delta u \, d\Omega + \int_\Omega e(\tilde{\mathbf{b}} \cdot \nabla u)\, d\Omega \\
&+ \tilde{\lambda}\int_\Omega eu \, d\Omega - \frac{1}{\gamma_0}\tilde{\varepsilon}\dot{\hat{\varepsilon}} - \frac{1}{\gamma_1}\tilde{\mathbf{b}} \cdot \dot{\hat{\mathbf{b}}} - \frac{1}{\gamma_2}\tilde{\lambda}\dot{\hat{\lambda}}.
\end{aligned} \tag{9.40}$$

With update laws

$$\dot{\hat{\varepsilon}} = -\gamma_0 \mathrm{Proj}_{\underline{\varepsilon}} \left\{ \int_\Omega \nabla u \cdot \nabla(u - \hat{u})\, d\Omega \right\}, \tag{9.41}$$

$$\dot{\hat{\mathbf{b}}} = \gamma_1 \int_\Omega (u - \hat{u}) \nabla u\, d\Omega, \tag{9.42}$$

$$\dot{\hat{\lambda}} = \gamma_2 \int_\Omega (u - \hat{u}) u\, d\Omega, \tag{9.43}$$

where $\gamma_0, \gamma_1, \gamma_2 > 0$ and $\mathrm{Proj}_{\underline{\varepsilon}} = \mathrm{Proj}_{[\underline{\varepsilon},\infty)}$ is defined by (8.64), we get

$$\dot{V} \leq -\underline{\varepsilon}\|\nabla e\|^2 - \gamma^2 \|e\|^2 \|\nabla u\|^2, \tag{9.44}$$

which implies $V(t) \leq V(0)$, so that $\tilde{\varepsilon}$, $|\tilde{\mathbf{b}}|$, $\tilde{\lambda}$, and $\|e\|$ are bounded. Integrating (9.44) with respect to time from zero to infinity we get the square integrability of $\|\nabla e\|$ and $\|e\|\|\nabla u\|$, which, together with the update laws (9.41)–(9.43), gives the square integrability of $|\dot{\hat{\mathbf{b}}}|$ and $\dot{\hat{\lambda}}$.

Lemma 9.1. *The identifier* (9.34)–(9.36) *with update laws* (9.42)–(9.43) *guarantees the following properties:*

$$\|\nabla e\|,\ \|e\|\|\nabla u\| \in \mathcal{L}_2, \quad \|e\| \in \mathcal{L}_\infty \cap \mathcal{L}_2, \tag{9.45}$$

$$\tilde{\varepsilon},\ \tilde{b}_1,\ \tilde{b}_2,\ \tilde{b}_3,\ \tilde{\lambda} \in \mathcal{L}_\infty, \tag{9.46}$$

$$\dot{\hat{b}}_1,\ \dot{\hat{b}}_2,\ \dot{\hat{b}}_3,\ \dot{\hat{\lambda}} \in \mathcal{L}_2. \tag{9.47}$$

We employ the controller

$$U(t) = -\int_0^1 \frac{\hat{\lambda} + c}{\hat{\varepsilon}} \xi e^{-\frac{\hat{b}_1(1-\xi)}{2\hat{\varepsilon}}} \frac{I_1\left(\sqrt{\frac{\hat{\lambda}+c}{\hat{\varepsilon}}(1-\xi^2)}\right)}{\sqrt{\frac{\hat{\lambda}+c}{\hat{\varepsilon}}(1-\xi^2)}} \hat{u}(\xi, y, z, t)\, d\xi \tag{9.48}$$

with $c \geq 0$, which is a straightforward generalization of the one proposed in the previous section.

Having established the stability of the identifier, we now turn to proving closed-loop stability. Unfortunately, it is very hard to prove the result in the case of unknown ε. This is because, while the identifier guarantees the properties (9.45) for $\|e\|$ and $\|\nabla e\|$, it does not provide any estimates for $\|\Delta e\|$, which are required in the case of unknown ε. Therefore, for the closed-loop result we assume that ε is known, and set $\hat{\varepsilon} = \varepsilon$ everywhere. The update law (9.41) nevertheless achieves closed-loop stability for unknown ε in simulations, as shown in Section 9.4.

THEOREM 9.2. *For any initial data* $u_0, \hat{u}_0 \in H^2(\Omega)$ *compatible with boundary conditions, the classical solution* $(\hat{\mathbf{b}}, \hat{\lambda}, u, \hat{u})$ *of the closed-loop system* (9.30)–(9.32), (9.34)–(9.36), (9.42), (9.43), (9.48) *is bounded for all* $(x, y, z) \in \Omega$, $t \geq 0$ *and*

$$\lim_{t\to\infty} \max_{(x,y,z)\in\Omega} |u(x, y, z, t)| = 0. \tag{9.49}$$

9.3 PROOF OF THEOREM 9.2

In the proof we will use Poincaré's and Agmon's inequalities (see, e.g., [115]):

$$\|u\| \leq d_1(\Gamma)\|\nabla u\|, \tag{9.50}$$

$$\max_{(x,y,z)\in\Omega} |u| \leq d_2(\Gamma)\|u\|_{H^1}^{1/2}\|u\|_{H^2}^{1/2}. \tag{9.51}$$

Here d_1 and d_2 are constants that depend only on Γ. The main difference in proving the result in the 3D case compared to the 1D case is that we need to show H^2 (instead of H^1) boundedness and H^1 (instead of L^2) regulation in order to have pointwise boundedness and regulation.

9.3.1 Target System

We use the transformation

$$\hat{w}(x, y, z, t) = \hat{u}(x, y, z, t) - \int_0^x \hat{k}(x, \xi)\hat{u}(\xi, y, z, t)\, d\xi, \tag{9.52}$$

$$\hat{k}(x, \xi) = -\frac{\hat{\lambda} + c}{\varepsilon}\xi e^{-\frac{\hat{b}_1(x-\xi)}{2\varepsilon}} \frac{I_1\left(\sqrt{\frac{\hat{\lambda}+c}{\varepsilon}(x^2 - \xi^2)}\right)}{\sqrt{\frac{\hat{\lambda}+c}{\varepsilon}(x^2 - \xi^2)}}, \tag{9.53}$$

which is a generalized version of the transformation presented in Chapter 3 for the case of known parameters.

LEMMA 9.2. *The transformation* (9.52) *maps* (9.34)–(9.36), (9.48) *into the target system*

$$\begin{aligned}\hat{w}_t = {} & \varepsilon \Delta \hat{w} + \hat{\mathbf{b}} \cdot \nabla \hat{w} - c\hat{w} + \dot{\hat{b}}_1 \Phi_1[\hat{w}] + \dot{\hat{\lambda}} \Phi_2[\hat{w}] \\ & + (\hat{\lambda} + \gamma^2\|\nabla u\|^2)e_1, \quad (x, y, z) \in \Omega,\end{aligned} \tag{9.54}$$

$$\hat{w} = 0, \quad (x, y, z) \in \partial\Omega, \tag{9.55}$$

where

$$\Phi_i[\hat{w}](x, y, z, t) = \int_0^x \varphi_i(x, \xi)\hat{w}(\xi, y, z, t)\, d\xi, \tag{9.56}$$

$$e_1(x, y, z, t) = e(x, y, z, t) - \int_0^x \hat{k}(x, \xi)e(\xi, y, z, t)\, d\xi, \tag{9.57}$$

and

$$\varphi_1(x,\xi) = \frac{x-\xi}{2\varepsilon}\hat{k}(x,\xi) + \frac{1}{2\varepsilon}\int_\xi^x (x-\sigma)\hat{k}(x,\sigma)\hat{l}(\sigma,\xi)\,d\sigma,$$

$$\varphi_2(x,\xi) = \frac{\xi}{2\varepsilon}e^{-\frac{\hat{b}_1}{2\varepsilon}(x-\xi)}. \tag{9.58}$$

Proof. Substituting (9.52) into (9.34) we get

$$\hat{w}_t = \varepsilon\Delta\hat{w} + \hat{\mathbf{b}}\cdot\nabla\hat{w} - c\hat{w} + (\hat{\lambda}+\gamma^2\|\nabla u\|^2)e_1 - \int_0^x\left(\dot{\hat{b}}_1\hat{k}_{\hat{b}_1}(x,\xi) + \dot{\hat{\lambda}}\hat{k}_{\hat{\lambda}}(x,\xi)\right)\hat{u}(\xi,y,z)\,d\xi. \tag{9.59}$$

To replace $\hat{u}$ with $\hat{w}$ we use an inverse transformation

$$\hat{u}(x,y,z,t) = \hat{w}(x,y,z,t) + \int_0^x \hat{l}(x,\xi)\hat{w}(\xi,y,z,t)\,d\xi, \tag{9.60}$$

$$\hat{l}(x,\xi) = -\frac{\hat{\lambda}+c}{\varepsilon}\xi e^{-\frac{\hat{b}_1(x-\xi)}{2\varepsilon}}\frac{J_1\left(\sqrt{\frac{\hat{\lambda}+c}{\varepsilon}(x^2-\xi^2)}\right)}{\sqrt{\frac{\hat{\lambda}+c}{\varepsilon}(x^2-\xi^2)}}. \tag{9.61}$$

We have

$$\int_0^x \hat{k}_{\hat{\lambda}}(x,\xi)\hat{u}(\xi,y,z,t)\,d\xi = \int_0^x\left(\hat{k}_{\hat{\lambda}}(x,\xi) + \int_\xi^x \hat{k}_{\hat{\lambda}}(x,\sigma)l(\sigma,\xi)\,d\sigma\right)\hat{w}(\xi,y,z,t)\,d\xi, \tag{9.62}$$

and similarly for $\hat{b}_1$. Computing the inner integrals with the help of [101], we get (9.54)–(9.58). □

We should mention that while the target system (9.54)–(9.55) is complicated, only the proof is affected by this complexity, and not the design (which is simple).

9.3.2 Boundedness and Regulation in L^2 and H^1

Let us denote the bounds on $|\hat{\mathbf{b}}|$, $\hat{\lambda}$ by b_0, λ_0. Since $\hat{k}$ and $\hat{l}$ and their derivatives with respect to parameters are bounded functions, we have the estimates

$$\|e_1\| \le M_1\|e\|, \quad \|\nabla u\| \le M_2\|\nabla\hat{w}\| + \|\nabla e\|, \tag{9.63}$$

where M_1, M_2 are some constants. The functions φ_1, φ_2 are also bounded; let us denote these bounds by $\bar{\varphi}_1$, $\bar{\varphi}_2$.

First we show the boundedness of the L^2-norm, starting with

$$\frac{1}{2}\frac{d}{dt}\|\hat{w}\|^2 = -\varepsilon\|\nabla\hat{w}\|^2 - c\|\hat{w}\|^2 + \dot{\hat{b}}_1\int_\Omega \hat{w}\Phi_1\,d\Omega + \dot{\hat{\lambda}}\int_\Omega \hat{w}\Phi_2\,d\Omega + (\hat{\lambda}+\gamma^2\|\nabla u\|^2)\int_\Omega e_1\hat{w}\,d\Omega. \tag{9.64}$$

Using the estimate

$$\dot{\hat{b}}_1 \int_\Omega \hat{w}\Phi_1 \, d\Omega \le |\dot{\hat{b}}_1||\bar{\varphi}_1|\|\hat{w}\|^2 \le \frac{\varepsilon}{8d_1^2}\|\hat{w}\|^2 + \frac{2}{\varepsilon}d_1^2|\dot{\hat{b}}_1|^2\bar{\varphi}_1^2\|\hat{w}\|^2$$

$$\le \frac{\varepsilon}{8}\|\nabla\hat{w}\|^2 + l_1\|\hat{w}\|^2, \tag{9.65}$$

and similarly for the term with $\dot{\hat{\lambda}}$, we get

$$\begin{aligned}\frac{1}{2}\frac{d}{dt}\|\hat{w}\|^2 \le & -\frac{3\varepsilon}{4}\|\nabla\hat{w}\|^2 + l_1\|\hat{w}\|^2 + M_1\lambda_0\|\hat{w}\|\|e\| \\ & + \gamma^2 M_1\|\nabla u\|(M_2\|\nabla\hat{w}\| + \|\nabla e\|)\|\hat{w}\|\|e\| \\ \le & -\frac{3\varepsilon}{4}\|\nabla\hat{w}\|^2 + l_1\|\hat{w}\|^2 + \frac{d_1^2}{\varepsilon}M_1^2\lambda_0^2\|e\|^2 \\ & + \frac{\varepsilon}{4d_1^2}\|\hat{w}\|^2 + \frac{\varepsilon}{4}\|\nabla\hat{w}\|^2 + \frac{\varepsilon}{4M_2^2}\|\nabla e\|^2 \\ & + \frac{2}{\varepsilon}\gamma^4 M_1^2 M_2^2\|\nabla u\|^2\|e\|^2\|\hat{w}\|^2 \\ \le & -\frac{\varepsilon}{4d_1^2}\|\hat{w}\|^2 + l_1\|\hat{w}\|^2 + l_1. \end{aligned} \tag{9.66}$$

Using Lemma D.3 we get $\|\hat{w}\| \in \mathcal{L}_\infty \cap \mathcal{L}_2$. Integrating (9.66) with respect to time from zero to infinity, we also get $\|\nabla\hat{w}\| \in \mathcal{L}_2$, and therefore $\|\nabla\hat{u}\|, \|\nabla u\| \in \mathcal{L}_2$.

Now let us show H^1 boundedness. In this case it is sufficient to consider e and $\hat{w}$ systems separately. First,

$$\begin{aligned}\frac{1}{2}\frac{d}{dt}\|\nabla e\|^2 = & \int_\Omega \nabla e_t \nabla e \, d\Omega = -\int_\Omega e_t \Delta e \, d\Omega \\ \le & -\varepsilon\|\Delta e\|^2 + b_0\|\Delta e\|\|\nabla e\| + |\tilde{\mathbf{b}}|\|\Delta e\|\|\nabla u\| \\ & + |\tilde{\lambda}|\|\Delta e\|\|u\| - \gamma^2\|\nabla e\|^2\|\nabla u\|^2 \\ \le & -\varepsilon\|\Delta e\|^2 + \frac{\varepsilon}{4}\|\Delta e\|^2 + \frac{b_0^2}{\varepsilon}\|\nabla e\|^2 + \frac{\varepsilon}{4}\|\Delta e\|^2 \\ & + \frac{|\tilde{\mathbf{b}}|^2}{\varepsilon}\|\nabla u\|^2 + \frac{\varepsilon}{4}\|\Delta e\|^2 + \frac{|\tilde{\lambda}|^2}{\varepsilon}\|u\|^2 \\ \le & -\frac{\varepsilon}{4d_1^2}\|\nabla e\|^2 + l_1. \end{aligned} \tag{9.67}$$

Using Lemma D.3 we get $\|\nabla e\| \in \mathcal{L}_\infty \cap \mathcal{L}_2$. Second,

$$\begin{aligned}
\frac{1}{2}\frac{d}{dt}\|\nabla \hat{w}\|^2 = &-\int_\Omega \hat{w}_t \Delta \hat{w}\, d\Omega \\
= &-\varepsilon\|\Delta\hat{w}\|^2 - c\|\nabla\hat{w}\|^2 - \int_\Omega \Delta\hat{w}\,(\hat{\mathbf{b}}\cdot\nabla\hat{w})\, d\Omega \\
&- \dot{\hat{b}}_1 \int_\Omega \Delta\hat{w}\Phi_1\, d\Omega - \dot{\hat{\lambda}} \int_\Omega \Delta\hat{w}\Phi_2\, d\Omega \\
&+ (\hat{\lambda} + \gamma^2\|\nabla u\|^2)\int_\Omega e\Delta\hat{w}\, d\Omega.
\end{aligned} \tag{9.68}$$

Using the estimates

$$\begin{aligned}
\int_\Omega \Delta\hat{w}\,(\hat{\mathbf{b}}\cdot\nabla\hat{w})\, d\Omega &\le b_0\|\Delta\hat{w}\|\|\nabla\hat{w}\| \le \frac{\varepsilon}{8}\|\Delta\hat{w}\|^2 + \frac{2b_0^2}{\varepsilon}\|\nabla\hat{w}\|^2 \\
&\le \frac{\varepsilon}{8}\|\Delta\hat{w}\|^2 + l_1, \\
\dot{\hat{b}}_1 \int_\Omega \Delta\hat{w}\Phi_1\, d\Omega &\le \frac{\varepsilon}{8}\|\Delta\hat{w}\|^2 + l_1\|\hat{w}\|^2,
\end{aligned} \tag{9.69}$$

and similarly for the term with $\dot{\hat{\lambda}}$, we get

$$\begin{aligned}
\frac{1}{2}\frac{d}{dt}\|\nabla\hat{w}\|^2 \le &-\frac{5\varepsilon}{8}\|\Delta\hat{w}\|^2 + l_1\|\hat{w}\|^2 + l_1 \\
&+ \gamma^2 M_1 M_2\|\nabla u\|\|\nabla\hat{w}\|\|\Delta\hat{w}\|\|e\| \\
&+ \gamma^2 M_1\|\nabla u\|\|\nabla e\|\|\Delta\hat{w}\|\|e\| + M_1\lambda_0\|\Delta\hat{w}\|\|e\| \\
\le &-\frac{5\varepsilon}{8}\|\Delta\hat{w}\|^2 + l_1 + \frac{\varepsilon\|\Delta\hat{w}\|^2}{4} + \frac{2M_1^2\lambda_0^2\|e\|^2}{\varepsilon} \\
&+ \frac{2}{\varepsilon}\gamma^4 M_1^2 M_2^2\|\nabla u\|^2\|e\|^2\|\nabla\hat{w}\|^2 \\
&+ \frac{2}{\varepsilon}\gamma^4 M_1^2\|\nabla u\|^2\|\nabla e\|^2\|e\|^2 + \frac{\varepsilon}{8}\|\Delta\hat{w}\|^2 \\
\le &-\frac{\varepsilon}{4d_1^2}\|\nabla\hat{w}\|^2 + l_1\|\nabla\hat{w}\|^2 + l_1.
\end{aligned} \tag{9.70}$$

Using Lemma D.3 we get $\|\nabla\hat{w}\| \in \mathcal{L}_\infty \cap \mathcal{L}_2$ and therefore $\|\nabla\hat{u}\|$, $\|\nabla u\| \in \mathcal{L}_\infty \cap \mathcal{L}_2$. Integrating (9.67), (9.70) over an infinite time interval we also get $\|\Delta e\|$, $\|\Delta\hat{w}\| \in \mathcal{L}_2$ and therefore $\|\Delta\hat{u}\|$, $\|\Delta u\| \in \mathcal{L}_2$.

Note that from the above properties and (9.67)–(9.70) it follows that $(d/dt)\|\nabla e\|^2$ and $(d/dt)\|\nabla\hat{w}\|^2$ are bounded. By Lemma D.2 we get $\|\nabla e\|$, $\|\nabla\hat{w}\| \to 0$, and therefore $\|\nabla\hat{u}\|$, $\|\nabla u\| \to 0$ as $t \to \infty$.

9.3.3 Pointwise Boundedness and Regulation

In order to prove pointwise boundedness in 3D, we need to show that the H^2-norms of the signals are bounded. It is more convenient to prove the boundedness of $\|\hat{w}_t\|$ and $\|e_t\|$ first and then use the equations (9.54), (9.37) to bound the H^2-norms. We start with

$$\begin{aligned}\frac{1}{2}\frac{d}{dt}\|e_t\|^2 =& \int_\Omega e_t e_{tt}\, d\Omega \\ \le& -\varepsilon\|\nabla e_t\|^2 + |\dot{\hat{\mathbf{b}}}|\|e_t\|\|\nabla e\| + |\dot{\hat{\mathbf{b}}}|\|e_t\|\|\nabla u\| + |\tilde{\mathbf{b}}|\|\nabla e_t\|\|u_t\| \\ &+ |\dot{\hat{\lambda}}|\|e_t\|\|u\| + |\tilde{\lambda}|\|e_t\|\|u_t\| + \gamma^2\|e\|\|e_t\|\left|\frac{d}{dt}\|\nabla u\|^2\right|. \end{aligned} \tag{9.71}$$

We first note that

$$\|u_t\|^2 \le 2(\varepsilon^2\|\Delta u\|^2 + |\mathbf{b}|^2\|\nabla u\|^2 + \lambda^2\|u\|^2) \le l_1 \tag{9.72}$$

and

$$\begin{aligned}\|e\|\|e_t\|\left|\frac{d}{dt}\|\nabla u\|^2\right| \le& 2\|e\|\|e_t\|(M_2^2\|\nabla\hat{w}\|\|\nabla\hat{w}_t\| + \|\nabla e\|\|\nabla e_t\|) \\ \le& l_1 + \frac{\varepsilon}{8}\|\nabla e_t\|^2 + c_1\|\nabla\hat{w}_t\|^2, \end{aligned} \tag{9.73}$$

where c_1 is an arbitrary constant. We get the following estimate:

$$\frac{1}{2}\frac{d}{dt}\|e_t\|^2 \le -\frac{\varepsilon}{2}\|\nabla e_t\|^2 + l_1\|e_t\|^2 + c_1\|\nabla\hat{w}_t\|^2 + l_1. \tag{9.74}$$

Now we estimate the time derivative of $\|\hat{w}_t\|^2$,

$$\begin{aligned}\frac{1}{2}\frac{d}{dt}\|\hat{w}_t\|^2 =& \int_\Omega \hat{w}_t\hat{w}_{tt}\, d\Omega \\ \le& -\varepsilon\|\nabla\hat{w}_t\|^2 + |\dot{\hat{\mathbf{b}}}|\|\hat{w}_t\|\|\nabla\hat{w}\| - c\|\hat{w}_t\|^2 \\ &+ |\ddot{\hat{b}}_1|\bar{\varphi}_1\|\hat{w}_t\|\|\hat{w}\| + |\dot{\hat{b}}_1|\|\hat{w}_t\dot{\Phi}_1\| \\ &+ |\ddot{\hat{\lambda}}|\bar{\varphi}_2\|\hat{w}_t\|\|\hat{w}\| + |\dot{\hat{\lambda}}|\|\hat{w}_t\dot{\Phi}_2\| \\ &+ (\lambda_0 + \gamma^2\|\nabla u\|^2)\|e_{1t}\|\|\hat{w}_t\| \\ &+ \left(|\dot{\hat{\lambda}}| + \gamma^2\left|\frac{d}{dt}\|\nabla u\|^2\right|\right)\|e\|M_1\|\hat{w}_t\|. \end{aligned} \tag{9.75}$$

Using the estimates

$$
\begin{aligned}
\|\hat{w}_t\dot{\Phi}_1\|^2 &\le \bar{\varphi}_1^2\|\hat{w}_t\|^2 + M_3\|\hat{w}_t\|^2\|\hat{w}\|^2,\\
\|\hat{w}_t\dot{\Phi}_2\|^2 &\le \bar{\varphi}_2^2\|\hat{w}_t\|^2 + M_4\|\hat{w}_t\|^2\|\hat{w}\|^2,\\
|\ddot{\hat{b}}_1|^2 &\le 2\gamma_1^2\|e_t\|^2\|\nabla u\|^2 + 2\gamma_1^2\|e\|^2\|\nabla u_t\|^2\\
&\le l_1\|e_t\|^2 + M_5(\|\nabla\hat{w}_t\|^2 + \|\nabla e_t\|^2),\\
|\ddot{\hat{\lambda}}|^2 &\le 2\gamma_2^2\|e_t\|^2\|u\|^2 + 2\gamma_2^2\|e\|^2\|u_t\|^2\\
&\le l_1\|e_t\|^2 + l_1,\\
\|e_{1t}\|^2 &\le 2M_1^2\|e_t\|^2 + M_6\|e\|^2,
\end{aligned}
\tag{9.76}
$$

we get

$$
\begin{aligned}
\frac{1}{2}\frac{d}{dt}\|\hat{w}_t\|^2 \le &-\varepsilon\|\nabla\hat{w}_t\|^2 + \frac{1}{2}|\dot{\hat{\mathbf{b}}}|^2\|\hat{w}_t\|^2 + \frac{1}{2}\|\nabla\hat{w}\|^2\\
&+ l_1\|e_t\|^2 + c_2(\|\nabla\hat{w}_t\|^2 + \|\nabla e_t\|^2)\\
&+ \frac{M_5\bar{\varphi}_1^2}{4c_2}\|\hat{w}\|^2\|\hat{w}_t\|^2 + \frac{2}{\varepsilon}d_1^2\bar{\varphi}_1^2|\dot{\hat{b}}_1|^2\\
&+ \frac{\varepsilon}{8d_1^2}\|\hat{w}_t\|^2 + l_1\|\hat{w}_t\|^2 + l_1\|e_t\|^2 + l_1\\
&+ l_1\|\hat{w}_t\|^2 + \frac{2}{\varepsilon}d_1^2\bar{\varphi}_2^2|\dot{\hat{\lambda}}|^2 + \frac{\varepsilon}{8d_1^2}\|\hat{w}_t\|^2\\
&+ l_1\|\hat{w}_t\|^2 + \frac{4\lambda_0^2M_1^2d_1^2}{\varepsilon}\|e_t\|^2 + \frac{\varepsilon}{8d_1^2}\|\hat{w}_t\|^2\\
&+ l_1\|e_t\|^2 + l_1\|\hat{w}_t\|^2 + \frac{\varepsilon}{8}\|\nabla\hat{w}_t\|^2 + c_3\|\nabla e_t\|^2 + l_1\|\hat{w}_t\|^2 + l_1\\
\le &-\left(\frac{\varepsilon}{4} - c_2\right)\|\nabla\hat{w}_t\|^2 + (c_2 + c_3)\|\nabla e_t\|^2 + l_1\\
&+ \frac{4\lambda_0^2M_1^2d_1^2}{\varepsilon}\|e_t\|^2 + l_1\|\hat{w}_t\|^2 + l_1\|e_t\|^2.
\end{aligned}
\tag{9.77}
$$

Combining (9.77) and (9.74) with a weighting constant A, we get

$$
\begin{aligned}
\frac{A}{2}\frac{d}{dt}\|e_t\|^2 + \frac{1}{2}\frac{d}{dt}\|\hat{w}_t\|^2 \le &-\left(\frac{\varepsilon}{4} - c_2 - c_1A\right)\|\nabla\hat{w}_t\|^2\\
&-\left(\frac{\varepsilon}{2}A - \frac{4\lambda_0^2M_1^2d_1^4}{\varepsilon} - c_2 - c_3\right)\|\nabla e_t\|^2\\
&+ l_1\|\hat{w}_t\|^2 + \|e_t\|^2 + l_1\,.
\end{aligned}
\tag{9.78}
$$

Choosing $A = 1 + 8\lambda_0^2M_1^2d_1^4\varepsilon^{-2}$, $c_1 = \varepsilon/(16A)$, and $c_2 = c_3 = \varepsilon/8$, we get

$$
\begin{aligned}
\frac{A}{2}\frac{d}{dt}\|e_t\|^2 + \frac{1}{2}\frac{d}{dt}\|\hat{w}_t\|^2 \le &-\frac{\varepsilon}{16d_1^2}\|\hat{w}_t\|^2 - \frac{\varepsilon}{4d_1^2}\|e_t\|^2\\
&+ l_1\|\hat{w}_t\|^2 + \|e_t\|^2 + l_1.
\end{aligned}
\tag{9.79}
$$

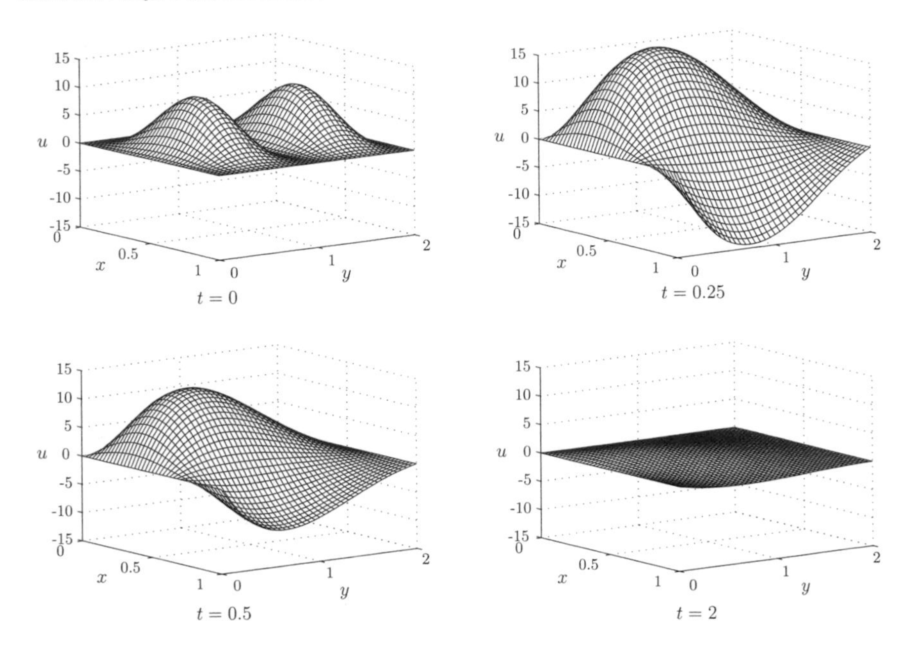

Figure 9.2 The closed-loop state for the plant (9.81) at different times.

By Lemma D.3, $\|\hat{w}_t\|, \|e_t\| \in \mathcal{L}_\infty \cap \mathcal{L}_2$, and therefore $\|\hat{u}_t\|, \|u_t\| \in \mathcal{L}_\infty \cap \mathcal{L}_2$. From (9.34) and (9.30) we get $\|\Delta\hat{u}\|, \|\Delta u\| \in \mathcal{L}_\infty \cap \mathcal{L}_2$. Using now Agmon's inequality (9.51) and the result of previous section ($\|u\|, \|\nabla u\| \to 0$ as $t \to \infty$), we get the regulation result:

$$\lim_{t\to\infty} \max_{(x,y,z)\in\Omega} |u(x, y, z, t)| \le d_2 \lim_{t\to\infty} \|u\|_{H^1}^{1/2} \|u\|_{H^2}^{1/2} = 0. \tag{9.80}$$

The proof of Theorem 9.2 is completed.

9.4 SIMULATIONS

We illustrate the design with a passive identifier on a 2D plant (which is easier to visualize than a 3D system) with four unknown parameters, ε, b_1, b_2, and λ:

$$u_t = \varepsilon(u_{xx} + u_{yy}) + b_1 u_x + b_2 u_y + \lambda u \tag{9.81}$$

on the rectangle $0 \le x \le 1$, $0 \le y \le L$ with actuation applied on the side with $x = 1$ and Dirichlet boundary conditions on the other three sides. Even though for the 2D case the actuation needs to be distributed, in practice only a limited number of actuators would be used, enough for the controller to perform well. The adaptive laws (9.41)–(9.43) are modified in a straightforward way from the 3D to the 2D setting. We set the simulation parameters to $\varepsilon = 1$, $b_1 = 1$, $b_2 = 2$, $\lambda = 22$, $L = 2$. With this choice the plant has two unstable eigenvalues at 8.4 and 1. Initial estimates are set to $\hat{\varepsilon}(0) = 2$, $\hat{b}_1(0) = 3$, $\hat{b}_2(0) = 0$, $\hat{\lambda}(0) = 5$, and the bound on $\hat{\varepsilon}$ from below is $\underline{\varepsilon} = 0.5$. The initial conditions for the plant and the observer are

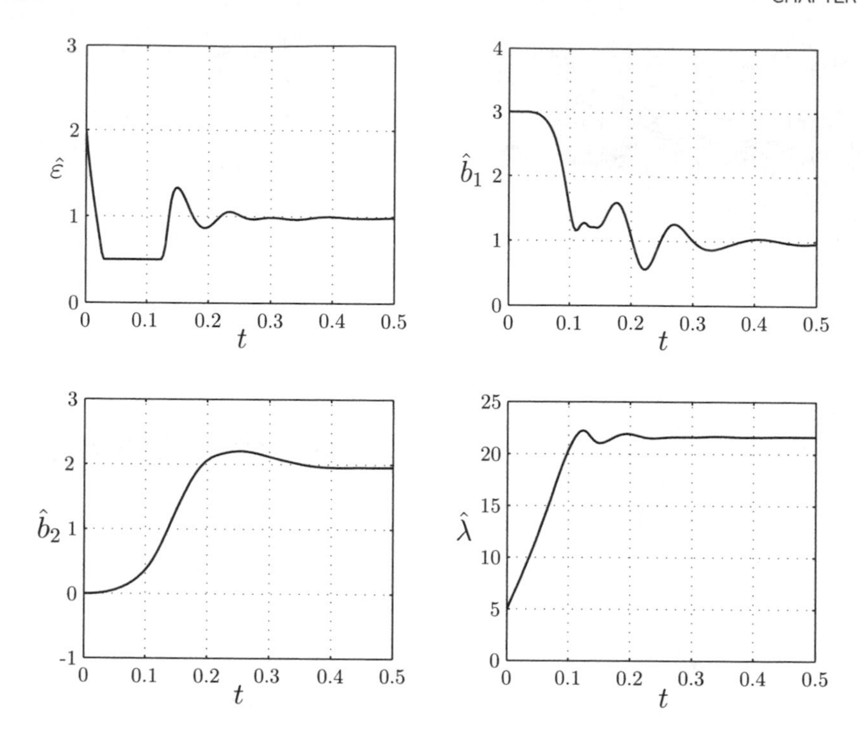

Figure 9.3 The parameter estimates for the plant (9.81) with adaptive controller based on a passive identifier.

$u(x, y, 0) = 10\sin^2(\pi x)\sin^2(\pi y)$ and $\hat{u}(x, y, 0) \equiv 0$. The simulation results are presented in Figure 9.2 (several snapshots of the state) and Figure 9.3 (estimates of the unknown parameters). One can see that projection keeps $\hat{\varepsilon} \geq \underline{\varepsilon} = 0.5$. All the estimates come close to the true values at approximately $t = 0.5$, and after that the controller stabilizes the system.

9.5 NOTES AND REFERENCES

The passivity-based approach (also known as the observer-based approach) presented in this chapter is the prevalent identification technique in nonbackstepping results on adaptive control for PDEs. For parabolic PDEs with *in-domain* actuation, passive identifiers were employed in [9, 11, 12, 32, 33, 34, 49, 93, 94, 114] (without the nonlinear damping). In these references the parameter estimates are shown to converge to the true parameters if the plant is excited by a sufficiently rich input. In this chapter we dealt with *boundary* actuation but were concerned only with stabilization, without the assumption of persistency of excitation.

Even though we considered only Dirichlet boundary conditions in this chapter, the designs can be easily extended to other cases (Dirichlet uncontrolled with Neumann actuation; Robin uncontrolled end with Dirichlet or Neumann actuation).

Note that, compared to the Lyapunov designs in Chapter 8, passivity-based design requires measurement of both the state and the first spatial derivative(s) of the state, except for the case of known diffusion and advection coefficients. This brings up the problem of sensitivity to noise in the passivity-based approach. This problem can be remedied to a certain extent by spatial "low-pass" filtering, in the same way that PD control is implemented in practice.

This chapter is notable for the control design for a 3D plant. While all the designs in the book (nonadaptive and adaptive) can be extended to higher-dimensional plants with constant coefficients (on rectangular 2D domains or cylinder-type 3D domains), the actual extension is presented only in this chapter.

Chapter Ten

Certainty Equivalence Design with Swapping Identifiers

In this chapter we present a second estimation-based approach, the design with swapping identifiers. This approach (often called simply the gradient or least squares method) is the most commonly used identification method in finite-dimensional adaptive control [51]. One converts a dynamic parametrization of the problem into a static one using filters of the regressor and of the measured part of the plant. The standard gradient and least squares estimation techniques are then applied.

In this chapter we deal with a 1D reaction-advection-diffusion plant with three unknown constant parameters. As in Chapter 9, there exists a fundamental difficulty in proving closed-loop stability when the diffusion coefficient is unknown. For this reason, we present the stability proof for the case of a known diffusion coefficient. However, stability of the identifier is established for the case of all three coefficients unknown, and the adaptive scheme is successful in simulations for that case, as shown in Section 10.3.

10.1 REACTION-ADVECTION-DIFFUSION PLANT

Consider the plant

$$u_t(x,t) = \varepsilon u_{xx}(x,t) + b u_x(x,t) + \lambda u(x,t), \tag{10.1}$$

$$u(0,t) = 0, \tag{10.2}$$

$$u(1,t) = U(t), \tag{10.3}$$

where ε, b, and λ are constant unknown parameters.

We need to employ four (the number of uncertain parameters plus one) filters. Let us first write the "estimation error" in the form

$$e = u - \varepsilon\psi - bp - \lambda v - \eta, \tag{10.4}$$

where v is the filter for u,

$$v_t(x,t) = \hat{\varepsilon} v_{xx}(x,t) + \hat{b} v_x(x,t) + u(x,t), \tag{10.5}$$

$$v(0,t) = 0, \tag{10.6}$$

$$v(1,t) = 0, \tag{10.7}$$

p is the filter for u_x,

$$p_t(x,t) = \hat{\varepsilon} p_{xx}(x,t) + \hat{b} p_x(x,t) + u_x(x,t), \tag{10.8}$$
$$p(0,t) = 0, \tag{10.9}$$
$$p(1,t) = 0, \tag{10.10}$$

ψ is the filter for u_{xx},

$$\psi_t(x,t) = \hat{\varepsilon} \psi_{xx}(x,t) + \hat{b} \psi_x(x,t) + u_{xx}(x,t), \tag{10.11}$$
$$\psi(0,t) = 0, \tag{10.12}$$
$$\psi(1,t) = 0, \tag{10.13}$$

and η is the following filter:

$$\eta_t(x,t) = \hat{\varepsilon} \eta_{xx}(x,t) + \hat{b} \eta_x(x,t) - \hat{b} u_x(x,t) - \hat{\varepsilon} u_{xx}(x,t), \tag{10.14}$$
$$\eta(0,t) = 0, \tag{10.15}$$
$$\eta(1,t) = U(t). \tag{10.16}$$

As usual, $\hat{\varepsilon} = \hat{\varepsilon}(t)$, $\hat{b} = \hat{b}(t)$, and $\hat{\lambda} = \hat{\lambda}(t)$ are time-varying estimates of unknown parameters. Note that in the case of known ε or b, the filter η can be modified by dropping the corresponding terms $\hat{\varepsilon} u_{xx}$ or $\hat{b} u_x$ in (10.14), so that there is no need to measure u_{xx} or u_x.

With the filters (10.5)–(10.16), the estimation error (10.4) satisfies the exponentially stable PDE

$$e_t(x,t) = \hat{\varepsilon} e_{xx}(x,t) + \hat{b} e_x(x,t), \tag{10.17}$$
$$e(0,t) = 0, \tag{10.18}$$
$$e(1,t) = 0. \tag{10.19}$$

We implement a "prediction error" as

$$\hat{e} = u - \hat{\varepsilon}\psi - \hat{b} p - \hat{\lambda} v - \eta, \tag{10.20}$$

which is related to the estimation error by

$$\hat{e} = e + \tilde{\varepsilon}\psi + \tilde{b} p + \tilde{\lambda} v. \tag{10.21}$$

To keep the parabolic character of the systems involved in the adaptive scheme, we must use projection to ensure $\hat{\varepsilon}(t) > \underline{\varepsilon} > 0$. The projection operator is defined in (8.64).

We choose gradient update laws with normalization,

$$\dot{\hat{\varepsilon}} = \gamma_1 \mathrm{Proj}_{\underline{\varepsilon}} \left\{ \frac{\int_0^1 \hat{e}(x,t)\psi(x,t)\,dx}{1 + \|\psi\|^2 + \|p\|^2 + \|v\|^2} \right\}, \tag{10.22}$$

$$\dot{\hat{b}} = \gamma_2 \frac{\int_0^1 \hat{e}(x,t) p(x,t)\,dx}{1 + \|\psi\|^2 + \|p\|^2 + \|v\|^2}, \tag{10.23}$$

$$\dot{\hat{\lambda}} = \gamma_3 \frac{\int_0^1 \hat{e}(x,t) v(x,t)\,dx}{1 + \|\psi\|^2 + \|p\|^2 + \|v\|^2}, \tag{10.24}$$

where γ_1, γ_2, and γ_3 are positive constants.

In contrast to the passive identifier, the normalization in the swapping identifier is employed in the update law. This makes time derivatives of the parameter estimates not only square integrable but also bounded, as stated in the following lemma.

Note that, unlike Lyapunov and passive approaches, the swapping approach requires a measurement of the second spatial derivative of the state when the diffusion coefficient is unknown.

Lemma 10.1. *The update laws* (10.22)–(10.24) *guarantee the following properties:*

$$\tilde{\varepsilon},\ \tilde{b},\ \tilde{\lambda} \in \mathcal{L}_\infty, \tag{10.25}$$

$$\dot{\hat{\varepsilon}},\ \dot{\hat{b}},\ \dot{\hat{\lambda}} \in \mathcal{L}_2 \cap \mathcal{L}_\infty, \tag{10.26}$$

$$\frac{\|\hat{e}\|}{\sqrt{1+\|\psi\|^2+\|p\|^2+\|v\|^2}} \in \mathcal{L}_2 \cap \mathcal{L}_\infty. \tag{10.27}$$

Proof. With the Lyapunov function

$$V = \frac{1}{2}\int_0^1 e^2(x,t)\,dx + \frac{1}{8\gamma_1}\tilde{\varepsilon}^2 + \frac{1}{8\gamma_2}\tilde{b}^2 + \frac{1}{8\gamma_3}\tilde{\lambda}^2 \tag{10.28}$$

we get

$$\begin{aligned}
\dot{V} &\le -\int_0^1 e_x^2(x,t)\,dx - \frac{\int_0^1 \hat{e}(x,t)(\tilde{\varepsilon}\psi(x,t) + \tilde{b}p(x,t) + \tilde{\lambda}v(x,t))\,dx}{4(1+\|\psi\|^2+\|p\|^2+\|v\|^2)} \\
&\le -\|e_x\|^2 - \frac{\int_0^1 \hat{e}^2(x,t)\,dx + \int_0^1 \hat{e}(x,t)e(x,t)\,dx}{4(1+\|\psi\|^2+\|p\|^2+\|v\|^2)} \\
&\le -\|e_x\|^2 - \frac{\|\hat{e}\|^2}{4(1+\|\psi\|^2+\|p\|^2+\|v\|^2)} + \frac{\|e_x\|\|\hat{e}\|}{2\sqrt{1+\|\psi\|^2+\|p\|^2+\|v\|^2}} \\
&\le -\frac{1}{2}\|e_x\|^2 - \frac{1}{8}\frac{\|\hat{e}\|^2}{1+\|\psi\|^2+\|p\|^2+\|v\|^2}.
\end{aligned} \tag{10.29}$$

This gives

$$\frac{\|\hat{e}\|}{\sqrt{1+\|\psi\|^2+\|p\|^2+\|v\|^2}} \in \mathcal{L}_2 \tag{10.30}$$

and the boundedness of $\tilde{\varepsilon}$, $\tilde{b}$, and $\tilde{\lambda}$. From (10.21) we get

$$\frac{\|\hat{e}\|}{\sqrt{1+\|\psi\|^2+\|p\|^2+\|v\|^2}} \in \mathcal{L}_\infty, \tag{10.31}$$

and from the update laws (10.22)–(10.24) the boundedness and square integrability of $\dot{\hat{\varepsilon}}$, $\dot{\hat{b}}$, and $\dot{\hat{\lambda}}$ follow. □

We use the controller

$$U(t) = -\int_0^1 \frac{\hat{\lambda}+c}{\hat{\varepsilon}} \xi e^{-\frac{\hat{b}(1-\xi)}{2\hat{\varepsilon}}} \frac{I_1\left(\sqrt{\frac{\hat{\lambda}+c}{\hat{\varepsilon}}(1-\xi^2)}\right)}{\sqrt{\frac{\hat{\lambda}+c}{\hat{\varepsilon}}(1-\xi^2)}} \times (\hat{\varepsilon}\psi(\xi,t) + \hat{b}p(\xi,t) + \hat{\lambda}v(\xi,t) + \eta(\xi,t))\, d\xi \tag{10.32}$$

with $c \geq 0$.

In the next section we prove the following result for a plant with known ε (i.e., we set $\hat{\varepsilon}(t) \equiv \varepsilon$ everywhere).

THEOREM 10.1. *For any initial data $v_0, p_0, \eta_0, u_0 \in H^2(0,1)$ compatible with boundary conditions, the classical solution $(\hat{b}, \hat{\lambda}, v, p, \eta, u)$ of the closed-loop system consisting of the plant* (10.1)–(10.3), *the controller* (10.32), *the filters* (10.5)–(10.10), (10.14)–(10.16), *and the update laws* (10.23)–(10.24) *is bounded for all $x \in [0,1]$, $t \geq 0$ and*

$$\lim_{t\to\infty} \max_{x\in[0,1]} |u(x,t)| = 0. \tag{10.33}$$

10.2 PROOF OF THEOREM 10.1

10.2.1 Target System

We use the transformation

$$\hat{w}(x,t) = \hat{b}p(x,t) + \hat{\lambda}v(x,t) + \eta(x,t) - \int_0^x \hat{k}(x,\xi)(\hat{b}p(\xi,t) + \hat{\lambda}v(\xi,t) + \eta(\xi,t))\, d\xi, \tag{10.34}$$

where $\hat{k}(x,\xi)$ is given by

$$\hat{k}(x,\xi) = -\frac{\hat{\lambda}+c}{\varepsilon} \xi e^{-\frac{\hat{b}(x-\xi)}{2\varepsilon}} \frac{I_1\left(\sqrt{\frac{\hat{\lambda}+c}{\varepsilon}(x^2-\xi^2)}\right)}{\sqrt{\frac{\hat{\lambda}+c}{\varepsilon}(x^2-\xi^2)}}. \tag{10.35}$$

The inverse transformation is defined as

$$\hat{b}p(x,t) + \hat{\lambda}v(x,t) + \eta(x,t) = \hat{w}(x,t) + \int_0^x \hat{l}(x,\xi)\hat{w}(\xi,t)\, d\xi \tag{10.36}$$

with the kernel

$$\hat{l}(x,\xi) = -\frac{\hat{\lambda}+c}{\varepsilon}\xi e^{-\frac{\hat{b}(x-\xi)}{2\varepsilon}} \frac{J_1\left(\sqrt{\frac{\hat{\lambda}+c}{\varepsilon}(x^2-\xi^2)}\right)}{\sqrt{\frac{\hat{\lambda}+c}{\varepsilon}(x^2-\xi^2)}}. \tag{10.37}$$

Lemma 10.2. *The transformation* (10.34) *produces the following target system:*

$$\hat{w}_t = \varepsilon\hat{w}_{xx} + \hat{b}\hat{w}_x - c\hat{w} + K[\dot{\hat{b}}p + \dot{\hat{\lambda}}v] + \hat{\lambda}K[\hat{e}] + \int_0^x (\dot{\hat{b}}\varphi_1(x,\xi) + \dot{\hat{\lambda}}\varphi_2(x,\xi))\hat{w}(\xi,t)\,d\xi, \tag{10.38}$$

$$\hat{w}(0,t) = 0, \tag{10.39}$$

$$\hat{w}(1,t) = 0, \tag{10.40}$$

where

$$K[v](x,t) = v(x,t) - \int_0^x \hat{k}(x,\xi)v(\xi,t)\,d\xi \tag{10.41}$$

and

$$\varphi_1(x,\xi) = \frac{x-\xi}{2\varepsilon}\hat{k}(x,\xi) + \frac{1}{2\varepsilon}\int_\xi^x (x-\sigma)\hat{k}(x,\sigma)\hat{l}(\sigma,\xi)\,d\sigma, \tag{10.42}$$

$$\varphi_2(x,\xi) = \frac{\xi}{2\varepsilon}e^{-\frac{\hat{b}}{2\varepsilon}(x-\xi)}. \tag{10.43}$$

Proof. Differentiating (10.34) with respect to x (twice) and t and using (10.5)–(10.10) and (10.14)–(10.16), we obtain

$$\hat{w}_t = \varepsilon\hat{w}_{xx} + \hat{b}\hat{w}_x - c\hat{w} + K[\dot{\hat{b}}p + \dot{\hat{\lambda}}v] + \hat{\lambda}K[\hat{e}] - \int_0^x (\dot{\hat{b}}\hat{k}_{\hat{b}}(x,\xi) + \dot{\hat{\lambda}}\hat{k}_{\hat{\lambda}}(x,\xi))(\hat{b}p + \hat{\lambda}v + \eta)\,d\xi. \tag{10.44}$$

Using the inverse transformation (10.36), we replace $(\hat{b}p + \hat{\lambda}v + \eta)$ in (10.44) by $\hat{w}$. Changing the order of the integration and computing the inner integral, we get (10.38). □

10.2.2 L^2 Boundedness

Let us use (10.20) and (10.36) to write the state u in filters (10.5)–(10.10) as

$$u(x,t) = \hat{e}(x,t) + \hat{w}(x,t) + \int_0^x \hat{l}(x,\xi)\hat{w}(\xi,t)\,d\xi. \tag{10.45}$$

We now have three interconnected systems for $\hat{w}$, v, and p with external driving signals $\hat{e}$, $\dot{\hat{b}}$, and $\dot{\hat{\lambda}}$ that go to zero in some sense, owing to the identifier properties (10.26)–(10.27).

The identifier properties imply that $\hat{k}$ and $\hat{l}$ are bounded and thus φ_1 and φ_2 are bounded. We denote these bounds by $\bar{\varphi}_1, \bar{\varphi}_2$. The bounds on $\hat{b}$, $\hat{\lambda}$ are denoted by b_0, λ_0, respectively.

We have the following estimates:

$$\int_0^1 \hat{w}(x,t)\int_0^x \varphi_i(x,\xi)\hat{w}(\xi,t)\,d\xi\,dx \le \bar{\varphi}_i\|\hat{w}\|^2, \tag{10.46}$$

$$\int_0^1 \hat{w}(x,t)K[\hat{e}](x,t)\,dx \le M_1\|\hat{w}\|\|\hat{e}\|, \tag{10.47}$$

$$\|u\| \le \|\hat{e}\| + M_2\|\hat{w}\|, \tag{10.48}$$

where M_1 and M_2 are some constants that depend on the bounds b_0 and λ_0.

We are now going to perform an L^2 Lyapunov analysis of the $(\hat{w}, v, p)$ system. We start with

$$\begin{aligned}\frac{1}{2}\frac{d}{dt}\|\hat{w}\|^2 \le &-\varepsilon\|\hat{w}_x\|^2 + \lambda_0 M_1\|\hat{w}\|\|\hat{e}\| + M_1\|\hat{w}\|\left(|\dot{\hat{b}}|\|p\| + |\dot{\hat{\lambda}}|\|v\|\right)\\ &+ \left(|\dot{\hat{b}}|\bar{\varphi}_1 + |\dot{\hat{\lambda}}|\bar{\varphi}_2\right)\|\hat{w}\|^2\\ \le &-\varepsilon\|\hat{w}_x\|^2 + \frac{\varepsilon}{4}\|\hat{w}\|^2 + \frac{\lambda_0^2 M_1^2}{\varepsilon}\|\hat{e}\|^2 + c_1(\|p\|^2 + \|v\|^2)\\ &+ \frac{M_1^2}{4c_1}\left(|\dot{\hat{b}}|^2 + |\dot{\hat{\lambda}}|^2\right)\|\hat{w}\|^2 + \frac{\varepsilon}{4}\|\hat{w}\|^2\\ &+ \frac{2}{\varepsilon}\left(|\dot{\hat{b}}|^2\bar{\varphi}_1^2 + |\dot{\hat{\lambda}}|^2\bar{\varphi}_2^2\right)\|\hat{w}\|^2.\end{aligned} \tag{10.49}$$

Here by c_1 we denote an arbitrary constant that will be defined later. Note that in the estimates we do not use the gain $c \ge 0$ to help stabilize the system.

Using properties (10.27) we have

$$\|\hat{e}\|^2 \le l_1\|p\|^2 + l_1\|v\|^2 + l_1, \tag{10.50}$$

so (10.49) can be written as

$$\begin{aligned}\frac{1}{2}\frac{d}{dt}\|\hat{w}\|^2 \le &-\frac{\varepsilon}{2}\|\hat{w}_x\|^2 + c_1(\|p\|^2 + \|v\|^2)\\ &+ l_1(\|\hat{w}\|^2 + \|p\|^2 + \|v\|^2) + l_1.\end{aligned} \tag{10.51}$$

We do a Lyapunov analysis for the filter v now:

$$\frac{1}{2}\frac{d}{dt}\|v\|^2 \le -\varepsilon\|v_x\|^2 + \int_0^1 v(x,t)u(x,t)\,dx. \tag{10.52}$$

Using (10.45) we have the estimate

$$\begin{aligned}\int_0^1 v(x,t)u(x,t)\,dx &\le M_2\|v\|\|\hat{w}\| + \|v\|\|\hat{e}\|\\ &\le \frac{\varepsilon}{4}\|v\|^2 + \frac{M_2^2}{\varepsilon}\|\hat{w}\|^2 + \frac{\varepsilon}{4}\|v\|^2 + l_1\|p\|^2 + l_1\|v\|^2 + l_1.\end{aligned} \tag{10.53}$$

With this estimate (10.52) becomes

$$\frac{1}{2}\frac{d}{dt}\|v\|^2 \leq -\frac{\varepsilon}{2}\|v_x\|^2 + \frac{M_2^2}{\varepsilon}\|\hat{w}\|^2 + l_1\|p\|^2 + l_1\|v\|^2 + l_1. \tag{10.54}$$

In a similar way one can show that for the filter p, we have:

$$\begin{aligned}\frac{1}{2}\frac{d}{dt}\|p\|^2 &\leq -\varepsilon\|p_x\|^2 + \int_0^1 p(x,t)u_x(x,t)\,dx \\ &\leq -\varepsilon\|p_x\|^2 + M_2\|p_x\|\|\hat{w}\| + \|p_x\|\|\hat{e}\| \\ &\leq -\varepsilon\|p_x\|^2 + \frac{\varepsilon}{2}\|p_x\|^2 + \frac{M_2^2}{\varepsilon}\|\hat{w}\|^2 + \frac{1}{\varepsilon}\|\hat{e}\|^2 \\ &\leq -\frac{\varepsilon}{2}\|p_x\|^2 + \frac{M_2^2}{\varepsilon}\|\hat{w}\|^2 + l_1\|p\|^2 + l_1\|v\|^2 + l_1. \end{aligned} \tag{10.55}$$

With a composite Lyapunov function

$$V = \frac{A}{2}\|\hat{w}\|^2 + \frac{1}{2}\|v\|^2 + \frac{1}{2}\|p\|^2, \tag{10.56}$$

where A is a constant yet to be defined, we get

$$\dot{V} \leq -\left(\frac{\varepsilon}{2}A - \frac{2M_2^2}{\varepsilon}\right)\|\hat{w}_x\|^2 - \left(\frac{\varepsilon}{2} - c_1 A\right)\left(\|v_x\|^2 + \|p_x\|^2\right) + l_1 V. \tag{10.57}$$

Choosing

$$A = 1 + \frac{4M_2^2}{\varepsilon^2} \tag{10.58}$$

and

$$c_1 = \frac{\varepsilon}{4A}, \tag{10.59}$$

we get

$$\dot{V} \leq -\frac{\varepsilon}{2A}V + l_1 V. \tag{10.60}$$

Using Lemma D.3 we get $V \in \mathcal{L}_\infty \cap \mathcal{L}_1$. Note that V depends on A, which depends on M_2, which depends on b_0 and λ_0, which in turn depend on the initial conditions of the system. However, $A \geq 1$, which implies that $\|\hat{w}\|^2$, $\|v\|^2$, $\|p\|^2 \leq 2V$, and hence $\|\hat{w}\|$, $\|v\|$, $\|p\| \in \mathcal{L}_\infty \cap \mathcal{L}_2$. Integrating (10.57), we also get $\|\hat{w}_x\|$, $\|v_x\|$, $\|p_x\| \in \mathcal{L}_2$.

10.2.3 Pointwise Boundedness and Regulation

We proceed now to H^1 analysis to establish pointwise boundedness and regulation. We start with

$$\begin{aligned}
\frac{1}{2}\frac{d}{dt}\|\hat{w}_x\|^2 &= \int_0^1 \hat{w}_x(x,t)\hat{w}_{xt}(x,t)\,dx \\
&= -\int_0^1 \hat{w}_{xx}(x,t)\hat{w}(x,t)\,dx \\
&\le -\varepsilon\|\hat{w}_{xx}\|^2 + b_0\|\hat{w}_x\|\|\hat{w}_{xx}\| + \lambda_0 M_1\|\hat{w}_{xx}\|\|\hat{e}\| \\
&\quad + M_1\|\hat{w}_{xx}\|\left(|\dot{\hat{b}}|\|p\| + |\dot{\hat{\lambda}}|\|v\|\right) \\
&\quad + \left(|\dot{\hat{b}}|\bar{\varphi}_1 + |\dot{\hat{\lambda}}|\bar{\varphi}_2\right)\|\hat{w}_{xx}\|\|\hat{w}\| \\
&\le -\varepsilon\|\hat{w}_{xx}\|^2 + \frac{\varepsilon}{8}\|\hat{w}_{xx}\|^2 + \frac{2b_0^2}{\varepsilon}\|\hat{w}_x\|^2 \\
&\quad + \frac{\varepsilon}{8}\|\hat{w}_{xx}\|^2 + \frac{2\lambda_0^2 M_1^2}{\varepsilon}\|\hat{e}\|^2 \\
&\quad + \frac{\varepsilon}{8}\|\hat{w}_{xx}\|^2 + \frac{4M_1^2}{\varepsilon}\left(|\dot{\hat{b}}|^2\|p\|^2 + |\dot{\hat{\lambda}}|^2\|v\|^2\right) \\
&\quad + \frac{\varepsilon}{8}\|\hat{w}_{xx}\|^2 + \frac{4}{\varepsilon}\left(|\dot{\hat{b}}|^2\bar{\varphi}_1^2 + |\dot{\hat{\lambda}}|^2\bar{\varphi}_2^2\right)\|\hat{w}\|^2 \\
&\le -\frac{\varepsilon}{2}\|\hat{w}_x\|^2 + l_1.
\end{aligned} \tag{10.61}$$

By Lemma D.3 we get $\|\hat{w}_x\| \in \mathcal{L}_\infty \cap \mathcal{L}_2$. For the filter v we have

$$\begin{aligned}
\frac{1}{2}\frac{d}{dt}\|v_x\|^2 &\le -\varepsilon\|v_{xx}\|^2 + b_0\|v_x\|\|v_{xx}\| + \|v_{xx}\|\|u\| \\
&\le -\varepsilon\|v_{xx}\|^2 + \frac{\varepsilon}{2}\|v_{xx}\|^2 + \frac{b_0^2}{\varepsilon}\|v_x\|^2 + \frac{1}{\varepsilon}\|u\|^2 \\
&\le -\frac{\varepsilon}{2}\|v_x\|^2 + l_1.
\end{aligned} \tag{10.62}$$

By Lemma D.3 we get $\|v_x\| \in \mathcal{L}_\infty \cap \mathcal{L}_2$. For the filter p we have

$$\begin{aligned}
\frac{1}{2}\frac{d}{dt}\|p_x\|^2 &\le -\varepsilon\|p_{xx}\|^2 + b_0\|p_x\|\|p_{xx}\| + \|p_{xx}\|\|u_x\| \\
&\le -\frac{\varepsilon}{2}\|p_{xx}\|^2 + \frac{b_0^2}{\varepsilon}\|p_x\|^2 + \frac{1}{\varepsilon}\|u_x\|^2.
\end{aligned} \tag{10.63}$$

Since

$$\begin{aligned}
\|u_x\|^2 &\le 2\|\hat{e}_x\|^2 + 2M_3\|\hat{w}_x\|^2 \\
&\le 4\|e_x\|^2 + 4|\tilde{b}|^2\|p_x\|^2 + 4|\tilde{\lambda}|^2\|v_x\|^2 \le l_1,
\end{aligned} \tag{10.64}$$

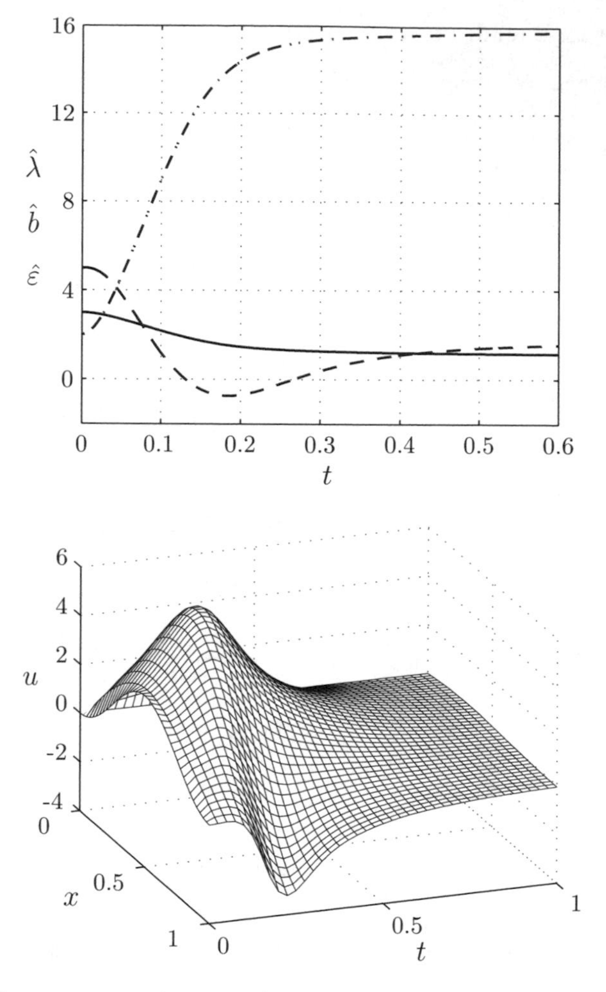

Figure 10.1 The parameter estimates and the closed-loop state for the plant (10.1)–(10.3) with adaptive controller based on swapping identifier (solid line, $\hat{\varepsilon}$; dashed line, $\hat{b}$; dashed-dotted line, $\hat{\lambda}$).

we get

$$\frac{1}{2}\frac{d}{dt}\|p_x\|^2 \leq -\frac{\varepsilon}{2}\|p_x\|^2 + l_1, \tag{10.65}$$

and by Lemma D.3, $\|p_x\| \in \mathcal{L}_\infty \cap \mathcal{L}_2$.

By Agmon's inequality we get the pointwise boundedness of signals $\hat{w}$, v, and p. From (10.36) we get the boundedness of η. Since $u = e + bp + \lambda v + \eta$, the state u is also bounded.

In order to prove regulation, we notice from (10.60) that

$$\dot{V} \leq \frac{\varepsilon}{4A}|V| + |l_1 V| < \infty, \tag{10.66}$$

where we used the fact that l_1 is a bounded function in this case. By Lemma D.2 we get $V \to 0$, and thus $\hat{w}$, v, $p \to 0$ as $t \to \infty$. From (10.36) we get $\eta \to 0$, and therefore (10.4) implies $u \to 0$ as $t \to \infty$. Using the boundedness of $\|u_x\|$, by Agmon's inequality we get

$$\lim_{t\to\infty} \max_{x\in[0,1]} |u(x,t)| \leq \lim_{t\to\infty} 2\|u\|^{1/2}\|u_x\|^{1/2} = 0. \tag{10.67}$$

The proof of Theorem 10.1 is completed. □

10.3 SIMULATIONS

We demonstrate the proposed adaptive scheme for the plant (10.1)–(10.3) with parameters $\varepsilon = 1$, $b = 2$, $\lambda = 15$. The plant has one unstable eigenvalue at $+4.1$. Initial estimates are set to $\hat{\varepsilon}(0) = 3$, $\hat{b}(0) = 5$, $\hat{\lambda}(0) = 2$. We use an implicit BTCS finite-difference scheme, with 100-step discretization in space. The results of the simulation are presented in Figure 10.1. Even though only the identifier properties (and not the closed-loop stabilization result) were proved in the case of an unknown diffusion coefficient, the adaptive controller successfully stabilizes the system. As expected for an adaptive regulation problem, the parameter estimates converge close to but not exactly to the true parameter values.

10.4 NOTES AND REFERENCES

The material in this chapter is based on [111]. The adaptive scheme presented in this chapter can be readily extended to other types of boundary conditions as well as to 2D and 3D plants with constant coefficients. Another straightforward extension not pursued here is the least squares estimation.

Chapter Eleven

State Feedback for PDEs with Spatially Varying Coefficients

In Chapters 8–10 we introduced three basic approaches for the design of parametric identifiers for boundary-controlled PDEs with constant coefficients: the Lyapunov, the passive, and the swapping approach. In this chapter we extend two of those designs, the Lyapunov and passive approaches, to plants with *spatially varying* parameters. The third approach (swapping) is not addressed here because it will be extended to plants with spatially varying coefficients in Chapter 13 in the output-feedback setting.

11.1 PROBLEM STATEMENT

We consider the plant

$$u_t(x,t) = \varepsilon(x)u_{xx}(x,t) + b(x)u_x(x,t) + \lambda(x)u(x,t)\,, \tag{11.1}$$

$$u(0,t) = 0\,, \tag{11.2}$$

$$u(1,t) = U(t)\,, \tag{11.3}$$

where the parameters $\varepsilon(x)$, $b(x)$, $\lambda(x)$ are unknown. We assume only that $\varepsilon(x)$ is positive for all $x \in [0,1]$ and that the parameters are sufficiently smooth for the classical solution of the closed-loop system to exist. The objective is to stabilize the zero equilibrium of this system using the boundary input $U(t)$.

We focus our attention on the case of unknown $\lambda(x)$. We do present adaptive designs for unknown $b(x)$ and $\varepsilon(x)$ in Section 11.5; however, the stability proofs are presented only for the case of unknown $\lambda(x)$ and known $b(x)$ and $\varepsilon(x)$. The main issues that arise in proving the stability of adaptive schemes for the case of unknown $b(x)$ and $\varepsilon(x)$ are related to the use of parameter projection. In addition to preventing the estimate of $\varepsilon(x)$ from becoming zero at some values of x, it turns out that one also needs to keep bounded the spatial H^1-norm of the estimate of $b(x)$ and the spatial H^2-norm of the estimate of $\varepsilon(x)$ (L^2-norms of all three parameter estimates are guaranteed to be a priori bounded by the choice of update laws). Parameter projection is a standard tool in adaptive control [51, 68]. We use it throughout this book to keep scalar parameters within certain bounds. It can also be employed to keep functional (spatially dependent) parameters within given bounds in an L^2 sense (as we do in this chapter) and in the L^∞ sense; however, it is not clear how this would be done in the H^1 and H^2 sense, though in a spatially discretized version of the problem keeping the resulting parameter vector within a predefined convex set that contains the true parameter would be standard. We also point out that even in the case of spatially constant parameters there exist considerable technical difficulties in proving stability for the case of an unknown diffusion coefficient. Certainly such challenges do not disappear when this coefficient is spatially varying.

11.2 NOMINAL CONTROL DESIGN

Before we proceed with the adaptive designs, we first present the nominal control design for the case of known parameters. This may seem unnecessary, given that the stabilizing controllers for this case were designed in Chapter 2. However, this was done with the help of the transformation (2.5)–(2.8), which avoids dealing with $b(x)$ and $\varepsilon(x)$ by transforming the plant into one without the advection coefficient and with a constant diffusion coefficient. When the parameters $\varepsilon(x)$ and $b(x)$ are unknown, the transformation (2.5)–(2.8) results in a plant with unmeasured state. Therefore, here we modify the control design presented in Chapter 2 by transforming the plant into the target system without the pretransformation (2.5)–(2.8).

Introducing the transformation

$$w(x,t) = u(x,t) - \int_0^x k(x,y)u(y,t)\,dy \tag{11.4}$$

and the controller

$$U(t) = \int_0^1 k(1,y)u(y,t)\,dy, \tag{11.5}$$

we map (11.1)–(11.3) into the target system

$$w_t(x,t) = \varepsilon(x)w_{xx}(x,t) + b(x)w_x(x,t) - cw(x,t), \tag{11.6}$$

$$w(0,t) = 0, \tag{11.7}$$

$$w(1,t) = 0. \tag{11.8}$$

Here c is an arbitrary nonnegative constant that allows the designer to change the desired decay rate of the closed-loop system.

One can show that the kernel $k(x,y)$ of the transformation (11.4) must satisfy the PDE

$$\begin{aligned}\varepsilon(x)k_{xx}(x,y) - (\varepsilon(y)k(x,y))_{yy} = {} & (\lambda(y)+c)k(x,y) \\ & -b(x)k_x(x,y) - (b(y)k(x,y))_y\end{aligned} \tag{11.9}$$

with boundary conditions

$$k(x,0) = 0, \tag{11.10}$$

$$k(x,x) = -\frac{1}{2\sqrt{\varepsilon(x)}}\int_0^x \frac{\lambda(y)+c}{\sqrt{\varepsilon(y)}}\,dy. \tag{11.11}$$

Next we establish stability of the target system and solve (11.9)–(11.11) by the method of successive approximations.

11.2.1 Stability of the Target System

Let us use the Lyapunov function

$$V = \frac{1}{2}\int_0^1 \rho(x) w^2(x)\,dx\,, \tag{11.12}$$

where $\rho(x)$ will be chosen later. The reason for using weighting in the Lyapunov function is that with the standard L^2-norm, one cannot prove the stability of (11.6)–(11.8) without imposing an additional condition on c. We have

$$\begin{aligned}
\dot V &= \int_0^1 \rho(x)(\varepsilon(x) w(x) w_{xx}(x) + b(x) w(x) w_x(x) - c w^2(x))\,dx \\
&= -\int_0^1 \rho(x)\varepsilon(x) w_x^2(x)\,dx - cV \\
&\quad + \int_0^1 w(x) w_x(x)\left(\rho(x) b(x) - \frac{d(\rho(x)\varepsilon(x))}{dx}\right) dx\,.
\end{aligned} \tag{11.13}$$

Choosing

$$\rho(x) = \frac{1}{\varepsilon(x)} \exp\left\{\int_0^x \frac{b(s)}{\varepsilon(s)}\,ds\right\}, \tag{11.14}$$

we get

$$\begin{aligned}
\dot V &= -\int_0^1 \rho(x)\varepsilon(x) w_x^2\,dx - cV \\
&\le -2\underline{\varepsilon} e^{-2B/\underline{\varepsilon}} V - cV\,,
\end{aligned} \tag{11.15}$$

where $B = \max_{x\in[0,1]} |b(x)|$ and $\underline{\varepsilon} = \min_{x\in[0,1]} \varepsilon(x)$. Therefore, the target system is exponentially stable at the origin for any $c \ge 0$.

11.2.2 Solution of the PDE for the Gain Kernel

Let us make the following change of variables in (11.9)–(11.11):

$$k(x,y) = \frac{\varepsilon^{\frac{1}{4}}(x)}{\varepsilon^{\frac{3}{4}}(y)} e^{-\int_y^x \frac{b(s)}{2\varepsilon(s)}\,ds} m(\xi,\eta)\,, \tag{11.16}$$

where ξ and η are defined as

$$\xi = \varphi(x) + \varphi(y)\,, \tag{11.17}$$

$$\eta = \varphi(x) - \varphi(y)\,, \tag{11.18}$$

$$\varphi(\zeta) = \int_0^\zeta \frac{1}{\sqrt{\varepsilon(s)}}\,ds\,. \tag{11.19}$$

It can be shown that the function m then satisfies the following PDE:

$$4 m_{\xi\eta}(\xi,\eta) = \lambda_a(\xi,\eta) m(\xi,\eta)\,, \tag{11.20}$$

$$m(\xi,\xi) = 0\,, \tag{11.21}$$

$$m(\xi,0) = -\frac{1}{4}\int_0^\xi \lambda_a(\sigma,0)\,d\sigma\,, \tag{11.22}$$

where

$$\lambda_a(\xi,\eta)=\lambda(y)+c+\frac{1}{2}(b'(x)-b'(y))-\frac{1}{4}(\varepsilon''(x)-\varepsilon''(y))$$
$$+\frac{1}{4}\left(\frac{b^2(x)}{\varepsilon(x)}-\frac{b^2(y)}{\varepsilon(y)}\right)+\frac{3}{16}\left(\frac{\varepsilon'^2(x)}{\varepsilon(x)}-\frac{\varepsilon'^2(y)}{\varepsilon(y)}\right)$$
$$-\frac{1}{2}\left(\frac{b(x)\varepsilon'(x)}{\varepsilon(x)}-\frac{b(y)\varepsilon'(y)}{\varepsilon(y)}\right). \tag{11.23}$$

Note that x and y in (11.23) are implicit functions of ξ and η.

The purpose of the transformation (11.16) is twofold. First, it removes all the coefficients in front of derivatives in (11.9) by "augmenting" λ to λ_a. Second, the resulting PDE can be converted into an integral equation and solved by the method of successive approximations.

As shown in Chapter 2, the PDE (11.20)–(11.22) has a unique, twice continuously differentiable solution that can be represented as

$$m(\xi,\eta)=\lim_{n\to\infty} m^n(\xi,\eta), \tag{11.24}$$

where

$$m^0(\xi,\eta)=-\frac{1}{4}\int_\eta^\xi \lambda_a(\sigma,0)\,d\sigma, \tag{11.25}$$

$$m^i(\xi,\eta)=m^0(\xi,\eta)+\frac{1}{4}\int_\eta^\xi\int_0^\eta \lambda_a(\sigma,s)\,m^{i-1}(\sigma,s)\,ds\,d\sigma \tag{11.26}$$

for $i=1,2,\dots,n$.

11.3 ROBUSTNESS TO ERROR IN GAIN KERNEL

When the coefficients ε, b, and λ are constant, the PDE (11.9)–(11.11) can be solved in closed form, and thus the explicitly parametrized controllers can be obtained. This remarkable fact was exploited in Chapters 8–10 to design explicit adaptive schemes that avoided numerical computation of the gain at each time step.

In case of spatially varying ε, b, and λ, the equation (11.9)–(11.11) has to be solved numerically. Therefore, robustness analysis is needed to ensure that the error in the kernel does not ruin the closed-loop stability of the adaptive scheme. The exact question to ask can be formulated as follows: How small should the difference between the approximate and exact solutions of (11.9)–(11.11) be to guarantee that the target system (11.6)–(11.8) is exponentially stable?

The robustness analysis (quantitatively) depends on the specific numerical method used to solve the PDE (11.9)–(11.11)—a finite-difference scheme, the successive approximation series, or some other method. Here we follow the successive approximation approach.

Let us use the following recursive formula to approximate $k(x,y)$:

$$k^n(x,y)=\frac{\varepsilon^{\frac{1}{4}}(x)}{\varepsilon^{\frac{3}{4}}(y)}e^{-\int_y^x\frac{b(s)}{2\varepsilon(s)}\,ds}m^n(\xi,\eta), \tag{11.27}$$

where $m^n(\xi,\eta)$ is given by (11.25)–(11.26).

Theorem 11.1. *There exists n^* such that for any initial data $u_0 \in H^1(0,1)$ compatible with boundary conditions, the system (11.1)–(11.3) with the controller*

$$U(t) = \int_0^1 k^n(1,y)u(y,t)\,dy, \tag{11.28}$$

where $n \geq n^*$ and k^n is given by (11.27), has a unique classical solution and is exponentially stable at the origin:

$$\|u(t)\|_{H^1} \leq Ce^{-(c+\delta)t}\|u_0\|_{H^1}, \tag{11.29}$$

where C is a positive constant independent of u_0 and $\delta = 2\underline{\varepsilon}e^{-2B/\underline{\varepsilon}}$.

11.3.1 Proof of Theorem 11.1

With an approximate transformation

$$w(x,t) = u(x,t) - \int_0^x k^n(x,y)u(y,t)\,dy, \tag{11.30}$$

we get the target system

$$\begin{aligned} w_t(x,t) &= \varepsilon(x)w_{xx}(x,t) + b(x)w_x(x,t) - cw(x,t) \\ &\quad + \int_0^x \delta k^n(x,y)u(y,t)\,dy, && (11.31) \\ w(0,t) &= 0, && (11.32) \\ w(1,t) &= 0, && (11.33) \end{aligned}$$

where $\delta k^n(x,y)$ is the error due to approximation,

$$\begin{aligned} \delta k^n(x,y) &= \varepsilon(x)k^n_{xx}(x,y) - (\varepsilon(y)k^n(x,y))_{yy} - (\lambda(y)+c)k^n(x,y) \\ &\quad + b(x)k^n_x(x,y) + (b(y)k^n(x,y))_y. \end{aligned} \tag{11.34}$$

When $n \to \infty$, we have $k^n \to k$, and so $\delta k^n \to 0$. Note that the integral term in (11.31) contains u instead of w. To express u in terms of w, we use the transformation that is inverse to (11.30):

$$u(x,t) = w(x,t) + \int_0^x l^n(x,y)w(y,t)\,dy. \tag{11.35}$$

The kernel of the inverse transformation l^n is related to k^n by the following equation:

$$l^n(x,y) = k^n(x,y) + \int_y^x k^n(x,\sigma)l^n(\sigma,y)\,d\sigma. \tag{11.36}$$

This relationship is easily obtained if we substitute (11.30) into (11.35), change the order of integration, and match the terms.

It is clear now that to prove stability of the system (11.31)–(11.33), we need to estimate δk^n, k^n, and l^n.

LEMMA 11.1. *The following bounds hold:*

$$|\delta k^n(x,y)| \le \alpha_n y(x^2-y^2)^n, \tag{11.37}$$

$$|k^n(x,y)| \le yK, \tag{11.38}$$

$$|l^n(x,y)| \le yL_n, \tag{11.39}$$

where

$$\alpha_n = \frac{2e^{\frac{B}{2\underline{\varepsilon}}}\bar{\lambda}_a(\bar{\varepsilon}/\underline{\varepsilon})^{1/4}}{n!(n+1)!}\left(\frac{\bar{\lambda}_a}{4\underline{\varepsilon}}\right)^{n+1}, \tag{11.40}$$

$$K = \frac{\bar{\varepsilon}^{1/4}}{\underline{\varepsilon}^{3/4}}e^{\frac{B}{2\underline{\varepsilon}}}\sqrt{\bar{\lambda}_a}I_1\left(\sqrt{\bar{\lambda}_a/\underline{\varepsilon}}\right), \tag{11.41}$$

$$L_n = K + \alpha_n\frac{4\underline{\varepsilon}e^{\frac{K-2B}{4\underline{\varepsilon}}}}{\lambda_a K^2}, \tag{11.42}$$

B, $\bar{\lambda}_a$ are bounds on $|b(x)|$, $|\lambda_a|$, and $\underline{\varepsilon}$, $\bar{\varepsilon}$ are upper and lower bounds on ε.

Proof. The proof of the three inequalities given in Lemma 11.1 is given in the next three subsections.

Bound on δk^n

If we substitute (11.27)–(11.23) into (11.34), after a long calculation we can show that

$$\delta k^n(x,y) = \lambda_a(\varphi(x)+\varphi(y),\varphi(x)-\varphi(y))[k^{n-1}(x,y)-k^n(x,y)]. \tag{11.43}$$

Using (11.25)–(11.26), it can be shown by induction that the following bound holds:

$$|m^n(\xi,\eta)-m^{n-1}(\xi,\eta)| \le \frac{(\xi-\eta)\xi^n\eta^n}{n!(n+1)!}\left(\frac{\bar{\lambda}_a}{4}\right)^{n+1}. \tag{11.44}$$

From (11.43), (11.27), and (11.44) we get

$$|\delta k^n(x,y)| \le \frac{\varepsilon^{\frac{1}{4}}(x)}{\varepsilon^{\frac{3}{4}}(y)}e^{-\int_y^x\frac{b(s)}{2\varepsilon(s)}ds}\varphi(y)\frac{(\varphi^2(x)-\varphi^2(y))^n}{n!(n+1)!}\left(\frac{\bar{\lambda}_a}{4}\right)^{n+1}. \tag{11.45}$$

Noting that $\varphi(y) \le y/\sqrt{\underline{\varepsilon}}$, we get (11.37).

Bound on k^n

Using the fact that $k^{-1} = m^{-1} = 0$ and (11.27), (11.44), we get the following estimate:

$$|k^n(x,y)| \le \sum_{i=0}^{n}\left|k^i(x,y)-k^{i-1}(x,y)\right|$$

$$\le y\sum_{i=0}^{n}\frac{2e^{\frac{B}{2\underline{\varepsilon}}}(\bar{\varepsilon}/\underline{\varepsilon})^{1/4}}{i!(i+1)!}\left(\frac{\bar{\lambda}_a}{4\underline{\varepsilon}}\right)^{i+1} \le yK, \tag{11.46}$$

where K, given by (11.41), is the sum of the series (11.46) when $n \to \infty$.

Note that (11.37) and (11.38) are tight bounds, since they become equalities when the parameters ε, b, and λ are constant.

Bound on l^n

Using (11.46), (11.36), and the Gronwall lemma, one can easily obtain the following bound on l^n:

$$|l^n(x, y)| \le yKe^{\frac{K}{2}(x^2-y^2)} \le yKe^{\frac{K}{2}}, \tag{11.47}$$

where K is given by (11.41). However, this bound is very conservative (roughly, it is an exponential of an exponential of $\sqrt{\bar{\lambda}_a}$). To obtain a much better bound (11.39), let us denote

$$\begin{aligned} \delta l^n(x, y) = {} & \varepsilon(x) l^n_{xx}(x, y) - (\varepsilon(y) l^n(x, y))_{yy} + (\lambda(x) + c) l^n(x, y) \\ & + b(x) l^n_x(x, y) + (b(y) l^n(x, y))_y . \end{aligned} \tag{11.48}$$

The reason for considering the expression (11.48) is that its right-hand side matches the PDE for the exact kernel of the inverse transformation when $n \to \infty$, so that $\delta l^n \to 0$.

If we apply the inverse transformation (11.35) to the target system (11.31)–(11.33) and require the result to be identical to the original plant, we get the following relationship between δl^n and δk^n:

$$\delta l^n - \int_y^x \delta l^n(x, \xi) k^n(\xi, y)\, d\xi = \delta k^n + \int_y^x l^n(x, \xi) \delta k^n(\xi, y)\, d\xi . \tag{11.49}$$

Using (11.30) and (11.35), we can rewrite (11.49) in the following form:

$$\begin{aligned} \delta l^n(x, y) = {} & \delta k^n(x, y) + \int_y^x \delta l^n(x, \xi) k^n(\xi, y)\, d\xi \\ & - \int_y^x k^n(x, \xi) \int_y^\xi \delta l^n(\xi, s) k^n(s, y)\, ds\, d\xi \\ & + \int_y^x k^n(x, \xi) \delta l^n(\xi, y)\, d\xi. \end{aligned} \tag{11.50}$$

This is an integral equation for δl^n with kernels that do not depend on l^n. Using the bounds (11.37) and (11.46), we get the following inequality:

$$\begin{aligned} |\delta l^n(x, y)| \le {} & \alpha_n y(x^2 - y^2) + K \int_y^x \left(y|\delta l^n(x, \xi)| + \xi |\delta l^n(\xi, y)| \right) d\xi \\ & + K^2 \int_y^x \int_y^\xi \xi y |\delta l^n(\xi, s)|\, ds\, d\xi. \end{aligned} \tag{11.51}$$

The sharpest bound for a solution of an integral inequality is given by a solution of the corresponding integral equation. We get the following estimate:

$$|\delta l^n(x, y)| \le \frac{\alpha_n y}{\sqrt{2}K}\left(e^{\frac{K(\sqrt{2}-1)}{2}(x^2-y^2)} - 1\right). \tag{11.52}$$

Since the PDEs (11.34) and (11.48) (without the errors δk^n and δl^n) differ only in the sign of the term without derivatives, we expect that the bound on l^n should be equal to the bound (11.46) on k^n plus some additional term due to δl^n. Applying to (11.48) the same technique that we used to obtain (11.46) and using the bound (11.52), we get the following estimate:

$$|l^n(x, y)| \leq y \left(K + \alpha_n \frac{4\underline{\varepsilon} e^{\frac{K-2B}{4\underline{\varepsilon}}}}{\lambda_a K^2} \right). \tag{11.53}$$

We can see that this bound is much better than the bound (11.47) since it approaches yK when n grows ($\alpha_n \to 0$) and not the exponential of K as (11.47) does. The proof of Lemma 11.1 is completed. □

Let us go back to the system (11.31)–(11.33) and consider the Lyapunov function

$$V = \frac{1}{2} \int_0^1 \rho(x) w^2(x)\, dx := \frac{1}{2} \|w\|_\rho^2, \tag{11.54}$$

where $\rho(x)$ is given by (11.14). We have

$$\begin{aligned} \dot{V} = &- \int_0^1 \rho(x) \varepsilon(x) w_x^2(x)\, dx - cV \\ &+ \int_0^1 \rho(x) w(x) \int_0^x \delta k^n(x, y) u(y)\, dy\, dx\,. \end{aligned} \tag{11.55}$$

The following lemma provides the estimate for the last term in (11.55).

Lemma 11.2. *There exists n^* such that for any $n \geq n^*$*

$$\int_0^1 \rho(x) w(x) \int_0^x \delta k^n(x, y) u(y)\, dy\, dx \leq \frac{1}{2} e^{-\frac{B}{\underline{\varepsilon}}} \|w_x\|^2. \tag{11.56}$$

Proof. Using (11.35), after the change of order of the integration we get

$$\begin{aligned} \int_0^1 \rho(x) w(x) & \int_0^x \delta k^n(x, y) u(y)\, dy\, dx \\ &\leq \int_0^1 \rho(x) w(x) \int_0^x w(y) \int_y^x \delta k^n(x, \sigma) l^n(\sigma, y)\, d\sigma\, dy \\ &\quad + \int_0^1 \rho(x) w(x) \int_0^x \delta k^n(x, y) w(y)\, dx\, dy \\ &\leq \int_0^1 \rho(x) |w(x)| \int_0^x \alpha_n y L_n \frac{(x^2 - y^2)^{n+1}}{2(n+1)} |w(y)|\, dy\, dx \\ &\quad + \int_0^1 \rho(x) |w(x)| \int_0^x \alpha_n y (x^2 - y^2)^n |w(y)|\, dy\, dx\,. \end{aligned} \tag{11.57}$$

Let us estimate the following integral:

$$\begin{aligned}
&\int_0^1 |w(x)| \int_0^x y(x^2-y^2)^n |w(y)|\,dy\,dx \\
&\le \|w\| \left(\int_0^1 \left(\int_0^x y(x^2-y^2)^n |w(y)|\,dy \right)^2 dx \right)^{\frac{1}{2}} \\
&\le \|w\| \left(\int_0^1 \frac{x^{2n+2}}{2(n+1)} \int_0^x \frac{(x^2-y^2)^{n+1}}{n+1} w w_y \,dy\,dx \right)^{\frac{1}{2}} \\
&\le \frac{2}{n+1} \|w_x\| \left(\int_0^1 \frac{1}{2} |w(x)||w_x(x)| \frac{1}{4n+5}\,dx \right)^{\frac{1}{2}} \\
&\le \frac{1}{(n+1)^{3/2}} \|w_x\|^2 .
\end{aligned} \tag{11.58}$$

Here we have used the Cauchy-Schwartz and Poincaré inequalities several times. Substituting the estimate (11.58) into (11.57), we get (11.56) if the following condition is satisfied:

$$\alpha_n \left(\frac{1}{(n+1)^{3/2}} + \frac{L_n}{2(n+1)(n+2)^{3/2}} \right) \le \frac{1}{2} e^{-\frac{B}{\underline{\varepsilon}}} . \tag{11.59}$$

Since $\alpha_n \to 0$ as $n \to \infty$ and L_n is uniformly bounded, it is clear that there exists n^* such that (11.59) holds for all $n > n^*$. □

Substituting (11.56) into (11.55) and using the Poincaré inequality, we get

$$\dot{V} \le -\underline{\varepsilon} e^{-\frac{2B}{\underline{\varepsilon}}} V - cV, \tag{11.60}$$

so that the zero equilibrium of the w-system is exponentially stable in L^2-norm for $c \ge 0$. It is straightforward to show with $\|w_x\|^2$ as the Lyapunov function that the condition (11.59) ensures H^1 stability as well. Since the kernel of the inverse transformation (11.35) is a twice continuously differentiable function, we get (11.29).

11.3.2 Practical Estimate of a Sufficient Number of Terms in the Series

The condition (11.59) produces a rather conservative estimate of n^*. The reason is that although the bounds α_n and L_n are tight, they provide a good estimate on δk^n and l^n for different parameter ranges. While the bound on δk^n is good when $\lambda_a \ge 0$, the bound on l^n is good when $\lambda_a \le 0$, and vice versa. Let us assume $\lambda_a \ge 0$ (which is normally the case for unstable plants). It then follows from (11.27)–(11.26) that k^n is negative and large. Thus, one can expect the kernel of the inverse transformation l^n to be "small," not of the size of k^n as (11.53) suggests. We give the following rule-of-thumb bound on l^n in this case:

$$|l^n(x,y)| \le y \frac{\bar{\lambda}_a}{2\underline{\varepsilon}} . \tag{11.61}$$

The motivation behind this bound is simple: since k^n in (11.36) is negative, one can assume that l^n will not grow beyond the "initial condition" $k^n(y,y)$, which is

Table 11.1 Minimum number of terms n required to satisfy the condition (11.59) for different $\bar{\lambda}$.

$\bar{\lambda}_a/\underline{\varepsilon}$	10	20	40	60
n^*, with a conservative bound (11.47)	6	16	64	204
n^*, with a tight bound (11.53)	4	7	25	74
n^*, with a "rule of thumb" bound (11.61)	4	6	8	10

bounded by $\bar{\lambda}_a y/(2\underline{\varepsilon})$. It is not possible to prove the estimate (11.61), but numerical computations seem to confirm that it holds.

To illustrate how the minimum number of terms n^* required to satisfy (11.59) depends on the parameters of the plant and on the selection of L_n, we computed the value of n^* for different values of $\bar{\lambda}_a/\underline{\varepsilon}$ (we set $b(x)\equiv 0$ for this computation, since the b-term does not contribute to instability and thus n^* is not affected much) using the different bounds on l^n (11.47), (11.53), and (11.61). The results are presented in Table 11.1. We can see that the bound (11.53) results in a much better estimate of n^* than the bound (11.47). It is also clear that in practice, only several terms are sufficient.

One can use the rule-of-thumb estimate for l^n (11.61) and the Stirling approximation $\ln n! \approx n\ln n - n$ to obtain the following approximation from (11.59):

$$n \geq \frac{e}{2}\sqrt{\frac{\bar{\lambda}_a}{\underline{\varepsilon}}} - 1 . \tag{11.62}$$

This is a very good practical estimate of the number of terms needed for implementation of the controller (11.27)–(11.23).

We are now ready to start the presentation of our adaptive schemes.

11.4 LYAPUNOV DESIGN

Consider the plant (11.1)–(11.3) with known $\varepsilon(x)$, $b(x)$ and unknown $\lambda(x)$. Given the nonadaptive control scheme (11.5)–(11.11), the natural approach (the so-called indirect adaptive control method) is to use the estimate $\hat{\lambda}(x)$ in (11.5)–(11.11) instead of $\lambda(x)$. The PDE for the control gain estimate $\hat{k}$ becomes:

$$\begin{aligned}\varepsilon(x)\hat{k}_{xx}(x,y) - (\varepsilon(y)\hat{k}(x,y))_{yy} &= (\hat{\lambda}(y)+c)\hat{k}(x,y) - b(x)\hat{k}_x(x,y)\\ &\quad - (b(y)\hat{k}(x,y))_y, \end{aligned}\tag{11.63}$$

$$\hat{k}(x,0)=0, \tag{11.64}$$

$$\hat{k}(x,x) = -\frac{1}{2\sqrt{\varepsilon(x)}}\int_0^x \frac{\hat{\lambda}(y)+c}{\sqrt{\varepsilon(y)}}\,dy. \tag{11.65}$$

As we pointed out in the previous section, one can follow different approaches to obtain the approximate solution of (11.63)–(11.65). Here we only prove closed-loop stability for the case when $\hat{k}$ is computed using the method of successive approximations.

The recursive scheme (11.25)–(11.27) is modified as follows:

$$\hat{k}^n(x,y) = \frac{\varepsilon^{\frac{1}{4}}(x)}{\varepsilon^{\frac{3}{4}}(y)} e^{-\int_y^x \frac{b(s)}{2\varepsilon(s)}\,ds} \hat{m}^n(\xi,\eta), \tag{11.66}$$

with $\hat{m}^n$ given by the series

$$\hat{m}^0(\xi,\eta) = -\frac{1}{4}\int_0^x \hat{\lambda}_a(\sigma,0)\,d\sigma, \tag{11.67}$$

$$\hat{m}^i(\xi,\eta) = \hat{m}^0(\xi,\eta) + \frac{1}{4}\int_\eta^\xi \int_0^\eta \hat{\lambda}_a(\sigma,s)\hat{m}^{i-1}(\sigma,s)\,ds\,d\sigma \tag{11.68}$$

for $i = 1, 2, \ldots, n$. The estimate $\hat{\lambda}_a$ is given by the expression

$$\begin{aligned}\hat{\lambda}_a(\xi,\eta) = {} & \hat{\lambda}(y) + c + \frac{1}{2}(b'(x) - b'(y)) - \frac{1}{4}(\varepsilon''(x) - \varepsilon''(y)) \\ & + \frac{1}{4}\left(\frac{b^2(x)}{\varepsilon(x)} - \frac{b^2(y)}{\varepsilon(y)}\right) + \frac{3}{16}\left(\frac{\varepsilon'^2(x)}{\varepsilon(x)} - \frac{\varepsilon'^2(y)}{\varepsilon(y)}\right) \\ & - \frac{1}{2}\left(\frac{b(x)\varepsilon'(x)}{\varepsilon(x)} - \frac{b(y)\varepsilon'(y)}{\varepsilon(y)}\right),\end{aligned} \tag{11.69}$$

where $\hat{\lambda}(x)$ is the estimate of the unknown function $\lambda(x)$. The variables x and y in (11.67) and (11.69) are expressed through ξ and η using (11.17)–(11.19). Note that all the variables with a "hat" depend on time, although we do not explicitly indicate that.

Let us introduce the projection operator

$$\mathrm{Proj}_{[\underline{\lambda},\bar{\lambda}]}\{\tau\} = \begin{cases} 0, & \|\hat{\lambda}\| = \underline{\lambda} \text{ and } \tau < 0 \\ 0, & \|\hat{\lambda}\| = \bar{\lambda} \text{ and } \tau > 0 \\ \tau, & \text{else.} \end{cases} \tag{11.70}$$

Note that we only keep $\|\hat{\lambda}\|$ a priori bounded. As explained in Chapter 8, it is also possible to make this operator continuous by introducing a small boundary layer instead of a hard switch.

Our main stability result is stated in the following theorem.

Theorem 11.2. *There exist n^* and γ^* such that, for all $n \geq n^*$ and $\gamma \in (0,\gamma^*)$, for any initial estimate $\hat{\lambda}_0 \in C^1[0,1]$, $\underline{\lambda} \leq \|\hat{\lambda}_0\| \leq \bar{\lambda}$, and for any initial condition $u_0 \in H^2(0,1)$ compatible with boundary conditions, the classical solution of the closed-loop system $(u,\hat{\lambda})$ consisting of the plant* (11.1)–(11.3), *update law*

$$\hat{\lambda}_t(x,t) = \gamma \mathrm{Proj}_{[\underline{\lambda},\bar{\lambda}]}\left\{\frac{\rho(x)w(x)u(x)}{1+\|w\|_\rho^2} - \frac{u(x)\int_x^1 \hat{k}^n(y,x)\rho(y)w(y)\,dy}{1+\|w\|_\rho^2}\right\}, \tag{11.71}$$

$$w(x) = u(x) - \int_0^x \hat{k}^n(x,y)u(y)\,dy \tag{11.72}$$

and the controller

$$U(t) = \int_0^1 \hat{k}^n(1, y)u(y, t)\,dy \tag{11.73}$$

is bounded for all $x \in [0, 1]$, $t \geq 0$ *and*

$$\lim_{t\to\infty} \max_{x\in[0,1]} |u(x, t)| = 0\,. \tag{11.74}$$

Proof. The transformation (11.72) converts the plant (11.1)–(11.3) into the following target system:

$$\begin{aligned} w_t = {} & \varepsilon(x)w_{xx} + b(x)w_x - cw - \int_0^x \hat{k}_t^n(x, y)u(y)\,dy \\ & + \int_0^x \delta\hat{k}^n(x, y)u(y)\,dy + \tilde{\lambda}(x)u(x) \\ & - \int_0^x \tilde{\lambda}(y)\hat{k}^n(x, y)u(y)\,dy, \end{aligned} \tag{11.75}$$

where u is related to w through the inverse tranformation

$$u(x) = w(x) + \int_0^x \hat{l}^n(x, y)w(y)\,dy, \tag{11.76}$$

and $\tilde{\lambda} = \lambda - \hat{\lambda}$.

Consider the Lyapunov function

$$V = \frac{1}{2}\log(1 + \|w\|_\rho^2) + \frac{1}{2\gamma}\|\tilde{\lambda}\|^2, \tag{11.77}$$

where $\|w\|_\rho$ and ρ are defined by (11.54) and (11.14), respectively. Computing $\dot{V}$ along the solutions we get

$$\begin{aligned} \dot{V} = {} & \frac{1}{1 + \|w\|_\rho^2}\left\{-\int_0^1 \rho(x)\varepsilon(x)w_x^2(x)\,dx - c\int_0^1 \rho(x)w^2(x)\,dx\right. \\ & + \int_0^1 \rho(x)\left(\tilde{\lambda}(x)u(x) - \int_0^x \tilde{\lambda}(y)\hat{k}^n(x, y)u(y)\,dy\right)dx \\ & - \int_0^1 \rho(x)w(x)\int_0^x \delta\hat{k}^n(x, y)u(y)\,dy\,dx \\ & \left.+ \int_0^1 \rho(x)w(x)\int_0^x \hat{k}_t^n(x, y)u(y)\,dy\,dx\right\} \\ & - \frac{1}{\gamma}\int_0^1 \tilde{\lambda}(y)\dot{\hat{\lambda}}(y)\,dy\,. \end{aligned} \tag{11.78}$$

With the update law (11.71) we obtain

$$\dot{V} = \frac{1}{1+\|w\|_\rho^2}\left\{-\int_0^1 \rho(x)\varepsilon(x)w_x^2(x)\,dx\right.$$
$$-\int_0^1 \rho(x)w(x)\int_0^x \delta\hat{k}^n(x,y)u(y)\,dy\,dx$$
$$\left.+\int_0^1 \rho(x)w(x)\int_0^x \hat{k}_t^n(x,y)u(y)\,dy\,dx\right\}. \tag{11.79}$$

The second term here is due to approximation of the control kernel $\hat{k}$ by (11.66)–(11.69) and the third term is due to the fact that $\hat{k}^n$ is time-varying.

By Lemma 11.2 there exists n^* such that for any $n \geq n^*$, the following holds:

$$\int_0^1 \rho(x)w(x)\int_0^x \delta\hat{k}^n(x,y)u(y)\,dy\,dx \leq \frac{1}{2}e^{-\frac{B}{\underline{\varepsilon}}}\|w_x\|^2. \tag{11.80}$$

To estimate the last term in (11.79), let us first note that $\hat{k}^n$ and $\hat{l}^n$ are bounded, which follows from (11.66)–(11.69) and the fact that $\|\hat{\lambda}\|$ is bounded by projection bounds. Let us denote these bounds by K and L, respectively. From (11.66)–(11.69) one can show that

$$|\hat{k}_t^n(x,y)| \leq M\|\hat{\lambda}_t\|, \tag{11.81}$$

where

$$M = \frac{\bar{\varepsilon}^{1/4}e^{B/2\underline{\varepsilon}}}{2\underline{\varepsilon}^{3/4}} I_0\left(\sqrt{\bar{\lambda}_a/\underline{\varepsilon}}\right), \tag{11.82}$$

and I_0 denotes the modified Bessel function of order zero. To see how this bound is obtained, note from (11.66)–(11.67) that $\hat{k}_t^0$ is bounded by $\|\hat{\lambda}_t\|$. Then from (11.68) we get that $\hat{k}^1$ is bounded by the maximum of $\hat{k}_t^0$ and $\|\hat{\lambda}_t\|$ and therefore is bounded by $\|\hat{\lambda}_t\|$. By induction we obtain a series of bounds, which can then be summed up to get (11.82).

From the update law (11.71) we get

$$\|\hat{\lambda}_t\| \leq \gamma e^{B/\underline{\varepsilon}}(1+L)(1+K). \tag{11.83}$$

Using (11.81) and (11.83) we obtain

$$\int_0^1 \rho(x)w(x)\int_0^x \hat{k}_t^n(x,y)u(y)\,dy\,dx \leq 2\gamma e^{B/\underline{\varepsilon}}M(1+L)(1+K)\|w_x\|^2. \tag{11.84}$$

Substituting the bounds (11.80) and (11.84) into (11.79), we get

$$\dot{V} \leq -\frac{e^{-B/\underline{\varepsilon}}(1-\gamma/\gamma^*)}{4(1+\|w\|_\rho^2)}\int_0^1 w_x^2(x)\,dx, \tag{11.85}$$

where

$$\gamma^* = \frac{\underline{\varepsilon}}{8}e^{-2B/\underline{\varepsilon}}\left(M(1+L)(1+K)\right)^{-1}. \tag{11.86}$$

Therefore, $\|w\|$ is bounded and $\|w_x\|^2$ is integrable over infinite time for $\gamma < \gamma^*$.

Let us now prove the boundedness of $\|w_x\|^2$. We start with

$$\begin{aligned}\frac{1}{2}\frac{d}{dt}\|w_x\|^2 \le & -\int_0^1 w_{xx}(x)w_t(x)\,dx \\ \le & -\int_0^1 \varepsilon(x)w_{xx}^2(x)\,dx - \int_0^1 b(x)w_x(x)w_{xx}(x)\,dx - c\|w_x\|^2 \\ & + \int_0^1 w_{xx}(x)\int_0^x \delta\hat{k}^n(x,y)u(y)\,dy \\ & - \int_0^1 w_{xx}(x)\int_0^x \hat{k}_t^n(x,y)u(y)\,dy\,. \end{aligned} \tag{11.87}$$

Using the bounds (11.80)–(11.83) we get

$$\begin{aligned}\frac{1}{2}\frac{d}{dt}\|w_x\|^2 \le & -\underline{\varepsilon}\|w_{xx}\|^2 + B\|w_x\|\|w_{xx}\| \\ & + \frac{\varepsilon}{2}\|w_{xx}\|\|w\| + \frac{\varepsilon}{4}\frac{\gamma}{\gamma^*}\|w_{xx}\|\|w\| - c\|w_x\|^2 \\ \le & -\underline{\varepsilon}\|w_{xx}\|^2 + \frac{\varepsilon}{4}\|w_{xx}\|^2 - c\|w_x\|^2 \\ & + B\|w_x\|^2 + \frac{\varepsilon}{2}\|w_{xx}\|^2 \\ & + \frac{\varepsilon}{8}\|w\|^2 + \frac{\varepsilon}{4}\|w_{xx}\|^2 + \frac{\varepsilon}{16}\|w\|^2 \\ \le & \left(\frac{3\varepsilon}{4} + \frac{1}{\underline{\varepsilon}}B^2 - c\right)\|w_x\|^2\,. \end{aligned} \tag{11.88}$$

Since $\|w_x\|^2$ is integrable over infinite time, by integrating (11.88) we get that $\|w_x\|$ is bounded. By Agmon's inequality, $w(x,t)$ is uniformly bounded for all $t \ge 0$ and for all $x \in [0,1]$.

In order to show regulation, we estimate

$$\begin{aligned}\left|\frac{1}{2}\frac{d}{dt}\|w\|^2\right| \le & \,\bar{\varepsilon}\|w_x\|^2 + B\|w\|\|w_x\| + c\|w\|^2 \\ & + \frac{\varepsilon}{4}\frac{\gamma}{\gamma^*}\|w_x\|^2 + \frac{\varepsilon}{2}\|w_x\|^2 \\ & + \gamma\bar{\lambda}e^{B/\underline{\varepsilon}}(1+L)(1+K)\,. \end{aligned} \tag{11.89}$$

The right-hand side of this inequality is bounded, so $|\frac{d}{dt}\|w\|^2|$ is bounded. Recalling that $\|w\|^2$ is bounded and integrable over infinite time, by Barbalat's lemma we get $\|w\| \to 0$ as $t \to \infty$. Using Agmon's inequality, we get $w(x,t) \to 0$ for all $x \in [0,1]$ as $t \to \infty$. The boundedness and regulation results for $u(x,t)$ follow from those for $w(x,t)$ owing to the boundedness of the kernel $\hat{l}^n$ of the transformation (11.76).

The proof of Theorem 11.2 is thus completed. □

11.5 LYAPUNOV DESIGN FOR PLANTS WITH UNKNOWN ADVECTION AND DIFFUSION PARAMETERS

To design the adaptive scheme when the parameters $\varepsilon(x)$ and $b(x)$ are unknown, let us start by introducing extra parameter estimates $\hat{b}_1(x)$, $\hat{\varepsilon}_1(x)$, and $\hat{\varepsilon}_2(x)$, which separately estimate $b'(x)$, $\varepsilon'(x)$, and $\varepsilon''(x)$, respectively. The purpose of this overparametrization is to avoid using state derivatives u_x or u_{xx} in the design (if the state is noisy, it is problematic to compute these derivatives).

Given six functional estimates $\hat{\lambda}$, $\hat{b}$, $\hat{b}_1$, $\hat{\varepsilon}$, $\hat{\varepsilon}_1$, $\hat{\varepsilon}_2$, the PDE for the gain kernel becomes

$$\begin{aligned}\hat{\varepsilon}(x)\hat{k}_{xx}(x,y) - \hat{\varepsilon}(y)\hat{k}_{yy}(x,y) &= (\hat{\lambda}(y) - \hat{b}_1(y) + \hat{\varepsilon}_2(y) + c)\hat{k}(x,y) \\ &\quad - \hat{b}(x)\hat{k}_x(x,y) - (\hat{b}(y) \\ &\quad - 2\hat{\varepsilon}_1(y))\hat{k}_y(x,y), \end{aligned} \tag{11.90}$$

$$\hat{k}(x,0) = 0, \tag{11.91}$$

$$\hat{k}(x,x) = -\frac{1}{2\sqrt{\hat{\varepsilon}(x)}}\int_0^x \frac{\hat{\lambda}(y)+c}{\sqrt{\hat{\varepsilon}(y)}}\,dy. \tag{11.92}$$

With the Lyapunov function

$$\begin{aligned}V &= \frac{1}{2}\log(1+\|w\|^2) + \frac{1}{2\gamma}(\|\tilde{\lambda}\|^2 + \|\tilde{b}\|^2 + \|\tilde{b}_1\|^2 \\ &\quad + \|\tilde{\varepsilon}\|^2 + \|\tilde{\varepsilon}_1\|^2 + \|\tilde{\varepsilon}_2\|^2),\end{aligned} \tag{11.93}$$

where w is the state of the target system (not written here explicitly), one gets the following update laws:

$$\hat{\lambda}_t(x) = \frac{\gamma}{1+\|w\|^2}\left(w(x)u(x) - u(x)\int_x^1 \hat{k}^n(\sigma,x)w(\sigma)\,d\sigma\right), \tag{11.94}$$

$$\hat{b}_t(x) = \frac{\gamma}{1+\|w\|^2}\left(w(x)\int_0^x \hat{k}^n_x(x,\sigma)w(\sigma)\,d\sigma + u(x)\int_x^1 \hat{k}^n_y(\sigma,x)w(\sigma)\,d\sigma\right), \tag{11.95}$$

$$\hat{b}_{1t}(x) = \frac{\gamma}{1+\|w\|^2}u(x)\int_x^1 \hat{k}^n(\sigma,x)w(\sigma)\,d\sigma, \tag{11.96}$$

$$\begin{aligned}\hat{\varepsilon}_t(x) &= \frac{\gamma}{1+\|w\|^2}\left(w(x)\int_0^x \hat{k}^n_{xx}(x,\sigma)u(\sigma)\,d\sigma - u(x)\int_x^1 \hat{k}^n_{yy}(\sigma,x)w(\sigma)\,d\sigma\right. \\ &\quad \left. + 2\frac{d}{dx}(\hat{k}^n(x,x))w(x)u(x)\right),\end{aligned} \tag{11.97}$$

$$\hat{\varepsilon}_{1t}(x) = \frac{\gamma}{1+\|w\|^2}\left(\hat{k}^n(x,x)w(x)u(x) - 2u(x)\int_x^1 \hat{k}^n_y(\sigma,x)w(\sigma)\,d\sigma\right), \tag{11.98}$$

$$\hat{\varepsilon}_{2t}(x) = -\frac{\gamma}{1+\|w\|^2}u(x)\int_x^1 \hat{k}^n(\sigma,x)w(\sigma)\,d\sigma. \tag{11.99}$$

One can show that

$$\dot{V} = \frac{1}{1+\|w\|^2}\left\{-\int_0^1 \varepsilon(x)w_x^2(x)\,dx - \int_0^1 \left(c + \frac{1}{2}b'(x)\right.\right.$$
$$\left. -\frac{1}{2}\varepsilon''(x)\right) w^2(x)\,dx - \int_0^1 w(x)\int_0^x \delta\hat{k}^n(x,y)u(y)\,dy\,dx$$
$$\left. + \int_0^1 w(x)\int_0^x \hat{k}_t^n(x,y)u(y)\,dy\,dx\right\}. \tag{11.100}$$

Assuming that

$$c > \frac{1}{2}\max_{x\in[0,1]} |\varepsilon''(x) - b'(x)|, \tag{11.101}$$

γ is sufficiently small, and n is sufficiently large, from (11.100) one gets boundedness and regulation in the same way as in the proof of Theorem 11.2. However, one has to use projection not only to bound L^2-norms of all the parameter estimates but also to bound $\|b'\|$, $\|\varepsilon'\|$, and $\|\varepsilon''\|$. This may seem surprising, since we use separate estimates for these derivatives. The reason to bound these norms is that the control gain kernel $\hat{k}^n$ obtained from (11.90)–(11.92) is bounded by a sum of the L^2-norms of all six parameter estimates, the H^1-norm of $\hat{b}$, and the H^2-norm of $\hat{\varepsilon}$.

The adaptive scheme presented in this section relies on the condition (11.101). It is possible to remove this condition by using a weighted L^2-norm of w in the Lyapunov function (11.93). However, since $\rho(x)$ given by (11.14) is unknown, one has to use its estimate, $\hat{\rho}(x)$. Unfortunately, with this approach the derivatives of the state u_x and u_{xx} appear in the update laws, and one loses the remarkable feature of the derivative-free scheme (11.94)–(11.99).

11.6 PASSIVITY-BASED DESIGN

We present now a certainty equivalence adaptive design with a passive identifier.

Consider the plant (11.1)–(11.3) with all three parameters unknown, and let us employ the following identifier:

$$\hat{u}_t = \hat{\varepsilon}(x)\hat{u}_{xx} + \hat{b}(x)u_x + \hat{\lambda}(x)u + \frac{1}{2}\hat{\varepsilon}''(x)(u-\hat{u})$$
$$- \gamma^2(u-\hat{u})\|u_x\|^2 - \gamma_0^2(u-\hat{u})\|u_{xx}\|^2, \tag{11.102}$$

$$\hat{u}(0,t) = 0, \tag{11.103}$$

$$\hat{u}(1,t) = U(t). \tag{11.104}$$

This system consists of a copy of the plant plus several additional terms. The term containing $\hat{\varepsilon}''(x)$ in (11.102) is needed to account for the spatial dependence of the parameter $\hat{\varepsilon}$ and produce a stable error system. The corresponding terms for

$\hat{\lambda}(x)$ and $\hat{b}(x)$ are not needed, since these estimates multiply u_x and u_{xx} instead of $\hat{u}_x$ and $\hat{u}_{xx}$. The last two terms in (11.102) act as nonlinear damping whose task is to ensure square integrability over infinite time of the spatial norms $\|\hat{\lambda}_t\|$, $\|\hat{b}_t\|$, and $\|\hat{\varepsilon}_t\|$, which serves as dynamic normalization. Note that when only $\lambda(x)$ is unknown, one can set $\gamma_0 = 0$ and replace the term $\gamma(u-\hat{u})\|u_x\|^2$ by $\gamma(u-\hat{u})\|u\|^2$ to avoid measuring tahe spatial derivatives of the state.

The error signal $e = u - \hat{u}$ satisfies the following PDE:

$$\begin{aligned} e_t &= \hat{\varepsilon}(x)e_{xx} + \tilde{\lambda}(x)u + \tilde{b}(x)u_x + \tilde{\varepsilon}(x)u_{xx} \\ &\quad - \frac{1}{2}e\hat{\varepsilon}''(x) - \gamma^2 e\|u_x\|^2 - \gamma_0^2 e\|u_{xx}\|^2, \end{aligned} \tag{11.105}$$

$$e(0,t) = 0, \tag{11.106}$$

$$e(1,t) = 0. \tag{11.107}$$

With the Lyapunov function

$$V = \frac{1}{2}\|e\|^2 + \frac{1}{2\gamma_1}\|\tilde{\lambda}\|^2 + \frac{1}{2\gamma_2}\|\tilde{b}\|^2 + \frac{1}{2\gamma_3}\|\tilde{\varepsilon}\|^2, \tag{11.108}$$

we get

$$\begin{aligned} \dot{V} &= \int_0^1 \hat{\varepsilon}(x)e(x)e_{xx}(x)\,dx + \int_0^1 (\tilde{\lambda}(x)u(x) + \tilde{b}(x)u_x(x) \\ &\quad + \tilde{\varepsilon}u_{xx}(x))e(x)\,dx - \frac{1}{2}\int_0^1 \hat{\varepsilon}''(x)e^2(x)\,dx - \gamma\int_0^1 e^2(x)\,dx\|u_x\|^2 \\ &\quad - \gamma_0\int_0^1 e^2(x)\,dx\|u_{xx}\|^2 - \int_0^1\left(\frac{1}{\gamma_1}\tilde{\lambda}(x)\hat{\lambda}_t(x) + \frac{1}{\gamma_2}\tilde{b}(x)\hat{b}_t(x) + \frac{1}{\gamma_3}\tilde{\varepsilon}(x)\hat{\varepsilon}_t(x)\right)dx \\ &= -\int_0^1 \hat{\varepsilon}(x)e_x^2(x)\,dx + \int_0^1 \tilde{\lambda}(x)\left(u(x)e(x) - \frac{1}{\gamma_1}\hat{\lambda}_t(x)\right)dx \\ &\quad + \int_0^1 \tilde{b}(x)\left(u_x(x)e(x) - \frac{1}{\gamma_2}\hat{b}_t(x)\right)dx \\ &\quad + \int_0^1 \tilde{\varepsilon}(x)\left(u_{xx}(x)e(x) - \frac{1}{\gamma_3}\hat{\varepsilon}_t(x)\right)dx \\ &\quad - \gamma^2\|e\|^2\|u_x\|^2 - \gamma_0^2\|e\|^2\|u_{xx}\|^2. \end{aligned} \tag{11.109}$$

Choosing the update laws

$$\hat{\lambda}_t = \gamma_1(u-\hat{u})u, \tag{11.110}$$

$$\hat{b}_t = \gamma_2(u-\hat{u})u_x, \tag{11.111}$$

$$\hat{\varepsilon}_t = \gamma_3(u-\hat{u})u_{xx}, \tag{11.112}$$

where $\gamma_1, \gamma_2, \gamma_3 > 0$, we get

$$\dot{V} \leq -\underline{\varepsilon}\|e_x\|^2 - \gamma^2\|e\|^2\|u_x\|^2 - \gamma_0^2\|e\|^2\|u_{xx}\|^2, \tag{11.113}$$

where $\underline{\varepsilon}$ is the lower bound on $\hat{\varepsilon}(x)$. This implies $V(t) \leq V(0)$, so that $\|\tilde{\varepsilon}\|$, $\|\tilde{b}\|$, $\|\tilde{\lambda}\|$, and $\|e\|$ are bounded. Integrating (11.113) with respect to time from zero to infinity we get the square integrability of $\|e_x\|$, $\|e\|\|u_x\|$, and $\|e\|\|u_{xx}\|$, which, together with the update laws (11.110)–(11.112), gives integrability of the L^2 spatial norms of the time derivatives of the parameter estimates.

LEMMA 11.3. *The identifier* (11.102)–(11.104) *with update laws* (11.110)–(11.112) *guarantees the following properties:*

$$\|\tilde{\varepsilon}\|, \|\tilde{b}\|, \|\tilde{\lambda}\|, \|e\| \in \mathcal{L}_\infty, \tag{11.114}$$

$$\|e_x\|, \|e\|\|u_x\|, \|e\|\|u_{xx}\| \in \mathcal{L}_2, \tag{11.115}$$

$$\|\hat{\lambda}_t\|, \|\hat{b}_t\|, \|\hat{\varepsilon}_t\| \in \mathcal{L}_2. \tag{11.116}$$

In contrast to the Lyapunov approach, here we do not require the adaptation gain to be small or the projection to be enforced for all the parameters and their derivatives. As a trade-off, the passivity-based design employs derivatives of the measured state.

Even though the identifier properties hold for unknown ε, b, and λ, for the reasons explained in Section 11.1 we state the stability result for known $\varepsilon(x)$ and $b(x)$. To this end, we set $\hat{\varepsilon} \equiv \varepsilon$, $\hat{b} \equiv b$, and $\gamma_0 = 0$, remove the term $(1/2)\varepsilon''(x)(u - \hat{u})$ from the identifier, and replace the term $\gamma^2(u - \hat{u})\|u_x\|^2$ by $\gamma^2(u - \hat{u})\|u\|^2$.

We employ the controller

$$U(t) = \int_0^1 \hat{k}^n(1, y)\hat{u}(y, t)\, dy, \tag{11.117}$$

where $\hat{k}^n$ is given by the successive approximations (11.66)–(11.69).

Our main result is given by the following theorem.

THEOREM 11.3. *There exists n^* such that, for all $n \geq n^*$, any initial estimate $\hat{\lambda}_0 \in C^1[0, 1]$ and any initial data $u_0, \hat{u}_0 \in H^2(0, 1)$ compatible with boundary conditions, the classical solution $(u, \hat{u}, \hat{\lambda})$ of the closed-loop system* (11.1)–(11.3), (11.102)–(11.104) *with the controller* (11.117) *and update law* (11.110) *is bounded for all $x \in [0, 1]$, $t \geq 0$ and*

$$\lim_{t\to\infty} \max_{x\in[0,1]} |u(x, t)| = 0. \tag{11.118}$$

Proof. Applying the transformation

$$\hat{w}(x) = \hat{u}(x) - \int_0^x \hat{k}^n(x, y)\hat{u}(y)\, dy := \hat{T}^n[\hat{u}](x) \tag{11.119}$$

to the identifier (11.102)–(11.104), we get the target system

$$\hat{w}_t = \varepsilon(x)\hat{w}_{xx} + b(x)\hat{w}_x - c\hat{w} + \hat{T}^n[\hat{\lambda}e + \gamma^2 e\|u\|^2] - \int_0^x \hat{k}_t^n(x,y)\hat{u}(y)\,dy - \int_0^x \delta\hat{k}^n(x,y)\hat{u}(y)\,dy, \quad (11.120)$$

$$\hat{w}(0) = 0, \quad (11.121)$$

$$\hat{w}(1) = 0. \quad (11.122)$$

To show the boundedness of $\|\hat{w}\|$, we start with

$$\begin{aligned}\frac{1}{2}\frac{d}{dt}\|\hat{w}\|_\rho^2 = &-\int_0^1 \rho(x)\varepsilon(x)\hat{w}_x^2(x)\,dx - c\|\hat{w}\|_\rho^2 \\ &+ \int_0^1 \rho(x)\hat{w}(x)\hat{T}^n[\hat{\lambda}e + \gamma^2 e\|u\|^2](x)\,dx \\ &- \int_0^1 \rho(x)\hat{w}(x)\int_0^x \delta\hat{k}^n(x,\xi)\hat{u}(y)\,dy \\ &- \int_0^1 \rho(x)\hat{w}(x)\int_0^x \hat{k}_t^n(x,y)\hat{u}(y)\,dy. \end{aligned} \quad (11.123)$$

Since $\|\hat{\lambda}\|$ is bounded (we denote this bound by $\bar{\lambda}$), the control gain $\hat{k}^n$ is also bounded, so that $\|\hat{T}^n[e]\| \le M_1\|e\|$, where M_1 is some constant. The kernel $\hat{l}^n$ of the inverse transformation

$$\hat{u}(x) = \hat{w}(x) + \int_0^x \hat{l}^n(x,y)\hat{w}(y)\,dy \quad (11.124)$$

is also bounded; let us denote this bound by L. We have

$$\begin{aligned}\int_0^1 &\rho(x)\hat{w}(x)\hat{T}^n[\hat{\lambda}e + \gamma^2 e\|u\|^2](x)\,dx \\ &\le \underline{\varepsilon}^{-1}e^{\frac{B}{\varepsilon}}\gamma^2 M_1\|u\|\|e\|\|\hat{w}\|(1+L)(\|\hat{w}\| + \|e\|) \\ &\quad + \underline{\varepsilon}^{-1}e^{\frac{B}{\varepsilon}} M_1\bar{\lambda}\|e\|\|\hat{w}\| \\ &\le \frac{1}{8}e^{-\frac{B}{\varepsilon}}\|\hat{w}\|^2 + 2\underline{\varepsilon}^{-2}e^{\frac{2B}{\varepsilon}}M_1^2\bar{\lambda}^2\|e\|^2 + \frac{1}{8}e^{-\frac{B}{\varepsilon}}\|\hat{w}\|^2 \\ &\quad + \frac{1}{8}e^{-\frac{B}{\varepsilon}}\|e\|^2 + 4\underline{\varepsilon}^{-2}e^{\frac{2B}{\varepsilon}}\gamma^4 M_1^2(1+L)^2\|e\|^2\|u\|^2\|\hat{w}\|^2 \\ &\le \frac{1}{4}e^{-\frac{B}{\varepsilon}}\|\hat{w}_x\|^2 + l_1(t)\|\hat{w}\|^2 + l_2(t), \end{aligned} \quad (11.125)$$

where $l_1(t)$ and $l_2(t)$ are integrable over infinite time functions.

The last term in (11.123) can be estimated as follows:

$$\begin{aligned}\int_0^1 \rho(x)\hat{w}(x)\int_0^x \hat{k}_t^n(x,y)\hat{u}(y)\,dy &\le \underline{\varepsilon}^{-1}e^{\frac{B}{\varepsilon}}(1+L)\|\hat{w}\|^2 M_2\|\hat{\lambda}_t\| \\ &\le \frac{1}{8}e^{-\frac{B}{\varepsilon}}\|\hat{w}_x\|^2 + l_3(t)\|\hat{w}\|^2, \end{aligned} \quad (11.126)$$

where M_2 is some constant and $l_3(t)$ is integrable, which follows from the identifier properties established in Lemma 11.3. Substituting the estimates (11.125)–(11.126) into (11.123) and using Lemma 11.2, we get

$$\frac{1}{2}\frac{d}{dt}\|\hat{w}\|^2 \leq -\frac{1}{8}e^{-\frac{B}{\underline{\varepsilon}}}\|\hat{w}_x\|^2 + (l_1(t)+l_3(t))\|\hat{w}\|^2 + l_2(t). \tag{11.127}$$

By Lemma D.3, $\|\hat{w}\|$ is bounded and $\|\hat{w}_x\|$ is square integrable over infinite time. Using the inverse transformation (11.124), we also get that $\|\hat{u}_x\|$ and $\|u_x\|$ are square integrable.

Now let us show the boundedness of the H^1-norm of the state. We start with

$$\begin{aligned}\frac{1}{2}\frac{d}{dt}\|e_x\|^2 =& -\int_0^1 \varepsilon(x)e_{xx}^2(x)\,dx - \int_0^1 \tilde{\lambda}(x)e_{xx}(x)u(x)\,dx - \gamma^2\|e\|^2\|u\|^2 \\ \leq& -\underline{\varepsilon}\|e_{xx}\|^2 + \|\tilde{\lambda}\|\|e_{xx}\|\|u\| \\ \leq& -\frac{\varepsilon}{4}\|e_{xx}\|^2 + \|\tilde{\lambda}\|^2\|u\|^2 + \|b\|^2\|u\|^2. \end{aligned}\tag{11.128}$$

Note that $\|\tilde{\lambda}\|^2\|u\|^2$ is an integrable function. From (11.128) we get that $\|e_x\|$ is bounded. The second calculation is

$$\begin{aligned}\frac{1}{2}\frac{d}{dt}\|\hat{w}_x\|^2 =& -\int_0^1 \hat{w}_{xx}(x)\hat{w}_t(x)\,dx \\ \leq& -\underline{\varepsilon}\|\hat{w}_{xx}\|^2 + \|b\|\|\hat{w}_x\|\|w_{xx}\| + M_1\bar{\lambda}\|e\|\|\hat{w}_{xx}\| \\ &+\gamma^2 M_1\|u\|^2\|e\|\|\hat{w}_{xx}\| + \frac{\varepsilon}{4}\|\hat{w}\|\|\hat{w}_{xx}\| \\ &+(1+L)\|\hat{w}\|\|\hat{w}_{xx}\|M_2\|\hat{\lambda}_t\| \\ \leq& -\frac{\varepsilon}{4}\|\hat{w}_x\|^2 + l_4(t)\|\hat{w}\|^2 + l_5(t), \end{aligned}\tag{11.129}$$

where $l_4(t)$ and $l_5(t)$ are integrable functions of time. Using Lemma D.3, we get the boundedness of $\|\hat{w}_x\|$. The boundedness of $\|\hat{u}_x\|$ and $\|u_x\|$ follows from (11.124) and the boundedness of $\|e_x\|$.

To show regulation, let us note that $\frac{d}{dt}\|e\|^2$ and $\frac{d}{dt}\|\hat{w}\|^2$ are bounded, which easily follows from (11.123). By Lemma D.2, we get $\|\hat{w}\| \to 0$ as $t \to \infty$. Using Agmon's inequality and the fact that $\|\hat{w}_x\|$ is bounded, we get $\hat{w}(x,t) \to 0$ for all $x \in [0,1]$ as $t \to \infty$. The boundedness and regulation results for $u(x,t)$ and $\hat{u}(x,t)$ follow from (11.124). The proof of Theorem 11.3 is completed. □

11.7 SIMULATIONS

The simulation results for the Lyapunov scheme for the plant (11.1)–(11.3) with unknown $b(x)$ and $\lambda(x)$ are shown in Figures 11.1–11.4.

We computed the series (11.66)–(11.69) using only eight terms at each time step, which allowed the scheme to be implementable in real time.

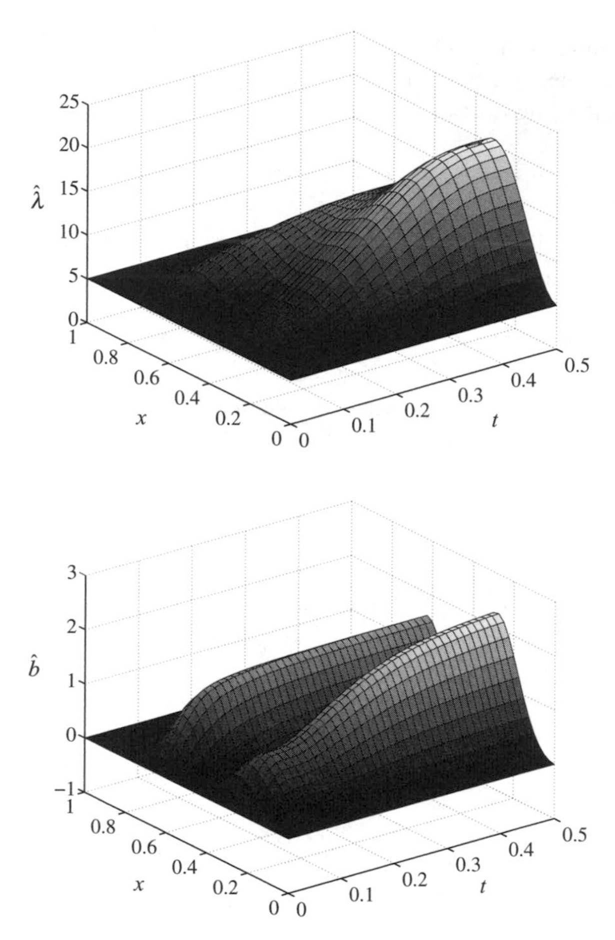

Figure 11.1 The parameter estimates $\hat{\lambda}(x, t)$ and $\hat{b}(x, t)$.

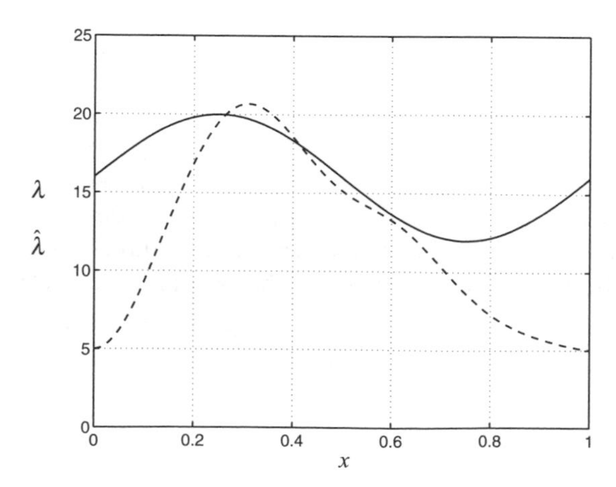

Figure 11.2 The final profile of the estimate $\hat{\lambda}(x)$ (dashed line) versus the true $\lambda(x)$ (solid line).

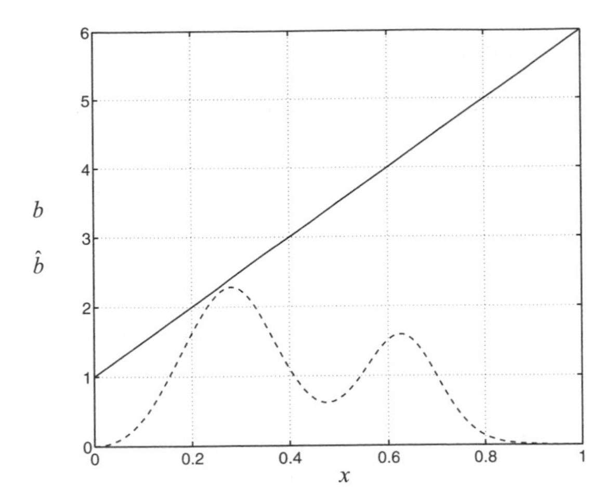

Figure 11.3 The final profile of the estimate $\hat{b}(x)$ (dashed line) versus the true $b(x)$ (solid line).

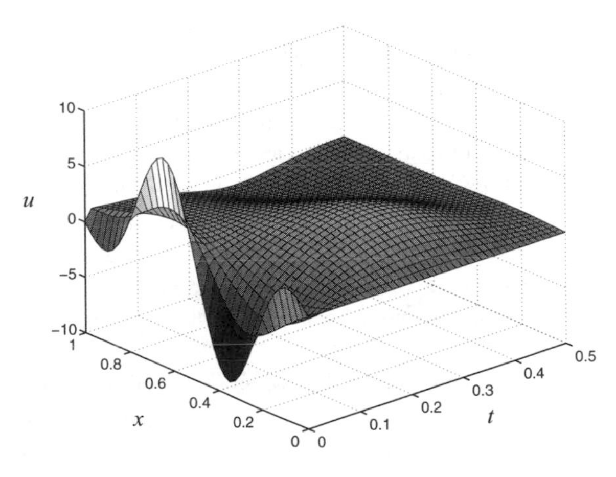

Figure 11.4 The closed-loop state $u(x, t)$.

The evolution of the parameter estimates $\hat{\lambda}(x)$ and $\hat{b}(x)$ is shown in Figure 11.1. In Figures 11.2 and 11.3 the final profiles of $\hat{\lambda}(x)$ and $\hat{b}(x)$ are shown in comparison with the true $\lambda(x)$ and $b(x)$. As expected, $\hat{\lambda}(x)$ and $\hat{b}(x)$ do not converge to the true functions. In Figure 11.4 the evolution of the state of the closed-loop system is shown. One can see that despite the parameter estimates not converging close to the true values, the state is regulated to zero.

11.8 NOTES AND REFERENCES

The designs presented in this chapter extend in a straightforward way to other types of boundary conditions (Robin uncontrolled end and Neumann controlled end).

Chapter Twelve

Closed-Form Adaptive Output-Feedback Controllers

In Chapters 8–11 we designed adaptive controllers for reaction-advection-diffusion systems under the assumption of availability of state measurements across the domain. In many applications, sensors are typically placed only at the boundary of the domain, for example, at the outlet of a chemical tubular reactor.

In this chapter we introduce output-feedback adaptive controllers for two parametrically uncertain, unstable parabolic PDEs controlled from the boundary. Both PDEs are motivated by a model of thermal instability in solid propellant rockets [13]. Perhaps even more important than the physical motivation behind those systems is the fact that they are conceptually the simplest possible benchmark problems for adaptive output-feedback control, so they are of pedagogical value. At the same time, these problems are nontrivial and retain many of the essential difficulties of parabolic PDEs with parametric uncertainties and boundary inputs and outputs that need to be addressed by the design, so they are of methodological value. Two key difficulties are emphasized. First, these benchmark plants cover two conceptually different cases of parametric uncertainties—in the domain and at the boundary. Second, both plants are of infinite relative degree, generally the most challenging case in adaptive control. Infinite relative degree typically arises in applications where actuators and sensors are on the "opposite sides" of the PDE domains.

There are several reasons that make the adaptive design for these problems particularly instructive. One reason is the availability of closed-form controllers, eliminating the need to compute the control gain online. Another reason is that these plants are already in the form in which unknown parameters multiply the output (the so-called observer canonical form) and one does not need to transform them into that form first. Finally, there is only one unknown parameter in each of the plants. With these difficulties out of the way (they are dealt with in the next chapter), we can focus on the designs and highlight difficult steps in the stability proofs.

We consider two schemes in this chapter—a Lyapunov-based scheme, with identifier and controller designed jointly, and a swapping-based scheme, with standard gradient and least squares identifiers. In both schemes, the control laws are adaptive versions of the explicit boundary control laws developed in Chapter 3, employing adaptive observers that are infinite-dimensional extensions of Kreisselmeier observers [68]. We should point out that the Lyapunov-based adaptive output-feedback design is presented only in this chapter. The general reaction-advection-diffusion problem in the next chapter will be handled using the scheme with a swapping identifier.

In general, swapping-based adaptive schemes have a considerably higher dynamic order than Lyapunov-based schemes (see the designs in Chapters 8 and 10). However, for the systems studied in this chapter we have been able to use the same set of Kreisselmeier filters both for designing an observer and for achieving a static parametrization from which a gradient update law is derived. Thus, the dynamic order for both output-feedback designs in this chapter is the same. The advantage of the swapping update laws is that they are considerably simpler and do not require a priori bounds on unknown parameters, whereas the advantage of the Lyapunov update laws is that they are derived from a complete Lyapunov function that incorporates the plant, the filters, and the update law, providing a tighter control over transient performance.

In addition to stabilization, in this chapter we also pursue parametric identification. It is in general a very challenging problem for distributed parameter systems because one needs to identify a functional uncertainty (in other words, infinitely many unknown parameters) using just a scalar (boundary) input and scalar (boundary) output. However, for the benchmark problems in this chapter, persistency of excitation can be achieved by simply augmenting feedback with a non-zero constant.

12.1 LYAPUNOV DESIGN—PLANT WITH UNKNOWN PARAMETER IN THE DOMAIN

Consider the plant

$$u_t(x,t) = u_{xx}(x,t) + gu(0,t), \tag{12.1}$$

$$u_x(0,t) = 0, \tag{12.2}$$

$$u(1,t) = U(t), \tag{12.3}$$

where g is a constant unknown parameter and $U(t)$ is the control input. We assume that only the boundary value $u(0,t)$ is available for measurement.

The system (12.2)–(12.3) is inspired by a model of thermal instability in solid propellant rockets [13]. For $U(t) \equiv 0$, this system is unstable for $g > 2$. The plant can be written in the frequency domain as a transfer function from the input $u(1,s)$ to the output $u(0,s)$ in the following way:

$$u(0,s) = \frac{s}{(s-g)\cosh\sqrt{s} + g}\, u(1,s). \tag{12.4}$$

We can see that it has no zeros (at $s = 0$ the transfer function is $2/(2-g)$) and has infinitely many poles, one of which is unstable and approximately equal to g as $g \to +\infty$. Thus, this is an infinite relative degree system.

In Chapter 3 we designed the following controller for (12.2)–(12.3) (assuming known g and measured $u(x,t)$):

$$U(t) = -\int_0^1 \sqrt{g}\sinh\left(\sqrt{g}(1-\xi)\right) u(\xi,t)\, d\xi. \tag{12.5}$$

Since the state $u(x,t)$ is not measured, we employ an adaptive observer that consists of the input filter

$$\eta_t(x,t) = \eta_{xx}(x,t), \tag{12.6}$$

$$\eta_x(0,t) = 0, \tag{12.7}$$

$$\eta(1,t) = U(t), \tag{12.8}$$

the output filter

$$v_t(x,t) = v_{xx}(x,t) + u(0,t), \tag{12.9}$$

$$v_x(0,t) = 0, \tag{12.10}$$

$$v(1,t) = 0, \tag{12.11}$$

and an estimate of $u(x,t)$ given by $\hat{g}v(x,t) + \eta(x,t)$. Replacing g and $u(x,t)$ in (12.5) with their estimates, we get the adaptive controller

$$U(t) = -\int_0^1 \sqrt{\hat{g}} \sinh\left(\sqrt{\hat{g}}(1-\xi)\right)\left(\hat{g}v(\xi,t) + \eta(\xi,t)\right) d\xi. \tag{12.12}$$

We propose the following update law, which is a modified version of the state-feedback update law (8.75):

$$\dot{\hat{g}} = \frac{\gamma}{1 + \|w\|^2 + a\|v\|^2} \times \operatorname{Proj}_{[2,\bar{g}]}\left\{ v(0,t) \int_0^1 \left(av(x,t) + \hat{g}\cosh\left(\sqrt{\hat{g}}x\right) w(x,t)\right) dx \right\}, \tag{12.13}$$

where a and γ are positive, sufficiently small normalization and adaptation gains and $w(x,t)$ is defined by

$$w(x,t) = \hat{g}v(x,t) + \eta(x,t) + \int_0^x \sqrt{\hat{g}} \sinh\left(\sqrt{\hat{g}}(x-\xi)\right)\left(\hat{g}v(\xi,t) + \eta(\xi,t)\right) d\xi. \tag{12.14}$$

THEOREM 12.1. *There exists $a^* > 0$ such that for all $a \in (0, a^*)$ there exists $\gamma^*(a) > 0$ (where both a^* and $\gamma^*(a)$ can be a priori estimated by the designer), such that for all $\gamma \in (0, \gamma^*)$ the following holds: for any initial data $u_0, v_0, \eta_0 \in H^2(0,1)$ compatible with boundary conditions, and any $\hat{g}(0) \in [2, \bar{g}]$, the classical solution $(u, v, \eta, \hat{g})$ of (12.2)–(12.3), (12.5)–(12.13) is bounded for all $x \in [0,1]$, $t \geq 0$ and*

$$\lim_{t\to\infty} \max_{x\in[0,1]} |u(x,t)| = 0. \tag{12.15}$$

Proof. Step 1 (Target system). We start the proof with the following result.

Lemma 12.1. *The transformation* (12.14), (12.23) *maps* (12.6)–(12.12) *into the system*

$$w_t(x,t) = w_{xx}(x,t) + \dot{\hat{g}} Q(x,t) + \hat{g}\cosh\left(\sqrt{\hat{g}}x\right)(e(0,t) + \tilde{g}v(0,t)), \quad (12.16)$$

$$w_x(0,t) = 0, \quad (12.17)$$

$$w(1,t) = 0, \quad (12.18)$$

where $\tilde{g} = g - \hat{g}$ is the parameter estimation error, signal Q is defined by

$$Q(x,t) = v(x,t) - \int_0^x (\hat{g}v(\xi,t) + w(\xi,t)) \frac{\sinh\left(\sqrt{\hat{g}}(x-\xi)\right)}{\sqrt{\hat{g}}} d\xi, \quad (12.19)$$

and $e(x,t)$ is the observer error, defined as

$$e(x,t) = u(x,t) - gv(x,t) - \eta(x,t). \quad (12.20)$$

Proof. It is easy to see that the boundary conditions (12.17), (12.18) are satisfied. To verify (12.16), we compute the derivatives of w:

$$\begin{aligned} w_{xx}(x,t) = {} & \hat{g}v_{xx}(x,t) + \eta_{xx}(x,t) + \hat{g}^2 v(x,t) + \hat{g}\eta(x,t) \\ & + \hat{g}\int_0^x \sqrt{\hat{g}}\sinh\left(\sqrt{\hat{g}}(x-\xi)\right)(\hat{g}v(\xi,t) + \eta(\xi,t))\, d\xi, \end{aligned} \quad (12.21)$$

$$\begin{aligned} w_t(x,t) = {} & \hat{g}v_t(x,t) + \eta_t(x,t) - \hat{g}u(0,t) + \hat{g}\cosh\left(\sqrt{\hat{g}}x\right)u(0,t) + \hat{g}^2 v(x,t) \\ & + \hat{g}\eta(x,t) - \hat{g}\cosh\left(\sqrt{\hat{g}}x\right)(\hat{g}v(0,t) + \eta(0,t)) + \dot{\hat{g}}v(x,t) \\ & + \dot{\hat{g}}\int_0^x \sqrt{\hat{g}}\sinh\left(\sqrt{\hat{g}}(x-\xi)\right)v(\xi,t)\, d\xi \\ & + \frac{\dot{\hat{g}}}{2}\int_0^x (x-\xi)\cosh\left(\sqrt{\hat{g}}(x-\xi)\right)(\hat{g}v(\xi,t) + \eta(\xi,t))\, d\xi \\ & + \dot{\hat{g}}\int_0^x \frac{\sinh\left(\sqrt{\hat{g}}(x-\xi)\right)}{2\sqrt{\hat{g}}}(\hat{g}v(\xi,t) + \eta(\xi,t))\, d\xi. \end{aligned} \quad (12.22)$$

In (12.22) we integrated by parts twice. Subtracting (12.21) from (12.22) and using the transformation inverse to (12.14),

$$\hat{g}v(x,t) + \eta(x,t) = w(x,t) - \hat{g}\int_0^x (x-\xi)w(\xi,t)\, d\xi, \quad (12.23)$$

we get

$$w_t(x,t) = w_{xx}(x,t) + \hat{g}\cosh\left(\sqrt{\hat{g}}x\right)(e(0,t) + \tilde{g}v(0,t)) + \dot{\hat{g}}v(x,t)$$
$$+ \dot{\hat{g}} \int_0^x \sqrt{\hat{g}} \sinh\left(\sqrt{\hat{g}}(x-\xi)\right) v(\xi,t)\, d\xi$$
$$+ \dot{\hat{g}} \int_0^x \left[\hat{g}(x-\xi)\cosh\left(\sqrt{\hat{g}}(x-\xi)\right) + \sqrt{\hat{g}}\sinh\left(\sqrt{\hat{g}}(x-\xi)\right)\right]$$
$$\times \left(w(\xi,t) - \hat{g}\int_0^\xi (\xi - y) w(y,t)\, dy\right) d\xi. \tag{12.24}$$

Note that we used the identity $u - \hat{g}v - \eta = e + \tilde{g}v$. Changing the order of integration in the double integral (last term in (12.24)), computing the internal integral, and gathering all the terms together, we obtain (12.16). □

Proof of Theorem 12.1 (continued).
Step 2 (Lyapunov analysis). Consider the Lyapunov function candidate

$$V = \frac{1}{2}\log\left(1 + \|w\|^2 + a\|v\|^2\right) + \frac{b}{2}\|e\|^2 + \frac{1}{2\gamma}\tilde{g}^2, \tag{12.25}$$

where $a \in (0,1)$ and b are positive constants yet to be defined. We note from (12.20) that

$$e_t(x,t) = e_{xx}(x,t), \tag{12.26}$$

$$e_x(0,t) = 0, \tag{12.27}$$

$$e(1,t) = 0 \tag{12.28}$$

and therefore

$$\frac{1}{2}\frac{d}{dt}\|e\|^2 = -\|e_x\|^2. \tag{12.29}$$

With (12.20), (12.14), and (12.9)–(12.11), we obtain

$$\frac{1}{2}\frac{d}{dt}\|v\|^2 = -\|v_x\|^2 + (w(0,t) + e(0,t) + \tilde{g}v(0,t))\int_0^1 v(x,t)\, dx. \tag{12.30}$$

With (12.25), (12.29), (12.30), and (12.16)–(12.18), we get

$$\dot{V} = -b\|e_x\|^2 + \frac{1}{1 + \|w\|^2 + a\|v\|^2}\Bigg\{ -\|w_x\|^2 - a\|v_x\|^2$$
$$+ aw(0,t)\int_0^1 v(x,t)dx + \dot{\hat{g}}\int_0^1 w(x,t)Q(x,t)\, dx$$
$$+ e(0,t)\int_0^1 \left(av(x,t) + \hat{g}\cosh\left(\sqrt{\hat{g}}x\right) w(x,t)\right) dx\Bigg\}, \tag{12.31}$$

which can be majorized by

$$\dot V \le \frac{1}{1+\|w\|^2+a\|v\|^2}\Bigg\{-(1-8a)\|w_x\|^2 - \frac{a}{2}\|v_x\|^2 - b\|e_x\|^2 + ae(0,t)\int_0^1 v(x,t)\,dx + \dot{\hat g}\int_0^1 w(x,t)Q(x,t)\,dx + e(0,t)\int_0^1 \hat g\cosh\left(\sqrt{\hat g}x\right)w(x,t)\,dx\Bigg\}. \tag{12.32}$$

By applying Young's inequality to the two cross-terms with $e(0,t)$, we get

$$\dot V \le \frac{1}{1+\|w\|^2+a\|v\|^2}\Bigg\{-\left(1-8a-\frac{2}{\mu_1}\right)\|w_x\|^2 - \left(\frac{a}{2}-\frac{2}{\mu_2}\right)\|v_x\|^2 + \dot{\hat g}\int_0^1 w(x,t)Q(x,t)\,dx - \left(b-2\mu_2 a^2 - 2\mu_1\bar g^2 e^{2\sqrt{\bar g}}\right)\|e_x\|^2\Bigg\}, \tag{12.33}$$

where μ_1 and μ_2 are positive constants that we can arbitrarily choose in our analysis. It can be shown that

$$\left|\int_0^1 w(x,t)Q(x,t)\,dx\right| \le 2e^{2\sqrt{\bar g}}\left(\|w\|^2+\|v\|^2\right), \tag{12.34}$$

which can then be used to prove that

$$\left|\dot{\hat g}\int_0^1 w(x)Q(x)\,dx\right| \le 2\frac{\gamma}{a}e^{2\sqrt{\bar g}}|v(0,t)|\left(a\|v\| + e^{2\sqrt{\bar g}}\|w\|\right).$$

With further calculations involving Young's, Poincaré's, and Agmon's inequalities, and using that fact that $a,\gamma\in(0,1)$, one arrives at a conservative bound

$$\left|\dot{\hat g}\int_0^1 w(x,t)Q(x,t)\,dx\right| \le 80\frac{\gamma}{a^2}e^{8\sqrt{\bar g}}\|v_x\|^2 + \frac{1}{4}\|w_x\|^2. \tag{12.35}$$

Substituting this bound into (12.33), we get

$$\dot V \le \frac{1}{1+\|w\|^2+a\|v\|^2}\Bigg\{-\left(\frac{3}{4}-8a-\frac{2}{\mu_1}\right)\|w_x\|^2 - \left(\frac{a}{2}-\frac{2}{\mu_2}-80\frac{\gamma}{a^2}e^{8\sqrt{\bar g}}\right)\|v_x\|^2 - \left(b-2\mu_2 a^2 - 2\mu_1\bar g^2 e^{2\sqrt{\bar g}}\right)\|e_x\|^2\Bigg\}. \tag{12.36}$$

Selecting now $\mu_1 = 16$, $\mu_2 = 16/a$, $b = 64(a+\bar g^2 e^{2\sqrt{\bar g}})$,

$$a^* = \frac{1}{16}, \qquad \gamma^* = \frac{a^3}{320}e^{-8\sqrt{\bar g}} \tag{12.37}$$

for $a \in (0, a^*]$ and $\gamma \in (0, \gamma^*]$, we obtain

$$\dot{V} \leq -\frac{1}{8} \frac{\|w_x\|^2 + a\|v_x\|^2 + 4b\|e_x\|^2}{1 + \|w\|^2 + a\|v\|^2}. \tag{12.38}$$

Step 3 (Boundedness and regulation).

From (12.38) one can conclude the boundedness of $\|w\|$, $\|v\|$ and the integrability in time of $\|w_x\|^2$, $\|v_x\|^2$ (and therefore, by Poincaré's inequality, the integrability of $\|w\|^2$, $\|v\|^2$). From this, one can conclude that $\|Q\|$ is bounded and, with Agmon's inequality, that $\dot{\hat{g}}$ is square integrable over infinite time, which implies that $\dot{\hat{g}}\|Q\|$ is square integrable. Agmon's inequality also guarantees that $\tilde{g}\cosh(\sqrt{\hat{g}}x)(e(0,t) + w(0,t))$, which appears in (12.16), is square integrable. Therefore, the target system (12.16)–(12.18) can be written as

$$w_t(x,t) = w_{xx}(x,t) + L(x,t), \tag{12.39}$$

$$w_x(0,t) = 0, \tag{12.40}$$

$$w(1,t) = 0, \tag{12.41}$$

where $\|L\|$ is square integrable over infinite time. Using the Lyapunov function

$$V = \frac{1}{2}\int_0^1 w_x(x,t)^2\,dt \tag{12.42}$$

we get

$$\begin{aligned}\dot{V} &= -\int_0^1 w_{xx}(x,t)^2\,dt - \int_0^1 L(x,t)w_{xx}(x,t)\,dt \\ &\leq -\frac{1}{4}V + \frac{1}{2}\|L\|^2.\end{aligned} \tag{12.43}$$

Using Lemma D.3, we get boundedness of $\|w_x\|$. Since $u(0,t) = w(0,t) + e(0,t) + \tilde{g}v(0,t)$ is square integrable over infinite time (which follows from the integrability of $\|w_x\|$, $\|v_x\|$ and Agmon's inequality), using the Lyapunov function $V = (1/2)\|v_x\|^2$ and Lemma D.3, we conclude that $\|v_x\|$ is bounded. The boundedness of $\|w_x\|$ and $\|v_x\|$ implies that $\tilde{g}\cosh(\sqrt{\hat{g}}x)(e(0,t) + w(0,t)) + \dot{\hat{g}}Q$ and $u(0,t)$ are bounded, and from (12.30) and (12.31) we get that the time derivatives of $\|w\|^2$, $\|v\|^2$ are bounded. By Barbalat's lemma (Lemma D.1), this implies the regulation of $\|w\|$, $\|v\|$ and, by Agmon's inequality, the regulation of $w(x,t)$, $v(x,t)$ for all $x \in [0,1]$. To obtain the corresponding boundedness and regulation results for u, we first use the inverse transformation (12.23), which establishes the boundedness and regulation of η, and then invoke (12.20). Theorem 12.1 is proved. □

It is clear that the conservative values of a^* and γ^* are for the purposes of the proof only. In an implementation one would be safe to choose higher values of a and γ.

12.2 LYAPUNOV DESIGN—PLANT WITH UNKNOWN PARAMETER IN THE BOUNDARY CONDITION

In this section we consider another benchmark system, a plant with parametric uncertainty in the boundary condition. Even though the plant appears to be quite similar to the plant considered in the previous section, the stability proof turns out to be substantially harder. To prove pointwise boundedness and regulation, we introduce a special technique, which will also prove useful in the analysis of more complex systems in Chapter 13.

Consider the plant

$$u_t(x,t) = u_{xx}(x,t), \tag{12.44}$$

$$u_x(0,t) = -qu(0,t), \tag{12.45}$$

$$u(1,t) = U(t), \tag{12.46}$$

where q is a constant unknown parameter and $U(t)$ is the control input. Only $u(0,t)$, the boundary value of $u(x,t)$ at $x=0$, is available for measurement. With $U(t)\equiv 0$, this PDE is unstable for $q>1$. The plant can be written in the frequency domain as a transfer function from input $u(1,s)$ to output $u(0,s)$:

$$u(0,s) = \frac{\sqrt{s}}{\sqrt{s}\cosh\sqrt{s} - q\sinh\sqrt{s}}\, u(1,s)\,. \tag{12.47}$$

Since this transfer function has infinitely many poles and no zeros (at $s=0$ the transfer function is $1/(1-q)$), this is an infinite relative degree system. One of the poles is unstable and is approximately equal to q^2 as $q\to+\infty$.

The nominal full-state feedback for (12.44)–(12.46) was designed in Chapter 3:

$$U(t) = -\int_0^1 qe^{q(1-y)}u(y,t)\,dy. \tag{12.48}$$

To estimate the state, we use the input filter (12.6)–(12.8) and the output filter

$$v_t(x,t) = v_{xx}(x,t), \tag{12.49}$$

$$v_x(0,t) = -u(0,t), \tag{12.50}$$

$$v(1,t) = 0. \tag{12.51}$$

Replacing q and u in (12.48) by their estimates $\hat{q}$ and $\hat{q}v+\eta$, we get the controller

$$U(t) = -\int_0^1 \hat{q}e^{\hat{q}(1-\xi)}\left(\hat{q}v(\xi,t)+\eta(\xi,t)\right)d\xi. \tag{12.52}$$

The update law is a modified version of the state-feedback update law (8.92):

$$\dot{\hat{q}} = \frac{\gamma}{1+\|w\|^2+a\|v\|^2} \times \mathrm{Proj}_{[1,\bar{q}]}\left\{v(0,t)\left(av(0,t)+\hat{q}w(0,t)+\hat{q}^2\int_0^1 e^{\hat{q}x}w(x,t)\,dx\right)\right\}, \tag{12.53}$$

where

$$w(x,t) = \hat{q}v(x,t) + \eta(x,t) + \int_0^x \hat{q}e^{\hat{q}(x-\xi)}\left(\hat{q}v(\xi,t) + \eta(\xi,t)\right)d\xi. \tag{12.54}$$

THEOREM 12.2. *There exists $a^* > 0$ such that for all $a \in (0, a^*)$, there exists $\gamma^*(a) > 0$ (where both a^* and $\gamma^*(a)$ can be a priori estimated by the designer), such that for all $\gamma \in (0, \gamma^*)$ the following holds: for any initial data $u_0, \eta_0, v_0 \in H^2(0,1)$ compatible with boundary conditions, and any $\hat{q}(0) \in [1, \bar{q}]$, the classical solution $(u, v, \eta, \hat{q})$ of* (12.44)–(12.46), (12.6)–(12.8), (12.49)–(12.54) *is bounded for all $x \in [0,1]$, $t \geq 0$ and*

$$\lim_{t\to\infty} \max_{x\in[0,1]} |u(x,t)| = 0. \tag{12.55}$$

Proof. Step 1 (Target system). We start by deriving the adaptive target system.

LEMMA 12.2. *The transformation* (12.54) *maps* (12.6)–(12.8), (12.49)–(12.52) *into*

$$w_t(x,t) = w_{xx}(x,t) + \dot{\hat{q}}\left\{v(x,t) + \int_0^x e^{\hat{q}(x-\xi)}\left(\hat{q}v(\xi,t) + w(\xi,t)\right)d\xi\right\}$$
$$+ \hat{q}^2 e^{\hat{q}x}\left(e(0,t) + \tilde{q}v(0,t)\right), \tag{12.56}$$

$$w_x(0,t) = -\hat{q}\left(e(0,t) + \tilde{q}v(0,t)\right), \tag{12.57}$$

$$w(1,t) = 0, \tag{12.58}$$

where $e(x,t)$ is an observer error defined as

$$e(x,t) = u(x,t) - qv(x,t) - \eta(x,t). \tag{12.59}$$

Proof. First, we verify the boundary condition (12.57):

$$\begin{aligned} w_x(0,t) &= \hat{q}v_x(0,t) + \eta_x(0,t) + \hat{q}(\hat{q}v(0,t) + \eta(0,t)) \\ &= -\hat{q}(e(0,t) + qv(0,t) + \eta(0,t)) + \hat{q}(\hat{q}v(0,t) + \eta(0,t)) \\ &= -\hat{q}\left(e(0,t) + \tilde{q}v(0,t)\right). \end{aligned} \tag{12.60}$$

The boundary condition (12.58) is obviously satisfied. To verify (12.56), we compute the derivatives of w:

$$\begin{aligned} w_{xx}(x,t) &= \hat{q}v_{xx}(x,t) + \eta_{xx}(x,t) + \hat{q}^2 v_x(x,t) \\ &\quad + \hat{q}\eta_x(x,t) + \hat{q}^3 v(x,t) + \hat{q}^2\eta(x,t) \\ &\quad + \int_0^x \hat{q}^3 e^{\hat{q}(x-\xi)}(\hat{q}v(\xi,t) + \eta(\xi,t))\,d\xi, \end{aligned} \tag{12.61}$$

$$
\begin{aligned}
w_t(x,t) = {} & \hat{q} v_t(x,t) + \eta_t(x,t) + \hat{q}^2 v_x(x,t) - \hat{q}^2 e^{\hat{q}x} v_x(0,t) \\
& + \hat{q}^3 v(x,t) + \hat{q}^2 \eta(x,t) + \hat{q}\eta_x(x,t) \\
& - \hat{q}^3 e^{\hat{q}x} v(0,t) - \hat{q}^2 e^{\hat{q}x} \eta(0,t) \\
& + \int_0^x \hat{q}^3 e^{\hat{q}(x-\xi)} (\hat{q} v(\xi,t) + \eta(\xi,t))\, d\xi \\
& + \dot{\hat{q}} v(x,t) + \dot{\hat{q}} \int_0^x \hat{q} e^{\hat{q}(x-\xi)} v(\xi,t)\, d\xi \\
& + \dot{\hat{q}} \int_0^x (1 + \hat{q}(x-\xi)) e^{\hat{q}(x-\xi)} (\hat{q} v(\xi,t) + \eta(\xi,t))\, d\xi. \qquad (12.62)
\end{aligned}
$$

In (12.62) we integrated by parts twice. Subtracting (12.61) from (12.62) and using the transformation inverse to (12.160),

$$
\hat{q} v(x,t) + \eta(x,t) = w(x,t) - \hat{q} \int_0^x w(y,t)\, dy, \qquad (12.63)
$$

we get

$$
\begin{aligned}
w_t(x,t) = {} & w_{xx}(x,t) + \hat{q}^2 e^{\hat{q}x} (e(0,t) + \tilde{q} v(0,t)) + \dot{\hat{q}} v(x,t) \\
& + \dot{\hat{q}} \int_0^x \hat{q} e^{\hat{q}(x-\xi)} v(\xi,t)\, d\xi \\
& + \dot{\hat{q}} \int_0^x (1 + \hat{q}(x-\xi)) e^{\hat{q}(x-\xi)} \left(w(\xi,t) - \hat{q} \int_0^\xi w(y,t)\, dy \right) d\xi.
\end{aligned}
\qquad (12.64)
$$

In the above we used the identity $u - \hat{q} v - \eta = e + \tilde{q} v$. Changing the order of integration in the double integral (last term in (12.64)) and computing the internal integral, we obtain (12.56). □

Step 2 (Boundedness in L^2). This step is quite similar to Step 2 in the proof of Theorem 12.1; therefore we just sketch it. We use the Lyapunov function

$$
V = \frac{1}{2} \log\left(1 + \|w\|^2 + a\|v\|^2\right) + \frac{b}{2}\|e\|^2 + \frac{1}{2\gamma}\tilde{q}^2, \qquad (12.65)
$$

where a and b are constants to be determined. Computing its time derivative and using (12.53) and Agmon's and Young's inequalities (as in (12.32)–(12.36)), for $a \in (0, a^*]$, $b = 2(a + \bar{q}^2 + 8\bar{q}^4 e^{\bar{q}})$, and $\gamma \in (0, \gamma^*]$, where

$$
a^* = \frac{1}{8}, \qquad \gamma^* = \frac{a^3}{128\bar{q}^2(1 + \bar{q}e^{\bar{q}})^4}, \qquad (12.66)
$$

one obtains

$$
\dot{V} \le -\frac{1}{8} \frac{\|w_x\|^2 + a\|v_x\|^2 + 4b\|e_x\|^2}{1 + \|w\|^2 + a\|v\|^2}. \qquad (12.67)
$$

From (12.67) and (12.65) we get the boundedness of $\|w\|$ and $\|v\|$ and the square integrability in time of $\|w_x\|$ and $\|v_x\|$ (which, by Poincaré's and Agmon's inequalities, also gives the square integrability of $w(0,t)$, $v(0,t)$, $\|w\|$, and $\|v\|$).

Step 3 (Pointwise boundedness and regulation). Using the Lyapunov function (12.42) does not help in proving boundedness here, because of the boundary terms appearing in $\dot{V}$ that cannot be majorized by an integrable function of time. To circumvent this difficulty, we propose using the following transformations:

$$\breve{w}(x,t) = w(x,t) + \hat{q}(1-x)\int_0^x (e(y,t) + \tilde{q}v(y,t))\,dy, \tag{12.68}$$

$$\breve{v}(x,t) = v(x,t) + (1-x)\int_0^x (e(y,t) + \tilde{q}v(y,t) + w(y,t))\,dy. \tag{12.69}$$

The purpose of these transformations is to make the boundary conditions of PDEs for $\breve{w}$ and $\breve{v}$ homogeneous[1] (note that $u(0,t) = w(0,t) + \tilde{q}v(0,t) + \eta(0,t)$). One can view (12.68), (12.69) as backstepping transformations that eliminate unwanted boundary terms, but only for the purpose of the proof, not as a design element.

With straightforward calculations, one can show that $\breve{w}$ and $\breve{v}$ satisfy the following PDEs:

$$\breve{w}_t(x,t) = \breve{w}_{xx}(x,t) + \dot{\hat{q}}M_1(x,t) + M_2(x,t), \tag{12.70}$$

$$\breve{w}_x(0,t) = 0, \tag{12.71}$$

$$\breve{w}(1,t) = 0, \tag{12.72}$$

and

$$\breve{v}_t(x,t) = \breve{v}_{xx}(x,t) + \dot{\hat{q}}N_1(x,t) + N_2(x,t), \tag{12.73}$$

$$\breve{v}_x(0,t) = 0, \tag{12.74}$$

$$\breve{v}(1,t) = 0, \tag{12.75}$$

where M_1, M_2, N_1, and N_2 are given by

$$M_1(x,t) = v(x,t) + \int_0^x e^{\hat{q}(x-y)}(\hat{q}v(y,t) + w(y,t))\,dy + \int_0^x (e(y,t) + \tilde{q}v(y,t) - \hat{q}v(y,t))\,dy, \tag{12.76}$$

$$M_2(x,t) = 2\hat{q}(e(x,t) + \tilde{q}v(x,t)) + (\hat{q}^2 e^{\hat{q}x} + \hat{q}\tilde{q}(1-x))(e(0,t) + \tilde{q}v(0,t)) + \hat{q}\tilde{q}(1-x)w(0,t) \tag{12.77}$$

[1]Note that a well-known method of moving nonhomogeneous terms from the boundary conditions to the domain using simple shifting of the form $\breve{w}(x,t) = w(x,t) - (1-x)\hat{q}(e(0,t) + \tilde{q}v(0,t))$ does not lead to any progress here. The reason is that the time derivatives $v_t(0,t)$ and $w_t(0,t)$ appear in the domain, and they cannot be majorized by an integrable signal.

and

$$N_1(x,t) = (1-x)\int_0^x (M_1(y,t) - v(y,t))\,dy, \tag{12.78}$$

$$\begin{aligned} N_2(x,t) = {} & 2(e(x,t) + \tilde{q}v(x,t) + w(x,t)) + (1-x)\tilde{q}w(0,t) \\ & + (1-x)(\tilde{q} - \hat{q})(e(0,t) + \tilde{q}v(0,t)) + (1-x)\int_0^x M_2(y,t)\,dy. \end{aligned} \tag{12.79}$$

Since $\|e\|$, $\|v\|$, and $\|w\|$ are bounded and $w(0,t)$, $v(0,t)$ are square integrable, we get that $\|M_1\|$, $\|M_2\|$, $\|N_1\|$, $\|N_2\|$ are square integrable functions of time and, in addition, $\|M_1\|$ and $\|N_1\|$ are bounded. The reason we separate the terms with and without $\dot{\hat{q}}$ is that, unlike in Section 12.1, here we do not get the square integrability of $\dot{\hat{q}}$ from L^2 analysis; from (12.53) we get only integrability.

With the Lyapunov function

$$V = \frac{1}{2}\int_0^1 \breve{w}_x(x,t)^2\,dx + \frac{1}{2}\int_0^1 \breve{v}_x(x,t)^2\,dx, \tag{12.80}$$

we obtain

$$\begin{aligned} \dot{V} = {} & -\int_0^1 \breve{w}_t(x,t)\breve{w}_{xx}(x,t)\,dx - \int_0^1 \breve{v}_t(x,t)\breve{v}_{xx}(x,t)\,dx \\ \leq {} & -\|\breve{w}_{xx}\|^2 - \|\breve{v}_{xx}\|^2 + |\dot{\hat{q}}|\|M_1\|\|\breve{w}_{xx}\| + \|M_2\|\|\breve{w}_{xx}\| \\ & + |\dot{\hat{q}}|\|N_1\|\|\breve{v}_{xx}\| + \|N_2\|\|\breve{v}_{xx}\|. \end{aligned} \tag{12.81}$$

Note from (12.68)–(12.69) that

$$\|w_x\| \leq \|\breve{w}_x\| + |\hat{q}|(\|e\| + |\tilde{q}|\|v\|) \leq \|\breve{w}_x\| + f_1(t), \tag{12.82}$$

$$\|v_x\| \leq \|\breve{v}_x\| + \|e\| + |\tilde{q}|\|v\| + \|w\| \leq \|\breve{v}_x\| + f_2(t), \tag{12.83}$$

where $f_1(t)$ and $f_2(t)$ are bounded and square integrable functions of time. From (12.53) and (12.82)–(12.83) we have that

$$\begin{aligned} |\dot{\hat{q}}| \leq {} & a\gamma\|v_x\|^2 + |\hat{q}|\gamma\|v_x\|\|w_x\| + |\hat{q}|^2 e^{|\hat{q}|}\|v_x\|\|w\| \\ \leq {} & f_3(t)(\|\breve{v}_x\| + \|\breve{w}_x\|), \end{aligned} \tag{12.84}$$

where $f_3(t)$ is square integrable. Going back to (12.81), with the help of Young's inequality we get

$$\begin{aligned} \dot{V} \leq {} & -\|\breve{w}_{xx}\|^2 - \|\breve{v}_{xx}\|^2 + 2f_3^2(t)\|M_1\|^2(\|\breve{v}_x\|^2 + \|\breve{w}_x\|^2) + \frac{1}{4}\|\breve{w}_{xx}\|^2 + \|M_2\|^2 \\ & + \frac{1}{4}\|\breve{w}_{xx}\|^2 + 2f_3^2(t)\|N_1\|^2(\|\breve{v}_x\|^2 + \|\breve{w}_x\|^2) + \frac{1}{4}\|\breve{v}_{xx}\|^2 + \|N_2\|^2 + \frac{1}{4}\|\breve{v}_{xx}\|^2 \\ \leq {} & -V + f_4(t)V + f_5(t), \end{aligned} \tag{12.85}$$

where $f_4(t)$ and $f_5(t)$ are integrable functions of time. Using Lemma D.3, we get that $\|\breve{w}_x\|$, $\|\breve{v}_x\|$ are bounded. From (12.82)–(12.83) we get the boundedness of

$\|w_x\|$ and $\|v_x\|$. Using Agmon's inequality, we get the boundedness of $w(x,t)$ and $v(x,t)$ for all $x \in [0,1]$. Using these properties, one can easily show that $\frac{d}{dt}(\|w\|^2)$ and $\frac{d}{dt}(\|v\|^2)$ are bounded. Since $\|w\|^2$ and $\|v\|^2$ are bounded and integrable, by Barbalat's lemma (Lemma D.1) we get $\|w\| \to 0$, $\|v\| \to 0$ for $t \to \infty$. By Agmon's inequality this also gives regulation of $w(x,t)$ and $v(x,t)$ to zero uniformly in x.

To obtain the corresponding boundedness and regulation results for u, we first use the inverse transformation (12.63), which establishes the boundedness and regulation of η, and then invoke (12.59). Theorem 12.2 is proved. □

12.3 SWAPPING DESIGN—PLANT WITH UNKNOWN PARAMETER IN THE DOMAIN

In this section and the next we design adaptive schemes with swapping identifiers for the same two plants considered in Sections 12.1–12.2. In addition to stabilization, here we are also concerned with parametric identification. We show that feedback plus a non-zero constant is sufficient to identify the true parameters.

Consider the plant

$$u_t(x,t) = u_{xx}(x,t) + gu(0,t), \quad 0 < x < 1, \tag{12.86}$$

$$u_x(0,t) = 0, \tag{12.87}$$

$$u(1,t) = U(t). \tag{12.88}$$

We start with the design of a swapping identifier and establish signal properties that are guaranteed independently of the feedback. In this section we use a standard gradient update law. In the next section, for the plant with parametric uncertainty in the boundary condition, we employ a least squares identifier.

In the following, by $\mathcal{L}_2$ and $\mathcal{L}_\infty$ we respectively denote spaces of square integrable and bounded functions of time for $t \geq 0$.

12.3.1 Adaptive Identifier Design

We propose the following gradient update law with normalization:

$$\dot{\hat{g}} = \gamma \frac{\hat{e}(0,t)v(0,t)}{1 + v^2(0,t)}, \tag{12.89}$$

where $\gamma > 0$ is the adaptation gain. The above identifier utilizes the prediction error

$$\hat{e}(x,t) = u(x,t) - \hat{g}v(x,t) - \eta(x,t), \tag{12.90}$$

evaluated at $x = 0$, and the filters

$$v_t(x,t) = v_{xx}(x,t) + u(0,t), \tag{12.91}$$

$$v_x(0,t) = 0, \tag{12.92}$$

$$v(1,t) = 0 \tag{12.93}$$

and

$$\eta_t(x,t) = \eta_{xx}(x,t), \tag{12.94}$$
$$\eta_x(0,t) = 0, \tag{12.95}$$
$$\eta(1,t) = U(t), \tag{12.96}$$

such that the boundary value problem

$$e_t(x,t) = e_{xx}(x,t), \tag{12.97}$$
$$e_x(0,t) = 0, \tag{12.98}$$
$$e(1,t) = 0, \tag{12.99}$$

written in terms of the error $e(x,t) = u(x,t) - gv(x,t) - \eta(x,t)$, is exponentially stable.

Lemma 12.3. *The adaptive law* (12.89) *guarantees the following properties*:

$$\frac{\hat{e}(0,t)}{\sqrt{1+v^2(0,t)}} \in \mathcal{L}_2 \cap \mathcal{L}_\infty,\ \tilde{g} \in \mathcal{L}_\infty,\ \dot{\hat{g}} \in \mathcal{L}_2 \cap \mathcal{L}_\infty. \tag{12.100}$$

Proof. Using the Lyapunov function

$$V = \frac{1}{2}\int_0^1 e^2(x,t)\,dx + \frac{1}{2\gamma}\tilde{g}^2 \tag{12.101}$$

we get

$$\begin{aligned}
\dot{V} &= -\int_0^1 e_x^2(x,t)\,dx - \frac{\tilde{g}\hat{e}(0,t)v(0,t)}{1+v^2(0,t)} \\
&\le -\int_0^1 e_x^2\,dx - \frac{\hat{e}^2(0,t)}{1+v^2(0,t)} + \frac{e(0,t)\hat{e}(0,t)}{1+v^2(0,t)} \\
&\le -\|e_x\|^2 - \frac{\hat{e}^2(0,t)}{1+v^2(0,t)} + \frac{\|e_x\||\hat{e}(0,t)|}{\sqrt{1+v^2(0,t)}} \\
&\le -\frac{1}{2}\|e_x\|^2 - \frac{1}{2}\frac{\hat{e}^2(0,t)}{1+v^2(0,t)}.
\end{aligned} \tag{12.102}$$

This gives the properties

$$\frac{\hat{e}(0,t)}{\sqrt{1+v^2(0,t)}} \in \mathcal{L}_2, \quad \tilde{g} \in \mathcal{L}_\infty. \tag{12.103}$$

Since

$$\frac{\hat{e}(0,t)}{\sqrt{1+v^2(0,t)}} = \frac{e(0,t)}{\sqrt{1+v^2(0,t)}} + \tilde{g}\frac{v(0,t)}{\sqrt{1+v^2(0,t)}} \tag{12.104}$$

and

$$\dot{\hat{g}} = \gamma\frac{\hat{e}(0,t)}{\sqrt{1+v^2(0,t)}}\frac{v(0,t)}{\sqrt{1+v^2(0,t)}}, \tag{12.105}$$

we get (12.100). □

The explicit bound on $\hat{g}$ in terms of initial conditions of all the signals can be obtained from (12.102):

$$\begin{aligned}\hat{g}^2(t) &= 2g^2 + 2\left(\tilde{g}(0)^2 + \gamma \int_0^1 e^2(x,0)\,dx\right)\\ &\leq 2g^2 + 2(g-\hat{g}(0))^2 + 2\gamma \int_0^1 (u(x,0) - gv(x,0) - \eta(x,0))^2\,dx.\end{aligned} \tag{12.106}$$

12.3.2 Stabilization and Identifiability

We propose the following boundary controller, which is a modified version of the controller (12.12):

$$U(t) = u_1 - \int_0^1 \sqrt{\hat{g}} \sinh\left(\sqrt{\hat{g}}(1-\xi)\right)(\hat{g}v(\xi,t) + \eta(\xi,t))\,d\xi. \tag{12.107}$$

A constant component $u_1 \neq 0$ makes the control signal sufficiently rich to persistently excite the system, yielding the desired parameter convergence.

Our main result is summarized as follows.

Theorem 12.3. *For any $\hat{g}(0)$ and any initial data $u_0, v_0, \eta_0 \in H^2(0,1)$ compatible with boundary conditions, the classical solution $(\hat{g}, u, v, \eta)$ of (12.86)–(12.88), (12.107), (12.91)–(12.96), (12.89) is bounded for all $x \in [0,1]$, $t \geq 0$ and*

$$\lim_{t\to\infty} \max_{x\in[0,1]} \left| u(x,t) - u_1\left(1 - g\frac{x^2}{2}\right)\right| = 0, \tag{12.108}$$

and, if in addition $u_1 \neq 0$, then

$$\lim_{t\to\infty} \hat{g}(t) = g. \tag{12.109}$$

Proof. Step 1 (Target system). Consider the transformation

$$w(x,t) = \hat{g}v(x,t) + \eta(x,t) + \int_0^x \sqrt{\hat{g}} \sinh\left(\sqrt{\hat{g}}(x-\xi)\right)(\hat{g}v(\xi,t) + \eta(\xi,t))\,d\xi. \tag{12.110}$$

Noting that $\hat{e}(0,t) = e(0,t) + \tilde{g}v(0,t)$, from Lemma 12.1 we know that w satisfies the following PDE:

$$\begin{aligned}w_t(x,t) &= w_{xx}(x,t) + \hat{g}\cosh\left(\sqrt{\hat{g}}x\right)\hat{e}(0,t) + \dot{\hat{g}}v(x,t)\\ &\quad + \dot{\hat{g}} \int_0^x \frac{\sinh\left(\sqrt{\hat{g}}(x-\xi)\right)}{\sqrt{\hat{g}}}\left(\hat{g}v(\xi,t) + w(\xi,t)\right)d\xi,\end{aligned} \tag{12.111}$$

$$w_x(0,t) = 0, \tag{12.112}$$

$$w(1,t) = u_1. \tag{12.113}$$

Using the fact that $u(0,t) = w(0,t) + \hat{e}(0,t)$, let us rewrite the filter (12.91)–(12.93) in the form

$$v_t(x,t) = v_{xx}(x,t) + w(0,t) + \hat{e}(0,t), \tag{12.114}$$
$$v_x(0,t) = 0, \tag{12.115}$$
$$v(1,t) = 0. \tag{12.116}$$

We now have two interconnected systems (12.111)–(12.116) for w and v, which are driven by three external signals: a constant u_1 and signals $\hat{e}(0,t)$, $\dot{\hat{g}}(t)$ with properties (12.100).

Step 2 (Boundedness). Our next goal is to demonstrate that the v-system and w-system are asymptotically stable around the limit points

$$w^{lp}(x) = u_1, \qquad v^{lp}(x) = u_1 \frac{1-x^2}{2}. \tag{12.117}$$

Let us introduce the error variables $\bar{w} = w - w^{lp}$, $\bar{v} = v - v^{lp}$. The equations for $\bar{w}$ and $\bar{v}$ are

$$\begin{aligned}
\bar{w}_t(x,t) = {} & \bar{w}_{xx}(x,t) + \hat{g}\cosh\left(\sqrt{\hat{g}}x\right)\hat{e}(0,t) + \dot{\hat{g}}\bar{v}(x,t) \\
& + \dot{\hat{g}} \int_0^x \frac{\sinh\left(\sqrt{\hat{g}}(x-\xi)\right)}{\sqrt{\hat{g}}} \left(\hat{g}\bar{v}(\xi,t) + \bar{w}(\xi,t)\right) d\xi, \\
& + \frac{u_1}{2}\dot{\hat{g}}\cosh(\sqrt{\hat{g}}x),
\end{aligned} \tag{12.118}$$
$$\bar{w}_x(0,t) = 0, \tag{12.119}$$
$$\bar{w}(1,t) = 0, \tag{12.120}$$

and

$$\bar{v}_t(x,t) = \bar{v}_{xx}(x,t) + \bar{w}(0,t) + \hat{e}(0,t), \tag{12.121}$$
$$\bar{v}_x(0,t) = 0, \tag{12.122}$$
$$\bar{v}(1,t) = 0. \tag{12.123}$$

Consider the Lyapunov function

$$V_1 = \frac{1}{2}\int_0^1 \bar{v}^2(x,t)\,dx + \frac{1}{2}\int_0^1 \bar{v}_x^2(x,t)\,dx. \tag{12.124}$$

Using Young's, Poincaré's, and Agmon's inequalities, we have[2]

$$
\begin{aligned}
\dot{V}_1 =& -\int_0^1 \bar{v}_x^2(x)\,dx + (\bar{w}(0)+\hat{e}(0))\int_0^1 \bar{v}(x)\,dx \\
& -\int_0^1 \bar{v}_{xx}^2(x)\,dx - (\bar{w}(0)+\hat{e}(0))\int_0^1 \bar{v}_{xx}(x)\,dx \\
\le& -\|\bar{v}_x\|^2 + \frac{1}{8}\|\bar{v}\|^2 + 4\frac{\hat{e}^2(0)}{1+v^2(0)}(1+v^2(0)) \\
& +4\|\bar{w}_x\|^2 - \|\bar{v}_{xx}\|^2 + \frac{1}{2}\|\bar{v}_{xx}\|^2 + \|\bar{w}_x\|^2 + \frac{\bar{e}^2(0)}{1+v^2(0)}(1+v^2(0)) \\
\le& -\frac{1}{2}\|\bar{v}_x\|^2 - \frac{1}{2}\|\bar{v}_{xx}\|^2 + 5\|\bar{w}_x\|^2 + 5\frac{\bar{e}^2(0)}{1+v^2(0)}\left(1+2\|\bar{v}_x\|^2+\frac{u_1^2}{2}\right) \\
\le& -\frac{1}{2}\|\bar{v}_x\|^2 - \frac{1}{2}\|\bar{v}_{xx}\|^2 + 5\|\bar{w}_x\|^2 + l_1\|\bar{v}_x\|^2 + l_1,
\end{aligned}
\tag{12.125}
$$

where by l_1 we denote a generic function of time in $\mathcal{L}_1$.

Using the following Lyapunov function for the $\bar{w}$-system,

$$
V_2 = \frac{1}{2}\int_0^1 \bar{w}^2(x)\,dx\,, \tag{12.126}
$$

we get

$$
\begin{aligned}
\dot{V}_2 =& -\int_0^1 \bar{w}_x^2\,dx + \hat{e}(0)\int_0^1 \hat{g}\cosh\left(\sqrt{\hat{g}}x\right)\bar{w}(x)\,dx \\
& +\dot{\hat{g}}\int_0^1 \bar{w}(x)(\bar{v}(x)+v^{lp}(x))\,dx \\
& +\dot{\hat{g}}\int_0^1 \bar{w}(x)\int_0^x \frac{\sinh\left(\sqrt{\hat{g}}(x-\xi)\right)}{\sqrt{\hat{g}}}(\hat{g}v^{lp}(y)+u_1+\hat{g}\bar{v}(y)+\bar{w}(y))\,dy\,dx.
\end{aligned}
\tag{12.127}
$$

Before we proceed, we note that $\tilde{g}$ is bounded and therefore $\hat{g}$ is also bounded; let us denote this bound by $\bar{g}$. With the help of Young's, Poincaré's, and Agmon's

[2]We drop the dependence on time in the proofs below to simplify the notation, that is, $v(0) \equiv v(0,t)$, and so on.

inequalities, we get the following estimate:

$$\begin{aligned}\dot{V}_2 \leq & -\|\bar{w}_x\|^2 + \frac{c_1}{2}\|\bar{w}\|^2 + \frac{|\dot{\hat{g}}|^2}{8c_1}u_1^2\cosh^2(\sqrt{\bar{g}}) \\ & + c_1\|\bar{w}\|^2 + \frac{|\dot{\hat{g}}|^2(1+\sinh(\sqrt{\bar{g}}))^2}{2c_1}\|\bar{v}\|^2 \\ & + \frac{\bar{g}^2\cosh^2(\sqrt{\bar{g}})}{2c_1}\frac{\hat{e}^2(0)}{1+v^2(0)}\left(1+\|\bar{v}_x\|^2+u_1^2\right) \\ & + c_1\|\bar{w}\|^2 + \frac{|\dot{\hat{g}}|^2\sinh^2(\sqrt{\bar{g}})}{2c_1\bar{g}}\|\bar{w}\|^2 \\ \leq & -(1-10c_1)\|\bar{w}_x\|^2 + l_1\|\bar{w}\|^2 + l_1\|\bar{v}_x\|^2 + l_1. \end{aligned} \tag{12.128}$$

Choosing $c_1 = 1/40$ and using the Lyapunov function $V = V_2 + (1/20)V_1$, we get

$$\begin{aligned}\dot{V} \leq & -\frac{1}{2}\|\bar{w}_x\|^2 - \frac{1}{40}\|\bar{v}_x\|^2 - \frac{1}{40}\|\bar{v}_{xx}\|^2 + l_1\|\bar{w}\|^2 + l_1\|\bar{v}_x\|^2 + l_1 \\ \leq & -\frac{1}{4}V + l_1 V + l_1, \end{aligned} \tag{12.129}$$

and by Lemma D.3 we obtain the boundedness and square integrability of $\|\bar{w}\|$, $\|\bar{v}\|$, and $\|\bar{v}_x\|$. Using these properties we can compute

$$\begin{aligned}\frac{1}{2}\frac{d}{dt}\|\bar{w}_x\|^2 \leq & -\|\bar{w}_{xx}\|^2 + \bar{g}\cosh^2(\sqrt{\bar{g}})|\hat{e}(0)|\|\bar{w}_{xx}\| \\ & + |\dot{\hat{g}}|\|\bar{w}_{xx}\|(1+\sqrt{\bar{g}}\sinh(\sqrt{\bar{g}}))\|\bar{v}\| \\ & + |\dot{\hat{g}}|\|\bar{w}_{xx}\|\left[\frac{\sinh(\sqrt{\bar{g}})}{\bar{g}}\|\bar{w}\| + |u_1|\frac{\cosh^2(\sqrt{\bar{g}})}{2}\right] \\ \leq & -\frac{1}{8}\|\bar{w}_x\|^2 + l_1, \end{aligned} \tag{12.130}$$

so that by Lemma D.3, $\|\bar{w}_x\| \in \mathcal{L}_2 \cap \mathcal{L}_\infty$. Using the fact that $\|\bar{v}_x\|$, $\|\bar{w}_x\|$ are bounded, it is easy to see that

$$\left|\frac{d}{dt}(\|\bar{v}\|^2 + \|\bar{w}\|^2)\right| < \infty. \tag{12.131}$$

By Lemma D.1 we get $\|\bar{w}\| \to 0$, $\|\bar{v}\| \to 0$. Using Agmon's inequality we get that $\bar{v}(x,t)$ is uniformly bounded and is regulated to zero as $t \to \infty$. By the same argument we get the boundedness and regulation of $\bar{w}(x,t)$. Thus, we have proved that the v-system and w-system are bounded and globally asymptotically stable around the limit points $v^{lp}(x)$ and $w^{lp}(x)$, respectively.

In order to show the boundedness of $\eta(x,t)$ and $u(x,t)$, we express η in terms of v and w with the inverse transformation to (12.110):

$$\hat{g}v(x,t) + \eta(x,t) = w(x,t) - \hat{g}\int_0^x (x-\xi)w(\xi,t)\,d\xi. \tag{12.132}$$

Since $v(x,t)$ and $w(x,t)$ are bounded, we see from (12.132) that $\eta(x,t)$ is also bounded. Finally, the boundedness of $u(x,t)$ is obtained from the relationship $u = e + gv + \eta$.

Step 3 (Parameter convergence and regulation). We showed that

$$\lim_{t\to\infty} v(0,t) = v^{lp}(0) = \frac{u_1}{2} \neq 0. \tag{12.133}$$

Thus, the update law (12.89), being rewritten in the form

$$\dot{\tilde{g}} = -\gamma \frac{(e(0,t) + \tilde{g}v(0,t))v(0,t)}{1 + v^2(0,t)} \tag{12.134}$$

and coupled to the parabolic system (12.97)–(12.99), turns out to be asymptotically autonomous. To complete the proof it remains to apply the invariance principle to the system (12.97)–(12.99), (12.134) and note that due to (12.133), the convergence of the solution to the maximal invariant set $\tilde{g}(t)v(0,t) = 0$ results in $\lim_{t\to\infty} \tilde{g}(t) = 0$.

Since $w(x,t) \to u_1$, from the inverse transformation (12.132) we get $\hat{g}v(x,t) + \eta(x,t) \to u_1 - gu_1x^2/2$. Since $u(x,t) \to gv(x,t) + \eta(x,t)$, we get (12.108). The proof of Theorem 12.3 is completed. □

12.4 SWAPPING DESIGN—PLANT WITH UNKNOWN PARAMETER IN THE BOUNDARY CONDITION

Consider the plant

$$u_t(x,t) = u_{xx}(x,t), \tag{12.135}$$

$$u_x(0,t) = -qu(0,t), \tag{12.136}$$

$$u(1,t) = U(t). \tag{12.137}$$

We start the adaptive design by employing the least squares adaptive identifier.

12.4.1 Least Squares Adaptive Identifier

First, we introduce input and output filters:

$$v_t(x,t) = v_{xx}(x,t), \tag{12.138}$$

$$v_x(0,t) = -u(0,t), \tag{12.139}$$

$$v(1,t) = 0, \tag{12.140}$$

$$\eta_t(x,t) = \eta_{xx}(x,t), \tag{12.141}$$

$$\eta_x(0,t) = 0, \tag{12.142}$$

$$\eta(1,t) = U(t). \tag{12.143}$$

The error $e = u - qv - \eta$ satisfies the exponentially stable heat equation (12.97)–(12.99). Let us define the prediction error as

$$\hat{e}(x,t) = u(x,t) - \hat{q}v(x,t) - \eta(x,t). \tag{12.144}$$

The least squares update law with normalization is

$$\dot{\hat{q}}(t) = \gamma(t)\frac{\hat{e}(0,t)v(0,t)}{1+\gamma(t)v^2(0,t)}, \tag{12.145}$$

$$\dot{\gamma}(t) = -\frac{\gamma^2(t)v^2(0,t)}{1+\gamma(t)v^2(0,t)}, \quad \gamma(0) > 0. \tag{12.146}$$

Lemma 12.4. *The identifier* (12.138)–(12.146) *guarantees the following*:

(i) $0 < \gamma(t) < \infty$, $|\dot{\gamma}(t)| < \infty$ *for all* $t \geq 0$.

(ii) $\tilde{q}(t)$ *is bounded.*

(iii) $\frac{\hat{e}(0)}{\sqrt{1+v^2(0,t)}} \in \mathcal{L}_2 \cap \mathcal{L}_\infty$ *and* $\dot{\hat{q}} \in \mathcal{L}_2 \cap \mathcal{L}_\infty$.

(iv) There exist γ_∞, q_∞ *such that*

$$\lim_{t\to\infty} \gamma(t) = \gamma_\infty, \quad \lim_{t\to\infty} q(t) = q_\infty. \tag{12.147}$$

Proof. : (i) Rewriting (12.146) as

$$\frac{d}{dt}(\gamma(t)^{-1}) = \frac{\gamma(t)^{-1}v^2(0,t)}{\gamma(t)^{-1}+v^2(0,t)}, \tag{12.148}$$

we can see that $\gamma(t)^{-1} \geq \gamma(0)^{-1} > 0$. Therefore, $\gamma(t)$ is bounded and positive. From (12.146) we get $|\dot{\gamma}(t)| \leq |\gamma(t)| < \infty$ for all $t \geq 0$.

(ii) Consider the Lyapunov function

$$V = \frac{1}{2}\|e\|^2 + \frac{1}{2\gamma(t)}\tilde{q}^2, \tag{12.149}$$

where $\tilde{q} = q - \hat{q}$. We get

$$\begin{aligned}
\dot{V} &= -\|e_x\|^2 + \frac{1}{2}\frac{\tilde{q}^2 v^2(0)}{1+\gamma v^2(0)} - \frac{\tilde{q}v(0)\hat{e}(0)}{1+\gamma v^2(0)} \\
&= -\|e_x\|^2 + \frac{1}{2}\frac{\tilde{q}v(0)}{1+\gamma v^2(0)}(\tilde{q}v(0) - 2\hat{e}(0)) \\
&= -\|e_x\|^2 + \frac{1}{2}\frac{(\hat{e}(0)-e(0))(-e(0)-\hat{e}(0))}{1+\gamma v^2(0)} \\
&= -\|e_x\|^2 + \frac{1}{2}\frac{e^2(0)-\hat{e}^2(0)}{1+\gamma v^2(0)} \\
&\leq -\frac{1}{2}\|e_x\|^2 - \frac{1}{2}\frac{\hat{e}^2(0)}{1+\gamma v^2(0)}.
\end{aligned} \tag{12.150}$$

Therefore, V is bounded, which in turn implies that $\tilde{q}$ is bounded.

(iii) Integrating (12.150) in time, we get $\frac{\hat{e}(0)}{\sqrt{1+\gamma(t)v^2(0,t)}} \in \mathcal{L}_2$. We also get

$$\frac{\hat{e}(0,t)}{\sqrt{1+v^2(0,t)}} = \frac{\hat{e}(0,t)}{\sqrt{1+\gamma(t)v^2(0,t)}} \frac{\sqrt{1+\gamma(t)v^2(0,t)}}{\sqrt{1+v^2(0,t)}}$$

$$\leq \frac{\hat{e}(0,t)\sqrt{1+\gamma(0)}}{\sqrt{1+\gamma(t)v^2(0,t)}} \in \mathcal{L}_2 \tag{12.151}$$

and

$$\frac{\hat{e}(0,t)}{\sqrt{1+v^2(0,t)}} = \frac{e(0,t)}{\sqrt{1+v^2(0,t)}} + \frac{\tilde{q}v(0,t)}{\sqrt{1+v^2(0,t)}} < \infty. \tag{12.152}$$

From (12.145) we have

$$\dot{\hat{q}} = \frac{\hat{e}(0)}{\sqrt{1+\gamma(t)v^2(0,t)}} \frac{\gamma(t)v(0,t)}{\sqrt{1+\gamma(t)v^2(0,t)}}$$

$$\leq \frac{\sqrt{\gamma(t)}\hat{e}(0)}{\sqrt{1+\gamma(t)v^2(0,t)}} \in \mathcal{L}_2 \cap \mathcal{L}_\infty. \tag{12.153}$$

(iv) Since $\gamma(t)$ is monotonically decreasing and is bounded from below, it has a limit: $\lim_{t\to\infty} \gamma(t) = \gamma_\infty$. We rewrite (12.145) as

$$\dot{\tilde{q}} = -\gamma \frac{v(0,t)(e(0,t) + \tilde{q}v(0,t))}{1+\gamma v^2(0,t)}$$

$$= -\gamma \frac{v(0,t)e(0,t)}{1+\gamma v^2(0,t)} + \tilde{q}\frac{\dot{\gamma}}{\gamma}. \tag{12.154}$$

The solution to this ODE is

$$\tilde{q}(t) = \frac{\tilde{q}(0)}{\gamma(0)}\gamma(t) - \gamma(t)\int_0^t \frac{v(0,\tau)e(0,\tau)}{1+\gamma(\tau)v^2(0,\tau)}\,d\tau. \tag{12.155}$$

Therefore,

$$\lim_{t\to\infty} \hat{q}(t) = q - \frac{\tilde{q}(0)}{\gamma(0)}\gamma_\infty + \gamma_\infty \int_0^\infty \frac{v(0,\tau)e(0,\tau)}{1+\gamma(\tau)v^2(0,\tau)}\,d\tau = q_\infty, \tag{12.156}$$

since the integral in (12.156) is obviously bounded. □

12.4.2 Stabilization and Identifiability

Our main result is stated in the following theorem.

Theorem 12.4. *For any $\hat{q}(0)$ and any initial data $u_0, v_0, \eta_0 \in H^2(0,1)$ compatible with boundary conditions, the classical solution $(\hat{q}, u, v, \eta)$ of the closed-loop system consisting of the plant* (12.135)–(12.137), *the controller*

$$U(t) = u_1 - \int_0^1 \hat{q}e^{\hat{q}(1-\xi)}(\hat{q}v(\xi,t) + \eta(\xi,t))\,d\xi\,, \tag{12.157}$$

the update law (12.145)–(12.146)*, and the filters* (12.138)–(12.143) *is bounded for all* $x \in [0, 1]$, $t \geq 0$ *and*

$$\lim_{t\to\infty} \sup_{x\in[0,1]} \left|u(x,t) - u_1\left(1 - \hat{q}x\right)\right| = 0, \tag{12.158}$$

and, if in addition $u_1 \neq 0$*, then*

$$\lim_{t\to\infty} \hat{q}(t) = q. \tag{12.159}$$

Proof. Step 1 (Target system). Let us introduce the transformation

$$w(x,t) = \hat{q}v(x,t) + \eta(x,t) + \int_0^x \hat{q}e^{\hat{q}(x-\xi)}(\hat{q}v(\xi,t) + \eta(\xi,t))\,d\xi, \tag{12.160}$$

where $\hat{q}v(x,t) + \eta(x,t)$ is the estimate of the state u. Using Lemma 12.2 and the relation $\hat{e}(0,t) = e(0,t) + \tilde{q}v(0,t)$, we obtain the following PDE for w:

$$w_t(x,t) = w_{xx}(x,t) + \hat{q}^2 e^{\hat{q}x}\hat{e}(0,t) + \dot{\hat{q}}v(x,t) + \dot{\hat{q}}\int_0^x e^{\hat{q}(x-\xi)}(\hat{q}v(\xi,t) + w(\xi,t))\,d\xi, \tag{12.161}$$

$$w_x(0,t) = -\hat{q}\hat{e}(0,t), \tag{12.162}$$

$$w(1,t) = u_1. \tag{12.163}$$

Noting that $u(0,t) = w(0,t) + \hat{e}(0,t)$, we rewrite the v-filter as

$$v_t(x,t) = v_{xx}(x,t), \tag{12.164}$$

$$v_x(0,t) = -w(0,t) - \hat{e}(0,t), \tag{12.165}$$

$$v(1,t) = 0. \tag{12.166}$$

We now have two interconnected systems for w and v by the constant boundary input u_1 and by the signal $\hat{e}(0,t)$ with properties established in Lemma 12.4.
Step 2 (L^2 boundedness). Our next goal is to demonstrate that these systems are exponentially stable around the limit points

$$w^{lp}(x) = u_1, \qquad v^{lp}(x) = u_1(1-x). \tag{12.167}$$

Introducing the error variables $\bar{w} = w - w^{lp}$, $\bar{v} = v - v^{lp}$, we get

$$\bar{w}_t(x,t) = \bar{w}_{xx}(x,t) + \hat{q}^2 e^{\hat{q}x}\hat{e}(0,t) + \dot{\hat{q}}\left[\bar{v}(x,t) + u_1 e^{\hat{q}x} + \int_0^x e^{\hat{q}(x-\xi)}(\hat{q}\bar{v}(\xi,t) + \bar{w}(\xi,t))\,d\xi\right], \tag{12.168}$$

$$\bar{w}_x(0,t) = -\hat{q}\hat{e}(0,t), \tag{12.169}$$

$$\bar{w}(1,t) = 0, \tag{12.170}$$

$$\bar{v}_t(x,t) = \bar{v}_{xx}(x,t), \tag{12.171}$$

$$\bar{v}_x(0,t) = -\bar{w}(0,t) - \hat{e}(0,t), \tag{12.172}$$

$$\bar{v}(1,t) = 0. \tag{12.173}$$

Consider the Lyapunov function

$$V=\frac{1}{2}\int_0^1 \bar{w}^2(x)\,dx+\frac{1}{2}\int_0^1 \bar{v}^2(x)\,dx. \tag{12.174}$$

We get

$$\begin{aligned}
\dot{V}=&\hat{q}\bar{w}(0)\hat{e}(0)-\int_0^1 \bar{w}_x^2(x)\,dx+\dot{\hat{q}}\int_0^1 \bar{w}(x)\bar{v}(x)\,dx+\hat{e}(0)\int_0^1 \hat{q}^2e^{\hat{q}x}\bar{w}(x)\,dx\\
&-\int_0^1 \bar{v}_x^2(x)\,dx+\dot{\hat{q}}\int_0^1 \bar{w}(x)\int_0^x e^{\hat{q}(x-\xi)}(\hat{q}\bar{v}(\xi)+\bar{w}(\xi))\,d\xi\,dx\\
&+\bar{v}(0)(\bar{w}(0)+\hat{e}(0))+\dot{\hat{q}}u_1\int_0^1 e^{\hat{q}x}\bar{w}(x)\,dx\\
\le&-\|\bar{w}_x\|^2+|\hat{e}(0)|(\bar{q}|\bar{w}(0)|+\bar{q}^2e^{\bar{q}}\|\bar{w}\|)+\frac{(1+\bar{q}e^{\bar{q}})^2|\dot{\hat{q}}|^2}{c_1}\|\bar{v}\|^2\\
&+\frac{e^{2\bar{q}}|\dot{\hat{q}}|^2}{c_1}\|\bar{w}\|^2+\frac{c_1}{2}\|\bar{w}\|^2-\|\bar{v}_x\|^2+\frac{1}{2}\|\bar{v}_x\|^2\\
&+\frac{1}{2}\|\bar{w}_x\|^2+|\bar{v}(0)||\hat{e}(0)|+\frac{c_1}{2}\|\bar{w}\|^2+\frac{|\dot{\hat{q}}|^2u_1^2e^{2\bar{q}}}{2c_1}\\
\le&-\left(\frac{1}{2}-4c_1\right)\|\bar{w}_x\|^2-\frac{1}{2}\|\bar{v}_x\|^2+l_1\|\bar{w}\|^2+l_1\|\bar{v}\|^2+l_1\\
&+\bar{q}|\hat{e}(0)||\bar{w}(0)|+\bar{q}^2e^{\bar{q}}|\hat{e}(0)|\|\bar{w}\|+|\bar{v}(0)||\hat{e}(0)|.
\end{aligned} \tag{12.175}$$

Here by $\bar{q}$ and $\bar{\gamma}$ we denoted the bounds on $\hat{q}$ and γ, respectively. We now separately estimate the last three terms of (12.175) using Poincaré's, Agmon's, and Young's inequalities:

$$\begin{aligned}
\bar{q}|\hat{e}(0)||\bar{w}(0)|\le&\,\bar{q}|\bar{w}(0)|\frac{\hat{e}(0)}{\sqrt{1+v^2(0)}}(1+|u_1|+|\bar{v}(0)|)\\
\le&\,c_2\|\bar{w}_x\|^2+\frac{\bar{q}^2(1+|u_1|)^2}{4c_2}\frac{\hat{e}^2(0)}{1+v^2(0)}\\
&+2\bar{q}\sqrt{\|\bar{w}\|\|\bar{w}_x\|\|\bar{v}\|\|\bar{v}_x\|}\frac{|\hat{e}(0)|}{\sqrt{1+v^2(0)}}\\
\le&\,c_2\|\bar{w}_x\|^2+l_1+\frac{\bar{q}|\hat{e}(0)|}{\sqrt{1+v^2(0)}}(\|\bar{w}\|\|\bar{w}_x\|+\|\bar{v}\|\|\bar{v}_x\|)\\
\le&\,c_2\|\bar{w}_x\|^2+l_1+c_3\|\bar{w}_x\|^2+c_4\|\bar{v}_x\|^2\\
&+\bar{q}^2\frac{\hat{e}^2(0)}{1+v^2(0)}\left(\frac{\|\bar{v}\|^2}{4c_3}+\frac{\|\bar{w}\|^2}{4c_4}\right)\\
\le&\,c_2\|\bar{w}_x\|^2+c_3\|\bar{w}_x\|^2+c_4\|\bar{v}_x\|^2+l_1\|\bar{v}\|^2+l_1\|\bar{w}\|^2+l_1,
\end{aligned} \tag{12.176}$$

$$\begin{aligned}\bar{q}^2 e^{\bar{q}}|\hat{e}(0)|\|\bar{w}\| &\leq \bar{q}^2 e^{\bar{q}}\|\bar{w}\|\frac{\hat{e}(0)}{\sqrt{1+v^2(0)}}(1+|u_1|+|v(0)|)\\ &\leq c_5\|\bar{w}\|^2+\frac{\bar{q}^4 e^{2\bar{q}}(1+|u_1|)^2}{4}\frac{\hat{e}^2(0)}{1+v^2(0)}\left(\frac{1}{c_5}+\frac{1}{c_6}\|\bar{w}\|^2\right)+c_6\|\bar{v}_x\|^2\\ &\leq c_5\|\bar{w}\|^2+c_6\|\bar{v}_x\|^2+l_1\|\bar{w}\|^2+l_1,\end{aligned} \tag{12.177}$$

and

$$\begin{aligned}|\bar{v}(0)||\hat{e}(0)| \leq& \frac{|\bar{v}(0)||\hat{e}(0)|}{1+v^2(0)}(1+2u_1^2+4\|\bar{v}\|\|\bar{v}_x\|)\\ \leq& \frac{c_7}{2}\|\bar{v}_x\|^2+\frac{(1+2u_1^2)^2}{2c_7}\frac{\hat{e}^2(0)}{1+v^2(0)}+\frac{8}{c_7}\left(\frac{|\bar{v}(0)||\hat{e}(0)|}{1+v^2(0)}\right)^2\|\bar{v}\|^2\\ &+\frac{c_7}{2}\|\bar{v}_x\|^2\\ \leq& c_7\|\bar{v}_x\|^2+l_1\|\bar{v}\|^2+l_1\,.\end{aligned} \tag{12.178}$$

Substituting (12.176), (12.177), and (12.178) into (12.175), we get

$$\begin{aligned}\dot{V} \leq& -\left(\frac{1}{2}-4c_1-c_2-c_3-4c_5\right)\|\bar{w}_x\|^2+l_1\|\bar{w}\|^2\\ &-\left(\frac{1}{2}-c_4-c_6-c_7\right)\|\bar{v}_x\|^2+l_1\|\bar{v}\|^2+l_1.\end{aligned} \tag{12.179}$$

Choosing $4c_1=c_2=c_3=4c_5=1/16$, $c_4=c_6=c_7=1/12$, we get

$$\dot{V} \leq -\frac{1}{8}V+l_1V+l_1, \tag{12.180}$$

and by Lemma D.3 we obtain $\|\bar{w}\|, \|\bar{v}\| \in \mathcal{L}_2 \cap \mathcal{L}_\infty$. By integrating (12.179) we also get $\|\bar{w}_x\|, \|\bar{v}_x\| \in \mathcal{L}_2$.

Step 3 (Parameter convergence). To show the parameter convergence, we note that

$$|\hat{e}(0,t)| \leq \frac{\hat{e}(0,t)}{\sqrt{1+v(0,t)^2}}(1+\|\bar{v}_x\|+|u_1|) \in \mathcal{L}_2, \tag{12.181}$$

because $\|\bar{v}_x\| \in \mathcal{L}_2$ and $\frac{\hat{e}(0,t)}{\sqrt{1+v(0,t)^2}} \in \mathcal{L}_2 \cap \mathcal{L}_\infty$ by Lemma 12.4. Using the definition of $\hat{e}$, we write

$$\hat{e}(0,t)=e(0,t)+\tilde{q}v(0,t)=e(0,t)+\tilde{q}\bar{v}(0,t)+\tilde{q}u_1, \tag{12.182}$$

from which we have for $u_1 \neq 0$:

$$|\tilde{q}| \leq \frac{|\hat{e}(0,t)|+|e(0,t)|+|\tilde{q}|\|\bar{v}_x\|}{|u_1|}. \tag{12.183}$$

Since $\tilde{q}$ is bounded and $\hat{e}(0,t)$, $e(0,t)$, and $\|\bar{v}_x\|$ are all square integrable, from (12.183) we conclude that $\tilde{q}$ is square integrable. By Lemma 12.4, $\tilde{q}$ and $\dot{\tilde{q}}$ are bounded, and therefore by Lemma D.1 we have $\lim_{t\to\infty}\tilde{q}(t)=0$.

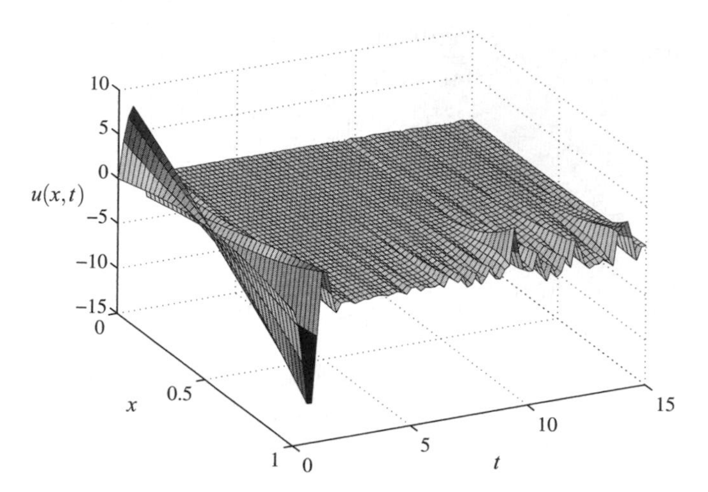

Figure 12.1 The state $u(x, t)$ for the plant (12.44)–(12.46) with the Lyapunov adaptive output-feedback controller (12.49)–(12.54).

Step 4 (Pointwise boundedness and regulation). We have already shown boundedness in the L^2-norm. Unlike in the identification proof for the g-plant, here it is considerably harder to show spatially uniform boundedness. The main difficulty is the presence of nonhomogeneous terms in the boundary conditions (12.169) and (12.172), which prevents us from using the H^1-norm as the standard Lyapunov function. To remove this difficulty, we use the approach introduced in Section 12.2 and consider the following transformation from $(\bar{w}, \bar{v})$ into $(\breve{w}, \breve{v})$:

$$\breve{w}(x, t) = \bar{w}(x, t) + \hat{q}(1 - x) \int_0^x \hat{e}(y, t)\, dy, \tag{12.184}$$

$$\breve{v}(x, t) = \bar{v}(x, t) + (1 - x) \int_0^x (\hat{e}(y, t) + \bar{w}(y, t))\, dy. \tag{12.185}$$

One can easily check that in the new variables we have $\breve{w}_x(0, t) = \breve{w}(1, t) = \breve{v}_x(0, t) = \breve{v}(1, t) = 0$. The right-hand side of the resulting PDEs for $\breve{w}$ and $\breve{v}$ is quite complicated but has a simple structure, with all the terms proportional either to $\dot{\hat{q}}$, $\hat{e}(0)$, or $\tilde{q}$, all of which are square integrable and bounded, as we have shown. After this crucial step, the rest of the proof closely follows the proof for the g-case. One first shows the boundedness of $\|\breve{w}_x\|$ and $\|\breve{v}_x\|$ with the Lyapunov function

$$V = \frac{1}{2}\left(\|\breve{w}\|^2 + \|\breve{v}\|^2 + \|\breve{w}_x\|^2 + \|\breve{v}_x\|^2\right). \tag{12.186}$$

By Agmon's inequality, this implies the boundedness of $\breve{w}(x, t)$ and $\breve{v}(x, t)$ for all $x \in [0, 1]$ and $t \geq 0$. Then from the transformation (12.184), (12.185) it follows that $\bar{w}$ and $\bar{v}$ are bounded (after substitution $\hat{e} = e + \tilde{q}\bar{v} + \tilde{q}u_1$). From the invertible transformation (12.160) it then follows that η is bounded. Since $u = e + qv + \eta$, we get the boundedness of $u(x, t)$ for all $x \in [0, 1]$ and $t \geq 0$.

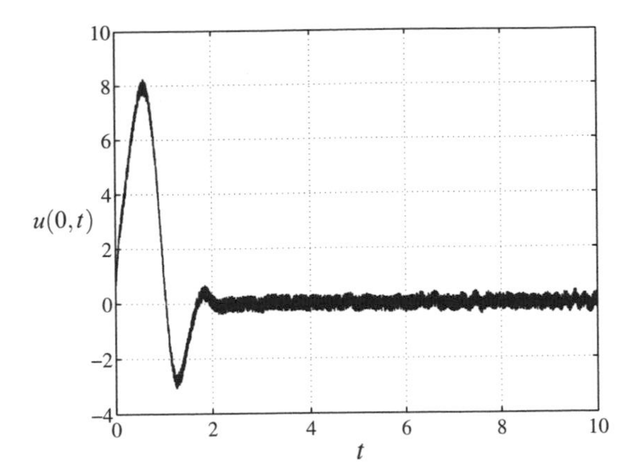

Figure 12.2 The output corrupted by noise for the plant (12.44)–(12.46) with the Lyapunov-based adaptive scheme.

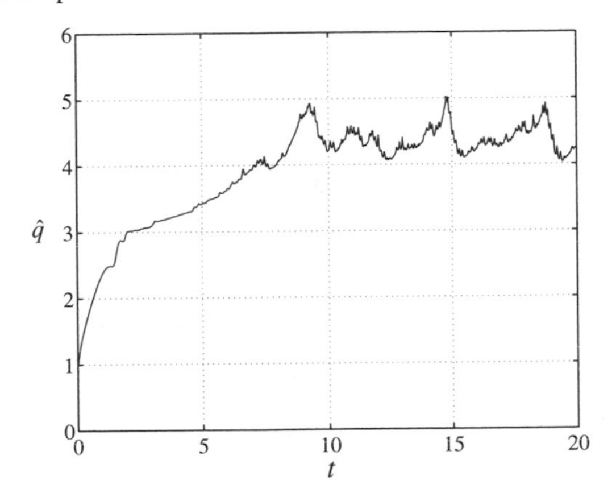

Figure 12.3 The parameter estimate for the plant (12.44)–(12.46) with the Lyapunov-based adaptive scheme.

Finally, since $w(x,t) \to u_1$ as $t \to \infty$, from the inverse transformation (12.63) we have $\hat{q}v(x,t)+\eta(x,t) \to u_1 - u_1\hat{q}x$ as $t \to \infty$. Because $u(x,t) \to qv(x,t)+\eta(x,t)$, we get (12.158). The proof of Theorem 12.4 is completed. □

12.5 SIMULATIONS

In this section we present the results of closed-loop simulations for both schemes developed in this chapter.

First we consider the Lyapunov design for the plant with parametric uncertainty in the boundary condition (12.44)–(12.46). The plant parameter is set to $q = 2$, which gives the unstable eigenvalue $\approx$3.6. The initial estimate is $\hat{q}(0) = 1$, and adaptation and normalization gains are $\gamma = 2.2$ and $a = 50$. We assume that the sensor measurements are noisy. The simulation results are presented in Figures 12.1–12.3. In Figure 12.1 (state response), we can see that although

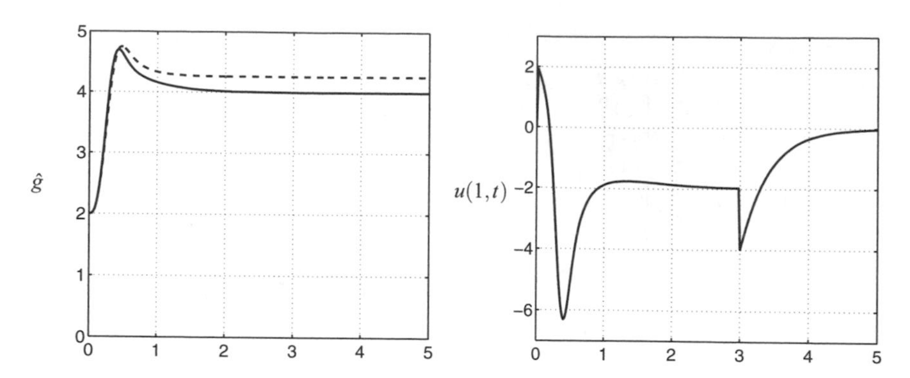

Figure 12.4 Simulation results for the plant (12.86)–(12.88) with the swapping-based adaptive scheme. Left: convergence of the parameter estimate $\hat{g}$ to the true value $g = 4$ with (solid line) and without (dashed line) additional constant boundary input. Right: the control effort; the additional constant input is turned off at $t = 3$.

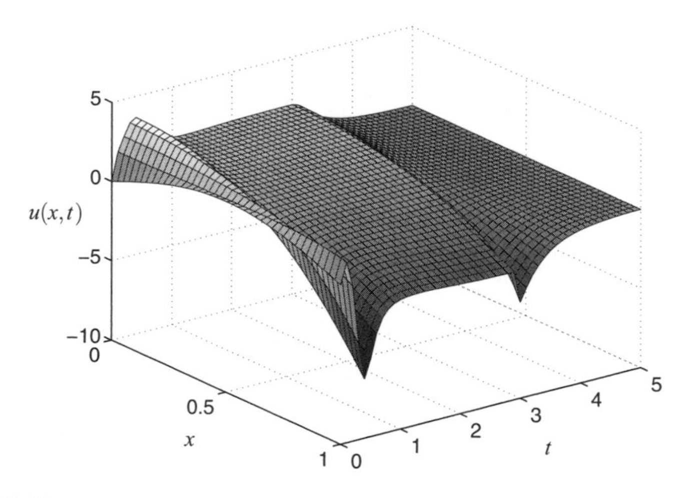

Figure 12.5 The closed-loop state $u(x, t)$ for the plant (12.86)–(12.88) with the swapping-based adaptive scheme.

instability occurs at the $x = 0$ boundary, the system is successfully regulated to zero by the control from the opposite boundary. The output corrupted by sensor noise is shown in Figure 12.2. The parameter estimate is shown in Figure 12.3. Because of noise and the lack of persistency of excitation, the identifier overestimates the parameter. The parameter estimate fluctuates (owing to the noise, there is no convergence to a constant value), but does not seem to drift.

The second simulation illustrates the swapping-based adaptive identification for the plant with parametric uncertainty in the domain (12.86)–(12.88). The unknown parameter is set to $g = 4$, with unstable eigenvalue ≈ 2.4. The initial estimate of the parameter is $\hat{g}(0) = 2$. A constant boundary input $u_1 = 2$ is added to the feedback. The simulation results are presented in Figures 12.4 and 12.5.

In Figure 12.4 the evolution of the parameter estimate is shown in comparison with the case $u_1 = 0$. We can see that the unknown parameter is successfully identified. In Figure 12.5 the closed-loop state is shown. Regulation to zero is achieved by turning off the additional constant input at $t = 3$, when the parameter estimate stops changing.

12.6 NOTES AND REFERENCES

The material presented in this chapter is based on [69], [112], and [113].

Both the Lyapunov and the swapping-based adaptive designs presented in this chapter are the first results reported in the literature for PDE systems of relative degree greater than one. Results for non-identifier-based adaptive controllers for infinite-dimensional systems of relative degree one are surveyed by Logemann and Townley [84].

The problem of parametric identification for a large class of distributed parameter systems has been considered in [9, 94] under the assumption that distributed actuation and measurements are available. Problems with boundary actuation have also been addressed in [94]; however, only open-loop identification has been considered, thus limiting the results to stable plants. In this chapter we have identified the parameters of open-loop unstable, boundary-actuated plants.

Chapter Thirteen

Output Feedback for PDEs with Spatially Varying Coefficients

In this chapter we consider the problem of output-feedback stabilization of reaction-advection-diffusion systems with uncertain parameters. We assume that sensing and actuation are performed at the opposite boundaries of the PDE domain. Thus, the open-loop plant has infinite relative degree, infinite-dimensional state, infinite-dimensional (functional) parametric uncertainties, and only scalar input and output.

We start the design by introducing the notion of an *observer canonical form* for PDEs. To estimate the state, we use the adaptive observers that are infinite-dimensional extensions of Kreisselmeier observers [68].

The identifiers are designed using the swapping approach. Typically the swapping method requires one filter per unknown parameter, and since we have functional parameters, infinitely many filters would seem to be needed. However, with a special algebraic representation of the filters, we reduce their number down to only two.

We also solve a problem of adaptive output *tracking* of a trajectory prescribed at the boundary. The solution is explicit and does not require any additional computations for most practical trajectories, such as combinations of exponentials, sinusoids, and polynomials.

In Sections 13.9–13.11 we extend the approach to complex-valued plants and Neumann measurements.

13.1 REACTION-ADVECTION-DIFFUSION PLANT

We consider the plant

$$u_t(x,t) = \varepsilon(x)u_{xx}(x,t) + b(x)u_x(x,t) + \lambda(x)u(x,t) + g(x)u(0,t) + \int_0^x f(x,y)u(y,t)\,dy \tag{13.1}$$

for $0 < x < 1$ with boundary conditions

$$u_x(0,t) = -qu(0,t), \tag{13.2}$$

$$u(1,t) = U(t), \tag{13.3}$$

where ε, b, λ, g, f, and q are unknown.

We assume that only the boundary value $u(0,t)$ is measured. The objective is to regulate the state of the plant to zero using the boundary input $U(t)$.

Assumption 13.1. *The scalar quantities*

$$\varepsilon(0) \quad \textit{and} \quad \int_0^1 \frac{1}{\sqrt{\varepsilon(x)}}\,dx \tag{13.4}$$

are known.

The need for this assumption will become clear later, in Section 13.2. Here we only point out that this assumption (which in most cases simply means that the diffusion coefficient $\varepsilon(x)$ is known) is critical for the output-feedback adaptive design. It is not made just to simplify the stability proof.

Before starting the adaptive design, we apply the gauge transformation (2.5)–(2.7) to eliminate the advection term $b(x)$ and to make diffusivity $\varepsilon(x)$ constant. Dropping the bars for convenience, we get

$$u_t(x,t) = u_{xx}(x,t) + \lambda(x)u(x,t) + g(x)u(0,t) + \int_0^x f(x,y)u(y,t)\,dy, \tag{13.5}$$

$$u_x(0,t) = -qu(0,t), \tag{13.6}$$

$$u(1,t) = \theta_2 U(t), \tag{13.7}$$

where $\lambda(x)$, $g(x)$, and $f(x,y)$ are new unknown continuously differentiable functions, and q and

$$\theta_2 = \left[\sqrt{\varepsilon(1)}\int_0^1 \frac{1}{\sqrt{\varepsilon(x)}}\,dx\right]^{-1/2} \exp\left\{\int_0^1 \frac{b(s)}{2\varepsilon(s)}\,ds\right\} > 0 \tag{13.8}$$

are new unknown constants.

Note that in (13.5) we set the constant diffusion coefficient to 1 with the appropriate time scaling (this is possible thanks to Assumption 13.1). Note also that Assumption 13.1 ensures that the new $u(0,t)$, which is the old $u(0,t)$ multiplied by $(\int_0^1 \sqrt{\varepsilon(0)/\varepsilon(x)}dx)^{-1/2}$, is measured.

13.2 TRANSFORMATION TO OBSERVER CANONICAL FORM

The key step in our design is the transformation of the plant (13.5)–(13.7) into a system in which unknown parameters multiply the measured output. This can be achieved with the help of a backstepping transformation very similar (but not identical) to the one used in Chapter 4 for observer design.

Consider the transformation

$$v(x,t) = u(x,t) - \int_0^x p(x,y)u(y,t)\,dy, \tag{13.9}$$

where $p(x,y)$ is a solution of the PDE

$$p_{xx}(x,y) = p_{yy}(x,y) + \lambda(y)p(x,y) - f(x,y) + \int_y^x p(x,\xi)f(\xi,y)\,d\xi, \tag{13.10}$$

$$p(1,y) = 0, \tag{13.11}$$

$$p(x,x) = \frac{1}{2}\int_x^1 \lambda(s)\,ds. \tag{13.12}$$

Straightforward computation shows that this transformation maps the system (13.5)–(13.7) into

$$v_t(x,t) = v_{xx}(x,t) + \theta(x)v(0,t), \tag{13.13}$$

$$v_x(0,t) = \theta_1 v(0,t), \tag{13.14}$$

$$v(1,t) = \theta_2 U(t), \tag{13.15}$$

where

$$\theta(x) = g(x) - p_y(x,0) - qp(x,0) - \int_0^x p(x,y)g(y)\,dy \tag{13.16}$$

and

$$\theta_1 = -p(0,0) - q \tag{13.17}$$

are the new unknown parameters.

The system (13.13)–(13.15) is the PDE analog of the observer canonical form and the transformation (13.9)–(13.12) is the similarity transformation. Note from (13.9) that

$$v(0,t) = u(0,t), \tag{13.18}$$

and therefore $v(0,t)$ is measured. It was shown in Chapter 4 (see (4.17)–(4.19)) that the PDE (13.10)–(13.12) has a unique, twice continuously differentiable solution. The transformation (13.9) is therefore invertible, and stability of the v-system implies stability of the u-system. Therefore, it is enough to design a stabilizing controller for the v-system and then use the condition (13.15) to obtain the controller for the original system.

We are going to directly estimate the unknown parameters $\theta(x)$, θ_1, and θ_2 instead of estimating the original parameters q, $b(x)$, $\lambda(x)$, $g(x)$, and $f(x,y)$. Thus, we do not need to solve the PDE (13.10)–(13.12) for the control scheme implementation.

From (13.13)–(13.15) the need for Assumption 13.1 becomes clear. Had there been an unknown constant in front of v_{xx} in (13.13), equations (13.13)–(13.15) would not be in the observer canonical form, since the second derivative of the state across the domain is not measured. Thus, Assumption 13.1 is crucial.

13.3 NOMINAL CONTROLLER

Let us first design a stabilizing controller for the case of known parameters. The general solution of this problem is given in Chapter 2; however, here we want to take advantage of the specific form of the plant (13.13)–(13.15).

As shown in Chapter 2, the backstepping transformation

$$w(x,t) = v(x,t) - \int_0^x k(x,y)v(y,t)\,dy, \tag{13.19}$$

with $k(x, y)$ satisfying the PDE

$$k_{xx}(x, y) = k_{yy}(x, y), \tag{13.20}$$

$$k_y(x, 0) = \theta_1 k(x, 0) + \theta(x) - \int_0^x k(x, y)\theta(y)\, dy, \tag{13.21}$$

$$k(x, x) = \theta_1, \tag{13.22}$$

maps the plant (13.13)–(13.15) into the exponentially stable target system

$$w_t(x, t) = w_{xx}(x, t), \tag{13.23}$$

$$w_x(0, t) = 0, \tag{13.24}$$

$$w(1, t) = 0. \tag{13.25}$$

The controller is

$$v(1, t) = \int_0^1 k(1, y) v(y, t)\, dy. \tag{13.26}$$

One can simplify the PDE (13.20)–(13.22) by setting

$$k(x, y) = \kappa(x - y), \tag{13.27}$$

which is a solution of (13.20). The function $\kappa(x)$ then has to satisfy the integro–differential equation

$$\kappa'(x) = -\theta_1 \kappa(x) - \theta(x) + \int_0^x \kappa(x - y)\theta(y)\, dy \tag{13.28}$$

with the initial condition

$$\kappa(0) = \theta_1. \tag{13.29}$$

Integrating (13.28) from 0 to x, we get the following Volterra integral equation:

$$\kappa(x) = \theta_1 - \int_0^x \theta(y)\, dy - \int_0^x \left[\theta_1 - \int_0^{x-y} \theta(s)\, ds\right] \kappa(y)\, dy. \tag{13.30}$$

It is considerably easier to solve this equation in *one variable* than the original PDE (13.20)–(13.22), which is very important for the adaptive control design because of the need to solve for the control gain online.

Remark 13.1. Note that it is possible to find the control gain explicitly for any *known* $\theta(x)$ and θ_1 by applying the Laplace transform in x to (13.30) to get

$$\kappa(s) = \frac{\theta(s) - \theta_1}{\theta(s) - s - \theta_1}. \tag{13.31}$$

However, this formula is not very helpful in the adaptive setting when the parameter estimates are known only numerically, and it is easier to solve (13.30) directly instead of computing the direct and inverse Laplace transforms.

13.4 FILTERS

The unknown parameters $\theta(x)$ and θ_1 enter the boundary condition and the domain of the v-system. Therefore we will need the following output filters:

$$\phi_t(x,t) = \phi_{xx}(x,t), \tag{13.32}$$

$$\phi_x(0,t) = u(0,t), \tag{13.33}$$

$$\phi(1,t) = 0 \tag{13.34}$$

and

$$\Phi_t(x,\xi,t) = \Phi_{xx}(x,\xi,t) + \delta(x-\xi)u(0,t), \tag{13.35}$$

$$\Phi_x(0,\xi,t) = 0, \tag{13.36}$$

$$\Phi(1,\xi,t) = 0. \tag{13.37}$$

Here the filter $\Phi = \Phi(x,\xi,t)$ is parametrized by $\xi \in [0,1]$ and $\delta(x-\xi)$ is the delta function. The reason for this parametrization is the presence of the functional parameter $\theta(x)$ in the domain. Therefore, loosely speaking, we need an infinite "array" of filters, one for each $x \in [0,1]$ (since the swapping design normally requires one filter per unknown parameter).

We also introduce the input filter

$$\psi_t(x,t) = \psi_{xx}(x,t), \tag{13.38}$$

$$\psi_x(0,t) = 0, \tag{13.39}$$

$$\psi(1,t) = U(t). \tag{13.40}$$

It is straightforward to show now that the error

$$\bar{e}(x,t) = v(x,t) - \theta_2\psi(x,t) - \theta_1\phi(x,t) - \int_0^1 \theta(\xi)\Phi(x,\xi,t)\,d\xi \tag{13.41}$$

satisfies the exponentially stable PDE

$$\bar{e}_t(x,t) = \bar{e}_{xx}(x,t), \tag{13.42}$$

$$\bar{e}_x(0,t) = 0, \tag{13.43}$$

$$\bar{e}(1,t) = 0. \tag{13.44}$$

The filters ϕ and Φ both have $u(0,t)$ as their input. Therefore, we can try to represent the state Φ algebraically through ϕ at each moment in time. The following lemma establishes this connection.

Lemma 13.1. *The signal*

$$e(x,t) = v(x,t) - \theta_2\psi(x,t) - \theta_1\phi(x,t) - \int_0^1 \theta(\xi)F(x,\xi,t)\,d\xi, \tag{13.45}$$

where $F(x, \xi, t)$ satisfies the PDE

$$F_{xx}(x, \xi, t) = F_{\xi\xi}(x, \xi, t), \tag{13.46}$$

$$F(0, \xi, t) = -\phi(\xi, t), \tag{13.47}$$

$$F_x(0, \xi, t) = 0, \tag{13.48}$$

$$F_\xi(x, 0, t) = 0, \tag{13.49}$$

$$F(x, 1, t) = 0 \tag{13.50}$$

satisfies the exponentially stable heat equation,

$$e_t(x, t) = e_{xx}(x, t), \tag{13.51}$$

$$e_x(0, t) = 0, \tag{13.52}$$

$$e(1, t) = 0. \tag{13.53}$$

The solution to (13.46)–(13.50) *is*

$$F(x, \xi, t) = -2 \sum_{n=0}^{\infty} \cos(\sigma_n x) \cos(\sigma_n \xi) \int_0^1 \cos(\sigma_n s)\, \phi(s, t)\, ds, \tag{13.54}$$

where $\sigma_n = \pi(n + 1/2)$.

Proof. The initial conditions ϕ_0 and Φ_0 for the filters ϕ and Φ are the design choice, so let us assume that they are continuous functions in x and ξ. We now write down the explicit solutions to the filters (see, e.g., [99]). The solution for the filter ϕ is

$$\begin{aligned} \phi(x, t) = {} & 2 \sum_{n=0}^{\infty} \cos(\sigma_n x)\, e^{-\sigma_n^2 t} \int_0^1 \phi_0(s) \cos(\sigma_n s)\, ds \\ & -2 \sum_{n=0}^{\infty} \cos(\sigma_n x) \int_0^t u(0, \tau) e^{-\sigma_n^2 (t-\tau)}\, d\tau, \quad \sigma_n = \pi(n + 1/2). \end{aligned} \tag{13.55}$$

The solution for the filter Φ is

$$\begin{aligned} \Phi(x, \xi, t) = {} & 2 \sum_{n=0}^{\infty} \cos(\sigma_n x) \int_0^1 \delta(s - \xi) \cos(\sigma_n s)\, ds \int_0^t u(0, \tau) e^{-\sigma_n^2 (t-\tau)}\, d\tau \\ & +2 \sum_{n=0}^{\infty} \cos(\sigma_n x)\, e^{-\sigma_n^2 t} \int_0^1 \Phi_0(s, \xi) \cos(\sigma_n s)\, ds \\ = {} & 2 \sum_{n=0}^{\infty} \cos(\sigma_n x) \cos(\sigma_n \xi) \int_0^t u(0, \tau) e^{-\sigma_n^2 (t-\tau)}\, d\tau \\ & +2 \sum_{n=0}^{\infty} \cos(\sigma_n x)\, e^{-\sigma_n^2 t} \int_0^1 \Phi_0(s, \xi) \cos(\sigma_n s)\, ds. \end{aligned} \tag{13.56}$$

Multiplying (13.55) by $\cos(\sigma_m x)$ and using the orthogonality of these functions on [0,1], we can rewrite (13.56) in the form

$$\Phi(x,\xi,t) = -2\sum_{n=0}^{\infty}\cos(\sigma_n x)\cos(\sigma_n \xi)\int_0^1 \cos(\sigma_n s)\,\phi(s,t)\,ds$$
$$+2\sum_{n=0}^{\infty}\cos(\sigma_n x)\,e^{-\sigma_n^2 t}$$
$$\times\int_0^1 \cos(\sigma_n s)\,(\phi_0(s)\cos(\sigma_n\xi)+\Phi_0(s,\xi))\,ds. \qquad (13.57)$$

Here the first term represents the explicit solution of the system (13.46)–(13.50) and the second term is the effect of the filters' initial conditions. Therefore, we can represent Φ as

$$\Phi(x,\xi,t) = F(x,\xi,t) + \Delta F(x,\xi,t), \qquad (13.58)$$

where ΔF satisfies

$$\Delta F_t(x,\xi,t) = \Delta F_{xx}(x,\xi,t), \qquad (13.59)$$
$$\Delta F_x(0,\xi,t) = 0, \qquad (13.60)$$
$$\Delta F(1,\xi,t) = 0, \qquad (13.61)$$

and using (13.42)–(13.44), we get (13.51)–(13.53). □

Lemma 13.1 allows us to avoid solving an infinite "array" of parabolic equations (13.35)–(13.37) by computing the solution of the standard wave equation (13.46)–(13.50) at each time step. Therefore, we have only two dynamic equations to solve (filters ϕ and ψ).

13.5 FREQUENCY DOMAIN COMPENSATOR WITH FROZEN PARAMETERS

In this section we present the frequency domain version of our control design when the parameter estimates are "frozen."

Applying the Laplace transform to (13.13)–(13.15) and solving the resulting ODE, we obtain the transfer function of the plant:

$$u(0,s) = \frac{\theta_2}{A(s)}\,u(1,s), \qquad (13.62)$$

$$A(s) = \theta_1\frac{\sinh\sqrt{s}}{\sqrt{s}} + \cosh\sqrt{s} - \int_0^1 \frac{\sinh\sqrt{s}(1-y)}{\sqrt{s}}\theta(y)\,dy. \qquad (13.63)$$

It is clear that the plant is of infinite relative degree.

For the filters, we obtain the following representation:

$$\psi(x,s) = \frac{\cosh\sqrt{s}x}{\cosh\sqrt{s}}\,u(1,s) \qquad (13.64)$$

for the input filter,

$$\phi(x,s)=\frac{\sinh\sqrt{s}(x-1)}{\sqrt{s}\cosh\sqrt{s}}u(0,s) \tag{13.65}$$

for the boundary output filter, and

$$F(x,\xi,s)=\frac{\sinh\sqrt{s}(1-\xi)\cosh(\sqrt{s}x)}{\sqrt{s}\cosh\sqrt{s}} -\frac{\sinh\sqrt{s}(x-\xi)}{\sqrt{s}}H(x-\xi)u(0,s) \tag{13.66}$$

for the in-domain output filter. (H is the Heavyside function.)

The adaptive compensator has the following form:

$$u(1,s)=\frac{Q(s)}{P(s)}u(0,s), \tag{13.67}$$

where

$$P(s)=\hat{\theta}_2\left(\cosh\sqrt{s}-\int_0^1\hat{\kappa}(1-y)\cosh(\sqrt{s}y)\,dy\right) \tag{13.68}$$

and

$$Q(s)=\frac{1}{\sqrt{s}}\int_0^1\hat{\kappa}(1-y)\Big[-\hat{\theta}_1\sinh\sqrt{s}(1-y) + \sinh\sqrt{s}(1-y)\int_0^y\hat{\theta}(\xi)\cosh(\sqrt{s}\xi)\,d\xi + \cosh(\sqrt{s}y)\int_y^1\hat{\theta}(\xi)\sinh\sqrt{s}(1-\xi)\,d\xi\Big]dy, \tag{13.69}$$

and where $\hat{\theta}_1$, $\hat{\theta}_2$, and $\hat{\theta}(x)$ are, respectively, the estimates of the unknown parameters θ_1, θ_2, and $\theta(x)$ and $\hat{\kappa}(x)$ is the solution of the integral equation

$$\hat{\kappa}(x)=\hat{\theta}_1-\int_0^x\hat{\theta}(y)\,dy-\int_0^x\left[\hat{\theta}_1-\int_0^{x-y}\hat{\theta}(s)\,ds\right]\kappa(y)\,dy. \tag{13.70}$$

13.6 UPDATE LAWS

To derive the update laws for generating the adaptive estimates $\hat{\theta}_1(t)$, $\hat{\theta}_2(t)$, and $\hat{\theta}(x,t)$, we take the following equation as a parametric model:

$$\begin{aligned} e(0,t)&=v(0,t)-\theta_2\psi(0,t)-\theta_1\phi(0,t)-\int_0^1\theta(\xi)F(0,\xi,t)\,d\xi \\ &=v(0,t)-\theta_2\psi(0,t)-\theta_1\phi(0,t)+\int_0^1\theta(\xi)\phi(\xi,t)\,d\xi\,. \end{aligned} \tag{13.71}$$

The estimation error is defined as

$$\hat{e}(0,t)=v(0,t)-\hat{\theta}_2(t)\psi(0,t)-\hat{\theta}_1(t)\phi(0,t)+\int_0^1\hat{\theta}(\xi,t)\phi(\xi,t)\,d\xi\,. \tag{13.72}$$

We pick the update laws based on the gradient algorithm with normalization:[1]

$$\hat{\theta}_t(x) = -\gamma(x)\frac{\hat{e}(0)\phi(x)}{1+\|\phi\|^2+\phi^2(0)+\nu\psi^2(0)}, \tag{13.73}$$

$$\dot{\hat{\theta}}_1 = \gamma_1\frac{\hat{e}(0)\phi(0)}{1+\|\phi\|^2+\phi^2(0)+\nu\psi^2(0)}, \tag{13.74}$$

$$\dot{\hat{\theta}}_2 = \gamma_2\frac{\hat{e}(0)\psi(0)}{1+\|\phi\|^2+\phi^2(0)+\nu\psi^2(0)}, \tag{13.75}$$

where $\gamma(x)$, γ_1, and γ_2 are positive adaptation gains and ν is a positive normalization constant.

Lemma 13.2. *The adaptive laws* (13.73)–(13.75) *guarantee the following properties:*

$$\|\tilde{\theta}\|,\ \tilde{\theta}_1,\ \tilde{\theta}_2 \in \mathcal{L}_\infty, \tag{13.76}$$

$$\|\hat{\theta}_t\|,\ \dot{\hat{\theta}}_1,\ \dot{\hat{\theta}}_2 \in \mathcal{L}_2\cap\mathcal{L}_\infty, \tag{13.77}$$

$$\frac{\hat{e}(0)}{\sqrt{1+\|\phi\|^2+\phi^2(0)+\nu\psi^2(0)}} \in \mathcal{L}_2\cap\mathcal{L}_\infty. \tag{13.78}$$

Proof. Using the Lyapunov function

$$V = \frac{1}{2}\|e\|^2 + \frac{1}{2\gamma_1}\tilde{\theta}_1^2 + \frac{1}{2\gamma_2}\tilde{\theta}_2^2 + \int_0^1 \frac{\tilde{\theta}^2(x)}{2\gamma(x)}\,dx \tag{13.79}$$

we get

$$\begin{aligned}
\dot{V} &= -\int_0^1 e_x^2\,dx + \frac{\int_0^1 \tilde{\theta}(x)\phi(x)\,dx - \tilde{\theta}_1\phi(0) - \tilde{\theta}_2\psi(0)}{1+\|\phi\|^2+\phi^2(0)+\nu\psi^2(0)}\hat{e}(0)\\
&\le -\|e_x\|^2 + \frac{e(0)\hat{e}(0)-\hat{e}^2(0)}{1+\|\phi\|^2+\phi^2(0)+\nu\psi^2(0)}\\
&\le -\|e_x\|^2 + \frac{\|e_x\|\,|\hat{e}(0)|}{\sqrt{1+\|\phi\|^2+\phi^2(0)+\nu\psi^2(0)}}\\
&\quad - \frac{\hat{e}^2(0)}{1+\|\phi\|^2+\phi^2(0)+\nu\psi^2(0)}\\
&\le -\frac{1}{2}\|e_x\|^2 - \frac{1}{2}\frac{\hat{e}^2(0)}{1+\|\phi\|^2+\phi^2(0)+\nu\psi^2(0)}.
\end{aligned} \tag{13.80}$$

This gives

$$\frac{\hat{e}(0)}{\sqrt{1+\|\phi\|^2+\phi^2(0)+\nu\psi^2(0)}} \in \mathcal{L}_2, \quad \|\tilde{\theta}\|,\ \tilde{\theta}_1,\ \tilde{\theta}_2 \in \mathcal{L}_\infty. \tag{13.81}$$

[1] From this point on, we suppress the time dependence whenever possible, that is, $\hat{e}(0) \equiv \hat{e}(0,t)$, etc.

The rest of the properties (13.77)–(13.78) follow from the relation

$$\hat{e}(0) = e(0) + \tilde{\theta}_2\psi(0) + \tilde{\theta}_1\phi(0) - \int_0^1 \tilde{\theta}(x)\phi(x)\,dx \tag{13.82}$$

and the update laws. □

13.7 STABILITY

To prove the stability of the closed-loop system, we need the following assumption.

Assumption 13.2. *The scalar quantities*

$$\varepsilon(1) \quad \textit{and} \quad \int_0^1 \frac{b(x)}{\varepsilon(x)}\,dx \tag{13.83}$$

are known.

This assumption essentially requires that the constant multiplying the input, θ_2, is known, which in most cases translates into the assumption of known $b(x)$. Unlike Assumption 13.1, the above assumption is needed only for the stability proof. The adaptive scheme can be designed even without knowledge of $b(x)$, as shown in the previous section. In Section 13.12 we present numerical results that show the adaptive regulation of the plant with unknown $b(x)$.

In light of Assumption 13.2, for the main closed-loop stability result, which we state next, we set $\hat{\theta}_2(0) = \theta_2$, $\gamma_2 = 0$, $\nu = 0$ in the update laws (13.73)–(13.75).

Theorem 13.1. *Consider the plant* (13.5)–(13.7) *under Assumptions 13.1 and 13.2 with the controller*

$$U(t) = \frac{1}{\theta_2}\int_0^1 \hat{\kappa}(1-y)\left[\theta_2\psi(y,t) + \hat{\theta}_1\phi(y,t) + \int_0^1 F(y,\xi,t)\hat{\theta}(\xi)\,d\xi\right]dy, \tag{13.84}$$

where $\hat{\kappa}(x)$ is the solution of the integral equation

$$\hat{\kappa}(x) = \hat{\theta}_1 - \int_0^x \hat{\theta}(y)\,dy - \int_0^x \left[\hat{\theta}_1 - \int_0^{x-y} \hat{\theta}(s)\,ds\right]\hat{\kappa}(y)\,dy, \tag{13.85}$$

the filters ϕ, F, and ψ are given by (13.32)–(13.34), (13.46)–(13.50), *and* (13.38)–(13.40), *and the update laws for $\hat{\theta}(x)$ and $\hat{\theta}_1$ are given by* (13.73)–(13.74). *For any initial data $\theta_1 \in \mathbb{R}$, $\hat{\theta}(\cdot,0) \in C^1[0,1]$, $u_0, \phi_0, \psi_0 \in H^2(0,1)$ compatible with boundary conditions, the classical solution $(u, \phi, \psi, \hat{\theta}, \hat{\theta}_1)$ is bounded for all $x \in [0,1]$, $t \geq 0$ and*

$$\lim_{t\to\infty} \max_{x\in[0,1]} |u(x,t)| = 0. \tag{13.86}$$

The proof consists of several steps, which are presented in the next four subsections.

13.7.1 Target System

Denote

$$h(x) = \theta_2\psi(x) + \hat{\theta}_1\phi(x) + \int_0^1 F(x,\xi)\hat{\theta}(\xi)\,d\xi, \tag{13.87}$$

and use the following backstepping transformation:

$$w(x) = h(x) - \int_0^x \hat{k}(x,y)h(y)\,dy := T[h](x). \tag{13.88}$$

Since we use the certainty equivalence approach, the kernel $\hat{k}(x,y)$ is obtained from (13.20)–(13.22) simply by substituting parameter estimates in place of the unknown parameters. Therefore it is given by $\hat{k}(x,y) = \hat{\kappa}(x-y)$, with $\hat{\kappa}(x)$ defined by (13.85).

One can show that the inverse transformation to (13.88) is

$$h(x) = w(x) + \int_0^x \hat{l}(x,y)w(y)\,dy, \tag{13.89}$$

where

$$\hat{l}(x,y) = \hat{\theta}_1 - \int_0^{x-y} \hat{\theta}(\xi)\,d\xi. \tag{13.90}$$

Using Lemma 13.1, the equations for the plant, filters ϕ and ψ, and the Volterra relationship between $\hat{l}$ and $\hat{k}$,

$$\hat{l}(x,y) = \hat{k}(x,y) + \int_y^x \hat{l}(x,\xi)\hat{k}(\xi,y)\,d\xi, \tag{13.91}$$

one can derive the following target system:

$$\begin{aligned} w_t &= w_{xx} + \hat{e}(0)\hat{k}_y(x,0) \\ &\quad - \int_0^x w(y)\left(\hat{l}_t(x,y) - \int_y^x \hat{k}(x,\xi)\hat{l}_t(\xi,y)\,d\xi\right)dy \\ &\quad + \dot{\hat{\theta}}_1 T[\phi] + T\left[\int_0^1 F(x,\xi)\hat{\theta}_t(\xi)\,d\xi\right], \end{aligned} \tag{13.92}$$

$$w_x(0) = \hat{\theta}_1\hat{e}(0), \tag{13.93}$$

$$w(1) = 0. \tag{13.94}$$

Compared to the nominal target system (13.23)–(13.25), two types of additional terms appear. First, there are terms with $\hat{e}(0)$ both in the equation and in the boundary condition. Second, there are terms that are proportional to time derivatives of the parameter estimates.

Let us rewrite the ϕ filter as follows:

$$\phi_t = \phi_{xx}, \tag{13.95}$$

$$\phi_x(0) = w(0) + \hat{e}(0), \tag{13.96}$$

$$\phi(1) = 0. \tag{13.97}$$

We now have an interconnection of two systems ϕ and w with forcing terms that have properties (13.78)–(13.76).

13.7.2 L^2 Boundedness

First, let us establish bounds on the gains $\hat{k}(x, y)$ and $\hat{l}(x, y)$. The boundedness of the parameter estimates $\hat{\theta}_1$ and $\|\hat{\theta}\|$ has been shown in Lemma 13.2. From (13.90) we get

$$|\hat{l}(x, y)| \leq \bar{\theta}_1 + \bar{\theta}, \tag{13.98}$$

where we denote $\bar{\theta}_1 = \max_{t\geq 0} |\hat{\theta}_1|$ and $\bar{\theta} = \max_{t\geq 0} \|\hat{\theta}\|$.

Using (13.91) and the Gronwall inequality, it is easy to get the following bound:

$$|\hat{k}(x, y)| \leq (\bar{\theta}_1 + \bar{\theta}) e^{\bar{\theta}_1 + \bar{\theta}} := K_1. \tag{13.99}$$

If we look at the right-hand side of the w-system, we can see that the estimates for $\hat{k}_y(x, 0)$ and $\hat{l}_t(x, y)$ are also needed. They are obtained from (13.85) and (13.90):

$$|\hat{k}_y(x, 0)| \leq (\bar{\theta}_1 + \bar{\theta}) K_1 + \bar{\theta} := K_2, \tag{13.100}$$

$$|\hat{l}_t(x, y)| \leq |\dot{\hat{\theta}}_1| + \|\hat{\theta}_t\|. \tag{13.101}$$

We are now ready to start the stability analysis of (13.92)–(13.97). Consider the Lyapunov function

$$V_1 = \frac{1}{2} \int_0^1 \phi^2(x)\, dx. \tag{13.102}$$

Computing its derivative along the solutions of the ϕ-system and using Young's, Poincaré's, and Agmon's inequalities, we get

$$\begin{aligned}
\dot{V}_1 = & -\phi(0) w(0) - \phi(0)\hat{e}(0) - \int_0^1 \phi_x^2(x)\, dx \\
\leq & \frac{1}{2} w^2(0) + \frac{1}{2}\phi^2(0) - \|\phi_x\|^2 \\
& + \frac{|\phi(0)\hat{e}(0)|}{1 + \|\phi\|^2 + \phi^2(0)} (1 + \|\phi\|^2 + 2\|\phi\|\|\phi_x\|) \\
\leq & -\frac{1}{2}\|\phi_x\|^2 + \frac{1}{2}\|w_x\|^2 + c_1\|\phi_x\|^2 \\
& + \frac{1}{4c_1} \frac{\hat{e}^2(0)}{1 + \|\phi\|^2 + \phi^2(0)} + c_1\|\phi\|^2 + c_1\|\phi_x\|^2 \\
& + \frac{3}{2c_1} \left(\frac{|\phi(0)\hat{e}(0)|}{1 + \|\phi\|^2 + \phi^2(0)} \right)^2 \|\phi\|^2 \\
\leq & -\left(\frac{1}{2} - 3c_1\right) \|\phi_x\|^2 + \frac{1}{2}\|w_x\|^2 + l_1\|\phi\|^2 + l_1.
\end{aligned} \tag{13.103}$$

Here c_1 is a positive constant that will be chosen later and l_1 denotes a generic integrable function of time.

With the Lyapunov function

$$V_2 = \frac{1}{2}\int_0^1 w^2(x)\,dx, \tag{13.104}$$

we get

$$\begin{aligned}\dot{V}_2 = &-\hat{\theta}_1 w(0)\hat{e}(0) + \hat{e}(0)\int_0^1 \hat{k}_y(x,0)w(x)\,dx\\ &-\int_0^1 w_x^2\,dx + \dot{\hat{\theta}}_1\int_0^1 w(x)T[\phi](x)\,dx\\ &+\int_0^1 w(x)T\left[\int_0^1 F(x,\xi)\hat{\theta}_t(\xi)\,d\xi\right]dx\\ &+\int_0^1 w(x)\int_0^x w(y)\left(\hat{l}_t(x,y) - \int_y^x \hat{k}(x,\xi)\hat{l}_t(\xi,y)\,d\xi\right)dy\,dx. \end{aligned}\tag{13.105}$$

We separately estimate each term on the right-hand side of (13.105).

Term 1:

$$\begin{aligned}|\hat{\theta}_1 w(0)\hat{e}(0)| \le &\frac{\bar{\theta}_1|w(0)||\hat{e}(0)|}{\sqrt{1+\|\phi\|^2+\phi^2(0)}}(1+\|\phi\|+|\phi(0)|)\\ \le &\, c_2\|w_x\|^2 + \frac{\bar{\theta}_1^2}{c_2}\frac{\hat{e}^2(0)(1+\|\phi\|^2)}{1+\|\phi\|^2+\phi^2(0)}\\ &+\frac{\bar{\theta}_1\hat{e}(0)(\|w\|\|w_x\|+\|\phi\|\|\phi_x\|)}{\sqrt{1+\|\phi\|^2+\phi^2(0)}}\\ \le &\, c_2\|w_x\|^2 + l_1(1+\|\phi\|^2) + c_3\|w_x\|^2\\ &+\frac{\bar{\theta}_1^2 l_1}{4c_3}\|w\|^2 + c_4\|\phi_x\|^2 + \frac{\bar{\theta}_1^2 l_1}{4c_4}\|\phi\|^2\\ \le &\,(c_2+c_3)\|w_x\|^2 + c_4\|\phi_x\|^2 + l_1\|w\|^2 + l_1\|\phi\|^2 + l_1. \end{aligned}\tag{13.106}$$

Term 2:

$$\begin{aligned}\hat{e}(0)\int_0^1 \hat{k}_y(x,0)w(x)\,dx \le &\, K_2\|w\|\frac{|\hat{e}(0)|(1+\|\phi\|+|\phi(0)|)}{\sqrt{1+\|\phi\|^2+\phi^2(0)}}\\ \le &\, c_5\|w\|^2 + \frac{K_2 l_1}{c_5}(1+\|\phi\|^2) + c_6\|\phi_x\|^2 + \frac{K_2 l_1}{2c_6}\|w\|^2\\ \le &\, c_5\|w_x\|^2 + c_6\|\phi_x\|^2 + l_1\|w\|^2 + l_1\|\phi\|^2 + l_1. \end{aligned}\tag{13.107}$$

Term 3:

$$\dot{\hat{\theta}}_1 \int_0^1 w(x)T[\phi](x)\,dx \le c_7\|w\|^2 + \frac{(1+K_1^2)}{2c_7}|\dot{\hat{\theta}}_1|^2\|\phi\|^2$$
$$\le c_7\|w_x\|^2 + l_1\|\phi\|^2. \tag{13.108}$$

Term 4:

$$\int_0^1 w(x)T\left[\int_0^1 F(x,\xi)\hat{\theta}_t(\xi)\,d\xi\right]dx$$
$$\le (1+K_1)\|w\|\|\hat{\theta}_t\|\left(\int_0^1 \max_{\xi} F^2(x,\xi)\,dx\right)^{1/2}$$
$$\le \frac{(1+K_1)^2}{c_8}\|w\|^2\|\hat{\theta}_t\|^2 + c_8\int_0^1 \left(F^2(x,1) + 2\|F(x,\cdot)\|\|F_\xi(x,\cdot)\|\right)dx$$
$$\le l_1\|w\|^2 + c_8\int_0^1\int_0^1 \left(F^2(x,\xi)+F_\xi^2(x,\xi)\right)d\xi\,dx$$
$$\le l_1\|w\|^2 + 2c_8\int_0^1 \phi^2(\xi)\,d\xi + 2c_8\int_0^1\int_0^1 \left(F_x^2(x,\xi)+F_\xi^2(x,\xi)\right)d\xi\,dx$$
$$\le l_1\|w\|^2 + 2c_8(\|\phi\|^2 + \|\phi_x\|^2). \tag{13.109}$$

Term 5:

$$\int_0^1 w(x)\int_0^x w(y)\left(\hat{l}_t(x,y) - \int_y^x \hat{k}(x,\xi)\hat{l}_t(\xi,y)\,d\xi\right)dy\,dx$$
$$\le 2c_9\|w\|^2 + \frac{1+K_1^2}{2c_9}|\dot{\hat{\theta}}_1|^2\|w\|^2 + \frac{1+K_2^2}{2c_9}\|\hat{\theta}_t\|^2\|w\|^2$$
$$\le 2c_9\|w_x\|^2 + l_1\|w\|^2. \tag{13.110}$$

Finally, we obtain

$$\dot{V}_2 \le (c_4+c_6+4c_8)\,\|\phi_x\|^2 + l_1\|w\|^2 + l_1\|\phi\|^2 + l_1$$
$$-(1-c_2-c_3-c_5-c_7-2c_9)\,\|w_x\|^2. \tag{13.111}$$

Using (13.103) and (13.111), for the Lyapunov function

$$V = V_1 + V_2 \tag{13.112}$$

we get

$$\dot{V} \le -\left(\frac{1}{2} - c_2 - c_3 - c_5 - c_7 - 2c_9\right)\|w_x\|^2$$
$$-\left(\frac{1}{2} - 3c_1 - c_4 - c_6 - 4c_8\right)\|\phi_x\|^2 + l_1\|w\|^2 + l_1\|\phi\|^2 + l_1. \tag{13.113}$$

Choosing $c_2 = c_3 = c_5 = c_7 = 2c_9 = 1/20$, $3c_1 = c_4 = c_6 = 4c_8 = 1/16$, we get

$$\dot{V} \le -\frac{1}{2}V + l_1 V + l_1, \tag{13.114}$$

and by Lemma D.3 we get $\|w\|, \|\phi\| \in \mathcal{L}_2 \cap \mathcal{L}_\infty$. From the transformation (13.89) we get $\|h\| \in \mathcal{L}_2 \cap \mathcal{L}_\infty$, and therefore $\|\psi\| \in \mathcal{L}_2 \cap \mathcal{L}_\infty$ follows from (13.87). From (13.45) and (13.9) we get $\|v\|, \|u\| \in \mathcal{L}_2 \cap \mathcal{L}_\infty$.

13.7.3 Pointwise Boundedness

In order to prove pointwise boundedness we need to show that $\|w_x\|$ and $\|\phi_x\|$ are bounded. Unfortunately, using the standard Lyapunov function of the form $V = \|w_x\|^2 + \|\phi_x\|^2$ does not lead to the desired result. The difficulty lies in the fact that systems (13.92)–(13.97) have the term $\hat{e}(0)$ in the boundary condition at $x = 0$. This term will produce the terms proportional to $\dot{\hat{e}}(0)$ in $\dot{V}$, and the properties of this signal are not known to us.

To convert (13.92)–(13.97) into a system with homogeneous boundary conditions, let

$$\bar{w}(x) = w(x) + \hat{\theta}_1 (x-1) \int_0^x \hat{e}(y)\, dy, \tag{13.115}$$

$$\bar{\phi}(x) = \phi(x) + (x-1) \int_0^x (\hat{e}(y) + w(y))\, dy, \tag{13.116}$$

where

$$\hat{e}(x) = e(x) + \tilde{\theta}_1 \phi(x) + \int_0^1 F(x, \xi, t) \tilde{\theta}(\xi)\, d\xi. \tag{13.117}$$

Straightforward computation shows that $\bar{w}$ and $\bar{\phi}$ satisfy

$$\bar{w}_t = \bar{w}_{xx} + f_1, \tag{13.118}$$

$$\bar{w}_x(0) = \bar{w}(1) = 0 \tag{13.119}$$

and

$$\bar{\phi}_t = \bar{\phi}_{xx} + f_2, \tag{13.120}$$

$$\bar{\phi}_x(0) = \bar{\phi}(1) = 0, \tag{13.121}$$

where $f_1(x,t)$ and $f_2(x,t)$ are composed of terms that are linear integral operators of w and ϕ, multiplied by $w(0)$, $\phi(0)$, $\hat{e}(0)$, or time derivatives of the parameter estimates:

$$f_1 = \hat{e}(0)\hat{k}_y(x,0) - \int_0^x w(y)\left(\hat{l}_t(x,y) - \int_y^x \hat{k}(x,\xi)\hat{l}_t(\xi,y)\,d\xi\right)dy$$

$$+ \dot{\hat{\theta}}_1 T[\phi] + T\left[\int_0^1 F(x,\xi)\hat{\theta}_t(\xi)\,d\xi\right] \tag{13.122}$$

$$f_2 = (1-x)(\tilde{\theta}_1 w(0) + \theta_1\hat{e}(0) + 2\hat{e}(x) + 2w(x)) + (1-x)\hat{e}(0)\int_0^x \hat{k}_y(\xi,0)\,d\xi$$

$$+ (1-x)\dot{\hat{\theta}}_1 \int_0^x\int_0^y \hat{k}(y,\xi)\phi(\xi)d\xi\,dy$$

$$+ (1-x)\int_0^x\int_0^y \hat{k}(y,\xi)\int_0^1 F(\xi,\eta)\hat{\theta}_t(\eta)d\eta\,d\xi\,dy$$

$$+ (1-x)\int_0^x\int_0^y w(\xi)\left(\hat{l}_t(y,\xi) - \int_\xi^y \hat{k}(y,\eta)\hat{l}_t(\eta,\xi)\,d\eta\right)d\xi\,dy. \tag{13.123}$$

Integrating (13.113) with respect to time, we get $\|w_x\|, \|\phi_x\| \in \mathcal{L}_2$, and therefore $w(0), \phi(0) \in \mathcal{L}_2$. From (13.82) we get $\hat{e}(0) \in \mathcal{L}_2$. Since $\|w\|, \|\phi\|$ are bounded, we have $\|f_1\|, \|f_2\| \in \mathcal{L}_2$. With the Lyapunov function

$$\bar{V} = \frac{1}{2}\|\overline{w}_x\|^2 + \frac{1}{2}\|\bar{\phi}_x\|^2, \tag{13.124}$$

we get

$$\dot{\bar{V}} \le -\frac{1}{4}\bar{V} + \frac{1}{2}\|f_1\|^2 + \frac{1}{2}\|f_2\|^2, \tag{13.125}$$

and therefore by Lemma D.3, $\|\overline{w}_x\|, \|\bar{\phi}_x\| \in \mathcal{L}_2 \cap \mathcal{L}_\infty$. Using the fact that $\|\hat{e}\| \in \mathcal{L}_2 \cap \mathcal{L}_\infty$, from (13.115)–(13.116) we get $\|w_x\|, \|\phi_x\| \in \mathcal{L}_2 \cap \mathcal{L}_\infty$. By Agmon's inequality, $w(x,t)$ and $\phi(x,t)$ are bounded for all $x \in [0,1]$. Since $\hat{e}(0,\cdot), \phi(x,\cdot) \in \mathcal{L}_2$, integrating (13.73) with respect to time we get the boundedness of $\hat{\theta}(x,t)$. From the transformation (13.89) we get the boundedness of $h(x,t)$, and from (13.87) the boundedness of $\psi(x,t)$ follows. From (13.45) and (13.9) we get the boundedness of $v(x,t)$ and $u(x,t)$ for all $x \in [0,1]$.

13.7.4 Regulation

It is easy to see from (13.114) that $\dot{V}$ is bounded from above. Using Lemma D.2 we get $V \to 0$, that is, $\|w\| \to 0$, $\|\phi\| \to 0$. From the transformation (13.89) we get $\|h\| \to 0$, and from (13.87) $\|\psi\| \to 0$ follows. From (13.45) and (13.9) we get $\|v\| \to 0$ and $\|u\| \to 0$. Using Agmon's inequality and the boundedness of $\|u_x\|$, we get $\max_{x\in[0,1]} |u(x,t)| \to 0$. The proof of Theorem 13.1 is completed.

13.8 TRAJECTORY TRACKING

Consider the plant (13.5)–(13.7), and suppose we want the output $u(0,t)$ to follow a prescribed trajectory $r(t)$ (we assume that it is analytic and bounded). After the transformation of the plant into the observer canonical form (13.13)–(13.15), we reformulate the objective as $v(0,t) \to r(t)$. If the parameters $\theta(x)$ and θ_1 were known, the design would consist of two steps. First, the trajectory of the full state $v^r(x,t)$ would have to be generated such that $v^r(0,t) = r(t)$, with $v^r(x,t)$ satisfying (13.13)–(13.14). The second step would be to stabilize the plant around that trajectory. However, in the case of unknown parameters the trajectory $v^r(x,t)$ cannot be generated. Therefore, our approach is to generate the trajectory $w^r(x,t)$ for the *nominal target system*. If the transformed plant follows $w^r(x,t)$, in particular, $w(0,t) \to r(t)$, then the output $v(0,t) = w(0,t)$ would also follow $r(t)$.

The nominal target system is defined by a Cauchy-Kovalevskaya problem given by the following three conditions:

$$w_t^r(x,t) = w_{xx}^r(x,t), \tag{13.126}$$

$$w_x^r(0,t) = 0, \tag{13.127}$$

$$w^r(0,t) = r(t). \tag{13.128}$$

The well-known solution to (13.126)–(13.128) is (see, e.g., [72])

$$w^r(x,t) = \sum_{n=0}^{\infty} \frac{r^{(n)}(t)x^{2n}}{(2n)!}, \tag{13.129}$$

which can be easily verified by direct substitution.

Remark 13.2. Note that the series (13.129) can be computed in closed form if $r(t)$ is a combination of exponentials, polynomials, and sinusoids, which covers most of the practically relevant trajectories. For example, for $r(t) = A\sin(\omega t)$ the reference trajectory is

$$w^r(x,t) = \frac{A}{2} e^{\sqrt{\frac{\omega}{2}}x} \sin\left(\omega t + \sqrt{\frac{\omega}{2}}x\right) + \frac{A}{2} e^{-\sqrt{\frac{\omega}{2}}x} \sin\left(\omega t - \sqrt{\frac{\omega}{2}}x\right). \tag{13.130}$$

THEOREM 13.2. *Consider the plant* (13.5)–(13.7) *with the controller*

$$U(t) = \frac{1}{\theta_2} \int_0^1 \hat{\kappa}(1-y) \left[\theta_2 \psi(y,t) + \hat{\theta}_1 \phi(y,t) + \int_0^1 F(y,\xi,t)\hat{\theta}(\xi)\, d\xi\right] dy + \frac{1}{\theta_2} \sum_{n=0}^{\infty} \frac{r^{(n)}(t)}{(2n)!}, \tag{13.131}$$

where $r(t)$ is a bounded analytic function for $t \geq 0$. Under the assumptions of Theorem 13.1, the classical solution $(u, \phi, \psi, \hat{\theta}, \hat{\theta}_1)$ of the closed-loop system is bounded for all $x \in [0,1]$, $t \geq 0$ and

$$\lim_{t\to\infty} [u(0,t) - r(t)] = 0. \tag{13.132}$$

Proof. The transformation (13.88) maps the closed-loop system into (13.92)–(13.94), with the boundary condition (13.94) changed to $w(1,t) = w^r(1,t)$. Introducing the error variable $\tilde{w} = w - w^r$, we get the following PDE for $\tilde{w}$:

$$\begin{aligned}\tilde{w}_t =& \tilde{w}_{xx} + \hat{e}(0)\hat{k}_y(x,0) + \dot{\hat{\theta}}_1 T[\tilde{\phi} + \phi^r] \\ &+ T\left[\int_0^1 (\tilde{F}(x,\xi) + F^r(x,\xi))\hat{\theta}_t(\xi)\,d\xi\right] \\ &- \int_0^x (\tilde{w}(y) + w^r(y))\left(\hat{l}_t(x,y) - \int_y^x \hat{k}(x,\xi)\hat{l}_t(\xi,y)\,d\xi\right)dy, \end{aligned} \tag{13.133}$$

$$\tilde{w}_x(0) = \hat{\theta}_1\hat{e}(0), \tag{13.134}$$

$$\tilde{w}(1) = 0. \tag{13.135}$$

Here we denote $\tilde{\phi} = \phi - \phi^r$, with ϕ^r satisfying

$$\phi^r_t(x,t) = \phi^r_{xx}(x,t), \tag{13.136}$$

$$\phi^r_x(0,t) = r(t), \tag{13.137}$$

$$\phi^r(1,t) = 0, \tag{13.138}$$

and $\tilde{F} = F - F^r$, with F^r given by

$$F^r(x,\xi,t) = -2\sum_{n=0}^{\infty} \cos(\sigma_n x)\cos(\sigma_n \xi)\int_0^1 \cos(\sigma_n s)\,\phi^r(s,t)\,ds. \tag{13.139}$$

One trajectory satisfying (13.136)–(13.138) is

$$\phi^r(x,t) = (x-1)\sum_{n=0}^{\infty} \frac{r^{(n)}(t)x^{2n}}{(2n+1)!}. \tag{13.140}$$

Using (13.95)–(13.97) and (13.136)–(13.138), we get the PDE for $\tilde{\phi}$:

$$\tilde{\phi}_t = \tilde{\phi}_{xx}, \tag{13.141}$$

$$\tilde{\phi}_x(0) = \tilde{w}(0) + \hat{e}(0), \tag{13.142}$$

$$\tilde{\phi}(1) = 0. \tag{13.143}$$

The interconnection of the systems (13.133)–(13.135) and (13.141)–(13.143) is very similar to that of (13.92)–(13.97). The only difference is in additional terms in (13.133), proportional to ϕ^r, F^r, and w^r (all bounded trajectories). Since these terms are multiplied by square integrable signals (time derivatives of the parameter estimates), the proofs of boundedness and regulation given in Section 13.7 apply without any changes; only the specific expression behind the generic l_1 function in (13.113) will be different. Therefore, $\tilde{w}(x,t)$ and $\tilde{\phi}(x,t)$ are bounded and converge to zero for all $x \in [0,1]$. This means that $w(x,t)$ and $\phi(x,t)$ are bounded. From (13.89) we get the boundedness of $\hat{h}$ and, as a result, the boundedness of $\psi(x,t)$. From (13.45) and (13.9) we get the boundedness of $v(x,t)$ and $u(x,t)$ for all $x \in [0,1]$.

To show tracking, we note that $w \to w^r$ as $t \to \infty$ (since $\tilde{w} \to 0$); in particular, $w(0,t) \to w^r(0,t) = r(t)$. Since $u(0,t) = w(0,t) + \hat{e}(0,t)$, we get

$u(0,t) \to r(t) + \hat{e}(0,t)$ as $t \to \infty$. It remains to show that $\hat{e}(0,t) \to 0$. Let us define an auxiliary signal $\hat{e}^r(0,t)$:

$$\begin{aligned}\hat{e}^r(0,t) &= e(0,t) + \tilde{\theta}_1 \phi^r(0,t) + \int_0^1 F^r(0,\xi,t)\tilde{\theta}(\xi)\,d\xi, \\ &= e(0,t) - \tilde{\theta}_1 r(t) - \int_0^1 \phi^r(\xi,t)\tilde{\theta}(\xi)\,d\xi. \end{aligned} \tag{13.144}$$

Using (13.82), we get

$$\hat{e}^r(0,t) = \hat{e}(0,t) - \tilde{\theta}_1 \tilde{\phi}(0,t) - \int_0^1 \tilde{F}(0,\xi,t)\tilde{\theta}(\xi)\,d\xi. \tag{13.145}$$

Since $\|\tilde{w}_x\|, \|\tilde{\phi}_x\| \in \mathcal{L}_2 \cap \mathcal{L}_\infty$, from (13.82) we get $\hat{e}(0,\cdot) \in \mathcal{L}_2 \cap \mathcal{L}_\infty$, and therefore (13.145) gives $\hat{e}^r(0,\cdot) \in \mathcal{L}_2 \cap \mathcal{L}_\infty$. Taking the time derivative of (13.144), we easily get the boundedness of $\hat{e}^r_t(0,t)$. By Barbalat's lemma, we have $\hat{e}^r(0,t) \to 0$. Since $\tilde{\phi}(0,t) \to 0$ and $\tilde{F}(0,\xi,t) = -\tilde{\phi}(\xi,t) \to 0$, from (13.145) we get $\hat{e}(0,t) \to \hat{e}^r(0,t) \to 0$ as $t \to \infty$. The proof is completed. □

13.9 THE GINZBURG-LANDAU EQUATION

Starting with this section, we present the extension of our approach to complex-valued plants and Neumann measurements.

Consider the Ginzburg-Landau equation

$$A_t(x,t) = a_0 A_{xx}(x,t) + a(x)A(x,t), \tag{13.146}$$

where A is a complex-valued function. The boundary conditions are

$$A(0,t) = 0, \tag{13.147}$$

$$A(1,t) = U(t). \tag{13.148}$$

We assume that only a boundary value $A_x(0,t)$ is available for measurement (this corresponds to pressure sensing in the vortex shedding problem considered in Chapter 6). The parameter a_0 is a known constant that satisfies $\mathrm{Re}(a_0) > 0$ and $a(x)$ is an unknown complex-valued continuously differentiable function. Without loss of generality, we assume that $\mathrm{Re}(a_0) = 1$ (one can always achieve that with appropriate scaling of time). The open-loop system is unstable when $a(x)/a_0$ is large in absolute value.

13.9.1 Observer Canonical Form

We start by transforming the plant into the observer canonical form. Consider the transformation

$$B(x,t) = A(x,t) - \int_0^x p(x,y)A(y,t)\,dy, \tag{13.149}$$

where the complex-valued function $p(x, y)$ is the solution of the PDE

$$p_{xx}(x, y) - p_{yy}(x, y) = \frac{a(y)}{a_0} p(x, y), \tag{13.150}$$

$$p(1, y) = 0, \tag{13.151}$$

$$p(x, x) = \frac{1}{2a_0} \int_x^1 a(s) ds. \tag{13.152}$$

It is straightforward to verify that the transformation (13.149) maps (13.146)–(13.148) into

$$B_t(x, t) = a_0 B_{xx}(x, t) + \theta(x) B_x(0, t), \tag{13.153}$$

$$B(0, t) = 0, \tag{13.154}$$

$$B(1, t) = U(t), \tag{13.155}$$

where the new unknown parameter is

$$\theta(x) = a_0\, p(x, 0). \tag{13.156}$$

Our design will be now pursued for the new system (13.153)–(13.155). Note that $B_x(0, t) = A_x(0, t)$ and $B(1, t) = A(1, t)$; that is, $B_x(0, t)$ is measured and the controller designed for (13.153)–(13.155) is applied to the original system without change. Since the PDE (13.150)–(13.152) has a twice continuously differentiable solution, the transformation (13.149) is invertible, and therefore asymptotic stability of B implies asymptotic stability of A. Since $\theta(x)$ will be estimated, there is no need to solve the PDE (13.150)–(13.152) for the controller implementation.

13.9.2 Nominal Controller

Before addressing the adaptive design we present the nominal control design on the assumption that $\theta(x)$ is known and the full state measurement is available.

Consider the transformation

$$W(x, t) = B(x, t) - \int_0^x k(x, y) B(y, t)\, dy. \tag{13.157}$$

One can show that if the kernel $k(x, y)$ is the solution of the complex-valued PDE

$$k_{xx}(x, y) - k_{yy}(x, y) = 0, \tag{13.158}$$

$$a_0\, k(x, x) = -\theta(0), \tag{13.159}$$

$$a_0\, k(x, 0) = -\theta(x) + \int_0^x k(x, y)\theta(y) dy, \tag{13.160}$$

then (13.157) maps the system (13.153)–(13.155) into the exponentially stable system

$$W_t(x, t) = a_0 W_{xx}(x, t), \tag{13.161}$$

$$W(0, t) = 0, \tag{13.162}$$

$$W(1, t) = 0. \tag{13.163}$$

The controller is

$$U(t) = \int_0^1 k(1, y)B(y, t)\, dy. \tag{13.164}$$

Exponential stability of (13.161)–(13.163), along with invertibility of the transformation (13.157), ensures that the closed-loop system (13.153)–(13.154), (13.164) is exponentially stable around zero equilibrium.

Note that the solution of (13.158)–(13.160) is easier to compute if we set $k(x, y) = \kappa(x - y)$ and then solve the integral equation in one variable,

$$a_0\, \kappa(x) = -\theta(x) + \int_0^x \kappa(x - y)\theta(y)dy. \tag{13.165}$$

13.10 IDENTIFIER FOR THE GINZBURG-LANDAU EQUATION

The next step is to design input and output filters that will provide the estimation of the system state and will also drive the update law for the parameter estimation.

Let us consider the following output filter, which can be viewed as a family of filters in the parameter $\xi \in [0,\ 1]$ (time dependence is suppressed):

$$\Phi_t(x, \xi) = a_0 \Phi_{xx}(x, \xi) + \delta(x - \xi) B_x(0), \tag{13.166}$$

$$\Phi(0, \xi) = 0, \tag{13.167}$$

$$\Phi(1, \xi) = 0 \tag{13.168}$$

and the input filter

$$\psi_t = a_0 \psi_{xx}, \tag{13.169}$$

$$\psi(0) = 0, \tag{13.170}$$

$$\psi(1) = B(1). \tag{13.171}$$

These two filters can be used to build an observer of the state, with error

$$e(x) = B(x) - \psi(x) - \int_0^1 \theta(\xi)\Phi(x, \xi)d\xi. \tag{13.172}$$

The error $e(x, t)$ of the observer is governed by the exponentially stable complex heat equation

$$e_t = a_0 e_{xx}, \tag{13.173}$$

$$e(0) = 0, \tag{13.174}$$

$$e(1) = 0, \tag{13.175}$$

as can be shown by direct substitution.

The observer error (13.172) provides us with the parametric model to build an estimator for $\theta(x)$. The parametric model is obtained by taking the spatial derivative of the error (13.172) on the boundary where the measurement is available:

$$e_x(0) = B_x(0) - \psi_x(0) - \int_0^1 \theta(\xi)\Phi_x(0, \xi)\, d\xi. \tag{13.176}$$

The estimation error

$$\hat{e}_x(0) = B_x(0) - \psi_x(0) - \int_0^1 \hat{\theta}(\xi)\Phi_x(0,\xi)\,d\xi \tag{13.177}$$

can then be used to drive the standard normalized gradient update law

$$\hat{\theta}_t(x) = \gamma(x)\frac{\hat{e}_x(0)\Phi_x^*(0,x)}{1+\|\Phi_x(0,\cdot)\|^2}, \tag{13.178}$$

where $\gamma(x)$ is a positive adaptation gain function.

Lemma 13.3. *The following properties hold:*

$$\|\hat{\theta}_t\|, \quad \frac{\hat{e}_x(0)}{\sqrt{1+\|\Phi_x(0,\cdot)\|^2}} \in \mathcal{L}_2 \cap \mathcal{L}_\infty, \tag{13.179}$$

$$|e_x(0)| \in \mathcal{L}_2, \tag{13.180}$$

$$\|\tilde{\theta}\|, \|e_x\| \in \mathcal{L}_\infty. \tag{13.181}$$

Proof. Consider the Lyapunov function

$$V = V_1 + V_2, \tag{13.182}$$

where

$$V_1 = \frac{1}{2}\int_0^1 \frac{|\tilde{\theta}(\xi)|^2}{\gamma(x)}\,d\xi, \tag{13.183}$$

$$V_2 = \frac{1}{2}\int_0^1 |e_x(\xi)|^2\,d\xi. \tag{13.184}$$

Their time derivatives are

$$\begin{aligned}
\dot{V}_1 &= \mathrm{Re}\left\{\int_0^1 \frac{\tilde{\theta}(\xi)^*\tilde{\theta}_t(\xi)}{\gamma(\xi)}\,d\xi\right\} \\
&= -\mathrm{Re}\left\{\frac{-\hat{e}_x(0)}{1+\|\Phi_x(0,\cdot)\|^2}\int_0^1 \tilde{\theta}^*\Phi_x^*(0,\xi)\,d\xi\right\} \\
&= -\mathrm{Re}\left\{\frac{-\hat{e}_x(0)}{1+\|\Phi_x(0,\cdot)\|^2}\left[\hat{e}_x^*(0) - e_x^*(0)\right]\right\} \\
&\le -\frac{|\hat{e}_x(0)|^2}{1+\|\Phi_x(0,\cdot)\|^2} + \frac{|\hat{e}_x(0)||e_x(0)|}{\sqrt{1+\|\Phi_x(0,\cdot)\|^2}}
\end{aligned}$$

and

$$\begin{aligned}
\dot{V}_2 &= \mathrm{Re}\left\{\int_0^1 e_x(\xi)e_{tx}^*(\xi)\,d\xi\right\} \\
&= \mathrm{Re}\left\{e_x(\xi)e_t^*(\xi)\Big|_0^1 - \int_0^1 e_{xx}(\xi)e_t^*(\xi)\,d\xi\right\} \\
&= -\mathrm{Re}(a)\|e_{xx}\|^2 = -\|e_{xx}\|^2.
\end{aligned}$$

Using the inequality $|e_x(0)| \leq \|e_{xx}\|$ and Young's inequality, we get

$$\dot{V}_1 + \dot{V}_2 \leq \frac{-|\hat{e}_x(0)|^2}{1 + \|\Phi_x(0, \cdot)\|^2} + \frac{|\hat{e}_x(0)||e_x(0)|}{\sqrt{1 + \|\Phi_x(0, \cdot)\|^2}} - \|e_{xx}\|^2$$

$$\leq -\frac{1}{2}\left[\frac{|\hat{e}_x(0)|^2}{1 + \|\Phi_x(0, \cdot)\|^2} + \|e_{xx}\|^2\right],$$

and from this we get

$$\frac{\hat{e}_x(0)}{\sqrt{1 + \|\Phi_x(0, \cdot)\|^2}} \in \mathcal{L}_2,$$

$$\|e_{xx}\| \in \mathcal{L}_2 \quad \Rightarrow \quad |e_x(0)| \in \mathcal{L}_2$$

$$\|e_x\| \in \mathcal{L}_\infty$$

$$\|\tilde{\theta}\| \in \mathcal{L}_\infty \quad \Rightarrow \quad \|\hat{\theta}\| \in \mathcal{L}_\infty.$$

Moreover, since $\hat{e}_x(0) = e_x(0) - \int_0^1 \tilde{\theta}(\xi)\Phi_x(0, \xi)\, d\xi$,

$$\frac{\hat{e}_x(0)}{\sqrt{1 + \|\Phi_x(0, \cdot)\|^2}} \in \mathcal{L}_\infty,$$

and the update law (13.178) gives $\|\hat{\theta}_t\| \in \mathcal{L}_\infty \cap \mathcal{L}_2$. □

13.11 STABILITY OF ADAPTIVE SCHEME FOR THE GINZBURG-LANDAU EQUATION

The stability result for the Ginzburg-Landau equation is given by the following theorem (note that functional spaces are complexified in this section).

Theorem 13.3. *Consider the system* (13.146)–(13.148) *with the controller*

$$U(t) = \int_0^1 \hat{k}(1, y)\left[\psi(y, t) + \int_0^1 \hat{\theta}(\xi)\Phi(y, \xi, t)\, d\xi\right] dy, \tag{13.185}$$

where the kernel $\hat{k}(x, y) = \hat{\kappa}(x - y)$ is the solution of

$$a_0\hat{\kappa}(x) = -\hat{\theta}(x) + \int_0^x \hat{\kappa}(x - y)\hat{\theta}(y)\, dy, \tag{13.186}$$

the filters ψ and Φ are defined by (13.166)–(13.168), (13.169)–(13.171), *and the estimate $\hat{\theta}$ is updated according to* (13.178). *For any initial data $\hat{\theta}(\cdot, 0) \in C^1[0, 1]$, $A(\cdot, 0)$, $\psi(\cdot, 0)$, $\Phi(\cdot, \xi, 0) \in H^2(0, 1)$ compatible with boundary conditions, the classical solution $(A, \psi, \Phi, \hat{\theta})$ is bounded for all $x \in [0, 1]$, $t \geq 0$ and*

$$\lim_{t\to\infty} \max_{x\in[0,1]} |A(x, t)| = 0. \tag{13.187}$$

Proof. Consider the backstepping transformation

$$w(x) = T[h](x) = h(x) - \int_0^x \hat{k}(x, y)h(y)\,dy \tag{13.188}$$

applied to the function

$$h(x) = \psi(x) + \int_0^1 \hat{\theta}(\xi)\Phi(x, \xi)\,d\xi. \tag{13.189}$$

The inverse transformation of (13.188) is

$$h(x) = w(x) + \int_0^x \hat{l}(x, y)w(y)dy, \tag{13.190}$$

where

$$\hat{l}(x, y) = \frac{-\hat{\theta}(x - y)}{a_0}. \tag{13.191}$$

The state w can be shown to satisfy the following PDE:

$$\begin{aligned} w_t =& a_0 w_{xx} - a_0\hat{k}(x, 0)\hat{e}_x(0) + T\left[\int_0^1 \hat{\theta}_t(\xi)\Phi(x, \xi)\,d\xi\right] \\ &- \int_0^x w(y)\left[\hat{l}_t(x, y) - \int_y^x \hat{l}_t(x, \xi)\hat{k}(\xi, y)\,d\xi\right]dy, \end{aligned} \tag{13.192}$$

$$w(0) = 0, \tag{13.193}$$

$$w(1) = 0. \tag{13.194}$$

Rewriting the filter (13.166)–(13.168) in the form

$$\Phi_t(x, \xi) = a_0\Phi_{xx}(x, \xi) + \delta(x - \xi)(w_x(0) + e_x(0)), \tag{13.195}$$

$$\Phi(0, \xi) = 0, \tag{13.196}$$

$$\Phi(1, \xi) = 0, \tag{13.197}$$

we obtain two interconnected systems driven by signals $e_x(0)$ and $\hat{\theta}_t$.

In the rest of the proof we are going to need the bounds on $\hat{k}$, $\hat{l}$, and $\hat{l}_t$. From (13.186) and (13.191) we have

$$\left|\hat{l}(x, y)\right| \le L_0, \qquad \left|\hat{k}(x, y)\right| \le K_0 = L_0 e^{L_0}, \tag{13.198}$$

where $L_0 = \frac{1}{|a_0|}\max_{t\ge 0}\|\hat{\theta}\|_\infty$. Moreover we have $\left|\hat{l}_t(x, y)\right| \le \left|\frac{\hat{\theta}_t(x-y)}{a_0}\right|$, which according to (13.179) is bounded and square integrable.

We start the stability analysis of the system (13.192)–(13.197) with the Lyapunov function

$$V_1 = \frac{1}{2}\|\Phi\|^2. \tag{13.199}$$

Its time derivative is

$$
\begin{aligned}
\dot{V}_1 &= \text{Re}\left\{\int_0^1\int_0^1 \left[a_0\Phi_{xx}(x,\xi)+\delta(x-\xi)[\hat{e}_x(0)+w_x(0)]\right]\Phi^*(x,\xi)dx\,d\xi\right\}\\
&= -\text{Re}\,(a_0)\|\Phi_x\|^2+\text{Re}\left\{[\hat{e}_x(0)+w_x(0)]\int_0^1\int_0^1 \delta(x-\xi)\Phi^*(x,\xi)dx\,d\xi\right\}\\
&\le -\|\Phi_x\|^2+\left|\hat{e}_x(0)+w_x(0)\right|\,\|\Phi_x\|\\
&\le -\|\Phi_x\|^2+\left[\frac{|\hat{e}_x(0)|(1+\|\Phi_x(0,\cdot)\|)}{\sqrt{1+\|\Phi_x(0,\cdot)\|^2}}+\|w_{xx}\|\right]\|\Phi_x\|\\
&\le -\left(1-c_1-\frac{1}{2c_3}\right)\|\Phi_x\|^2+l_1+l_1\|\Phi_x\|^2+c_2\|\Phi_{xx}\|^2+\frac{c_3}{2}\|w_{xx}\|^2,
\end{aligned}
$$

where l_1 is a generic function of time in $\mathcal{L}_1$. In the above estimate we used the inequality $\int_0^1 \Phi(x,x)dx \le \|\Phi_x\|$, as well as the Poincaré inequality and properties (13.179)–(13.181). Here and later, by c_i we denote arbitrary positive constants that will be chosen at the end of the proof.

Consider another Lyapunov function,

$$
V_2 = \frac{1}{2}\|w\|^2. \tag{13.200}
$$

Its time derivative $\dot{V}_2 = \text{Re}\int_0^1 w_t(x)w^*(x)dx$ can be evaluated separately for the different terms of $w_t(x)$ in (13.192):

Term 1:

$$
\text{Re}\left\{\int_0^1 a_0 w_{xx}(x)w^*dx\right\} = -\|w_x\|^2.
$$

Term 2:

$$
\begin{aligned}
&\text{Re}\left\{\int_0^1 -a_0\hat{k}(x,0)\hat{e}_x(0)w^*(x)dx\right\}\\
&\le \int_0^1 \left|a_0\hat{k}(x,0)\hat{e}_x(0)w^*(x)\right|dx\\
&\le |a_0|K_0|\hat{e}_x(0)|\int_0^1 |w^*(x)|dx\\
&\le |a_0|K_0|\hat{e}_x(0)|\|w\|\frac{1+\|\Phi_x(0,\cdot)\|}{\sqrt{1+\|\Phi_x(0,\cdot)\|^2}}\\
&\le |a_0|K_0\frac{|\hat{e}_x(0)|}{\sqrt{1+\|\Phi_x(0,\cdot)\|^2}}\|w\|[1+\|\Phi_{xx}\|]\\
&\le c_4\|w\|^2+c_5\|\Phi_{xx}\|^2+l_1\|w\|^2+l_1.
\end{aligned}
$$

Term 3:

$$
\begin{aligned}
&\mathrm{Re}\left\{\int_0^1 -\int_0^x w(y)\left[\hat{l}_t(x,y)-\int_y^x \hat{l}_t(x,\xi)\hat{k}(\xi,y)\,d\xi\right]dy\,w^*(x)dx\right\}\\
&\leq \int_0^1\int_0^x |w(y)|\left|\hat{l}_t(x,y)-\int_y^x \hat{l}_t(x,\xi)\hat{k}(\xi,y)\,d\xi\right|dy\,|w|dx\\
&\leq \frac{1}{4c_6}\left\|\int_0^1 |w(y)|\left[\left|\hat{l}_t(x,y)\right|+\int_y^x\left|\hat{l}_t(x,\xi)\hat{k}(\xi,y)\right|\,d\xi\right]dy\right\|^2+c_6\|w\|^2\\
&\leq \frac{1}{4c_6}\left[\int_0^1 |w(y)|\left(\frac{\|\hat{\theta}_t\|}{|a_0|}+\frac{\|\hat{\theta}_t\|}{|a_0|}K_1\right)dy\right]^2+c_6\|w\|^2\\
&\leq \frac{1}{4c_6}\frac{(1+K_1)^2}{|a_0|^2}\|\hat{\theta}_t\|^2\|w\|^2+c_6\|w\|^2\\
&\leq l_1\|w\|^2+c_6\|w\|^2.
\end{aligned}
$$

Term 4:

$$
\begin{aligned}
&\mathrm{Re}\left\{\int_0^1 T\left[\int_0^1 \hat{\theta}_t(\xi)\Phi(x,\xi)\,d\xi\right]w^*(x)\,dx\right\}\\
&\leq \left\|T\left[\int_0^1 \hat{\theta}_t(\xi)\Phi(x,\xi)\,d\xi\right]\right\|\|w\|\\
&\leq \left\{\left\|\int_0^1 \hat{\theta}_t(\xi)\Phi(x,\xi)\,d\xi\right\|+\left\|\int_0^x |\hat{k}(x,y)|\left|\int_0^1 \hat{\theta}_t(\xi)\Phi(y,\xi)\,d\xi\right|dy\right\|\right\}\|w\|\\
&\leq \left\{\|\hat{\theta}_t\|\|\Phi\|+K_0\|\hat{\theta}_t\|\|\Phi\|\right\}\|w\|\\
&\leq (1+K_0)\|\hat{\theta}_t\|\|\Phi\|\|w\|\\
&\leq l_1\|w\|^2+c_7\|\Phi\|^2.
\end{aligned}
$$

We have

$$
\dot{V}_2 \leq -\|w_x\|^2+(c_4+c_6)\|w\|^2+c_7\|\Phi\|^2+c_5\|\Phi_{xx}\|^2+l_1\|w\|^2+l_1. \quad (13.201)
$$

Using the third Lyapunov function,

$$
V_3=\frac{1}{2}\|\Phi_x\|^2, \quad (13.202)
$$

we get

$$\dot{V}_3 = \mathrm{Re}\left\{\int_0^1\int_0^1 \Phi_{xt}(x,\xi)\Phi_x^*(x,\xi)dx\,d\xi\right\}$$

$$= -\mathrm{Re}\left\{\int_0^1\int_0^1 [a_0\Phi_{xx}(x,\xi) + \delta(x,\xi)[\hat{e}_x(0) - w_x(0)]]\Phi_{xx}^*(x,\xi)dx\,d\xi\right\}$$

$$\leq -\mathrm{Re}(a_0)\|\Phi_{xx}\|^2 + |\hat{e}_x(0) - \hat{w}_x(0)|\|\Phi_{xx}\|$$

$$\leq -\mathrm{Re}(a_0)\|\Phi_{xx}\|^2 + \left[\frac{\hat{e}_x(0)(1+\|\Phi_x(0,\cdot)\|)}{\sqrt{1+\|\Phi_x(0,\cdot)\|^2}} + \|w_{xx}\|\right]\|\Phi_{xx}\|$$

$$\leq -\|\Phi_{xx}\|^2 + \frac{\hat{e}_x(0)(1+\|\Phi_x(0,\cdot)\|)}{\sqrt{1+\|\Phi_x(0,\cdot)\|^2}}\|\Phi_{xx}\| + \frac{c_{14}}{2}\|\Phi_{xx}\|^2 + \frac{1}{2c_{14}}\|w_{xx}\|^2.$$

Using Agmon's inequality and the result (13.179), we have

$$\frac{\hat{e}_x(0)(1+\|\Phi_x(0,\cdot)\|)}{\sqrt{1+\|\Phi_x(0,\cdot)\|^2}}\|\Phi_{xx}\| \leq l_1 + c_{10}\|\Phi_{xx}\|^2$$

$$+ \frac{\hat{e}_x(0)}{\sqrt{1+\|\Phi_x(0,\cdot)\|^2}}2\sqrt{\|\Phi_x\|\|\Phi_{xx}\|}\|\Phi_{xx}\|$$

$$\leq l_1 + c_{10}\|\Phi_{xx}\|^2 + \frac{l_1 4\|\Phi_x\|\|\Phi_{xx}\|}{2c_{11}} + \frac{c_{11}}{2}\|\Phi_{xx}\|^2$$

$$\leq l_1 + c_{10}\|\Phi_{xx}\|^2 + \frac{l_1}{2c_{11}}\left(\frac{8\|\Phi_x\|^2}{c_{12}}\right.$$

$$\left. + \|\Phi_{xx}\|^2\frac{c_{12}}{2}\right) + \frac{c_{11}}{2}\|\Phi_{xx}\|^2$$

$$\leq l_1 + \left(c_{10} + \frac{c_{11}}{2} + c_{13}\right)\|\Phi_{xx}\|^2 + l_1\|\Phi_x\|^2,$$

and then

$$\dot{V}_3 \leq -\left[1 - c_{10} - \frac{c_{11}}{2} - c_{13} - \frac{c_{14}}{2}\right]\|\Phi_{xx}\|^2 + l_1\|\Phi_x\|^2 + \frac{1}{2c_{14}}\|w_{xx}\|^2. \tag{13.203}$$

The fourth and last Lyapunov function we need is

$$V_4 = \frac{1}{2}\|w_x\|^2. \tag{13.204}$$

Its time derivative is

$$\begin{aligned}
\dot{V}_4 =& \operatorname{Re}\left\{\int_0^1 w_{xt}(x) w_x^*(x)\,dx\right\} \\
=& -\operatorname{Re}\left\{\int_0^1 \left[a_0 w_{xx}(x) - a_0 \hat{k}(x,0)\hat{e}_x(0)\right.\right. \\
& - \int_0^x w(y)\left[\hat{l}_t(x,y) - \int_y^x \hat{l}_t(x,\xi)\hat{k}(\xi,y)\,d\xi\right] dy \\
& \left.\left. + T\left[\int_0^1 \hat{\theta}_t(\xi)\Phi(x,\xi)\,d\xi\right]\right] w_{xx}^*(x) dx\right\}.
\end{aligned}$$

It can be evaluated similarly to the V_2 function, apart from the second term, which is

$$\begin{aligned}
& \operatorname{Re}\left\{\int_0^1 -a_0 \hat{k}(x,0)\hat{e}_x(0) w_{xx}^*\,dx\right\} \\
& \le |a_0| K_0 |\hat{e}_x(0)| \|w_{xx}\| \frac{1 + \|\Phi_x(0,\cdot)\|}{\sqrt{1 + \|\Phi_x(0,\cdot)\|^2}} \\
& \le |a_0| K_0 \left[\frac{1 + \|\Phi_x(0,\cdot)\|}{\sqrt{1 + \|\Phi_x(0,\cdot)\|^2}} \|w_{xx}\| + \right. \\
& \quad \left. + \frac{1 + \|\Phi_x(0,\cdot)\|}{\sqrt{1 + \|\Phi_x(0,\cdot)\|^2}} \|\Phi_x(0,\cdot)\| \|w_{xx}\|\right] \\
& \le |a_0| K_0 \left[l_1 + c_{15} \|w_{xx}\|^2 + l_1 \right. \\
& \quad \left. + \left(c_{16} + \frac{c_{17}}{2}\right) \|w_{xx}\|^2 + c_{18} \|\Phi_{xx}\|^2 + l_1 \|\Phi_x\|^2\right].
\end{aligned}$$

That gives

$$\begin{aligned}
\dot{V}_4 \le& -\left[1 - |a_0| K_0 \left(c_{15} + c_{16} + \frac{c_{17}}{2}\right) - c_{19} - c_{20}\right] \|w_{xx}\|^2 \\
& + |a_0| K_0 c_{18} \|\Phi_{xx}\|^2 + l_1 \|\Phi_x\|^2 + l_1 \|w\|^2 + l_1 \|\Phi\|^2 + l_1.
\end{aligned}$$

Taking the total Lyapunov function as

$$V = V_1 + V_2 + V_3 + \alpha_4 V_4,$$

with $\alpha_4 > 0$, and rewriting the time derivatives of the four terms in a more compact form, we have

$$\begin{aligned}
\dot{V}_1 + \dot{V}_2 \le& -\|w_x\|^2 - (1 - c_1 - d_1)\|\Phi_x\|^2 + d_2 \|w\|^2 \\
& + c_7 \|\Phi\|^2 + l_1 \|w\|^2 + l_1 \|\Phi_x\|^2 \\
& + l_1 + \frac{1}{4d_1} \|w_{xx}\|^2 + d_3 \|\Phi_{xx}\|^2, \\
\dot{V}_3 \le& -\left[1 - d_4 - \frac{c_{14}}{2}\right] \|\Phi_{xx}\|^2 + l_1 \|\Phi_x\|^2 + \frac{1}{2c_{14}} \|w_{xx}\|^2, \\
\alpha_4 \dot{V}_4 \le& -\alpha_4 \left[1 - d_5\right] \|w_{xx}\|^2 + d_6 \|\Phi_{xx}\|^2 + \\
& + l_1 \|w\|^2 + l_1 \|\Phi\|^2 + l_1 \|\Phi_x\|^2 + l_1,
\end{aligned}$$

where

$$
\begin{aligned}
d_1 &= \frac{1}{2c_3},\\
d_2 &= c_4 + c_6,\\
d_3 &= c_2 + c_5,\\
d_4 &= c_{10} + \frac{c_{11}}{2} + c_{13},\\
d_5 &= |a_0| K_0 \left(c_{15} + c_{16} + \frac{c_{17}}{2} \right) + c_{19} + c_{20},\\
d_6 &= \alpha_4 |a_0| K_0 c_{18}.
\end{aligned}
$$

Choosing $d_1 = 1/3$, $d_2 = 1/9$, $d_3 = d_4 = d_5 = d_6 = c_7 = 1/18$, $c_1 = 1/9$, and $c_{14} = \alpha_4 = 4/3$, we obtain

$$\dot{V} \leq -\frac{1}{9} V + l_1 V + l_1. \tag{13.205}$$

13.11.1 Boundedness

Given the inequality (13.205), by Lemma D.3 we conclude that all the signals $\|w\|$, $\|\Phi\|$, $\|w_x\|$, and $\|\Phi_x\|$ belong to $\mathcal{L}_2 \cap \mathcal{L}_\infty$. Using the inverse transformation (13.190) and the bound (13.198), we get that $\|h\|$ too belongs to $\mathcal{L}_2 \cap \mathcal{L}_\infty$. From the definition (13.189) of h, this implies that the state of the input filter ψ is also in $\mathcal{L}_2 \cap \mathcal{L}_\infty$. Finally, from the equation of the observer (13.172) and from the transformation (13.149), boundedness of $B(x)$ and $A(x)$ follows. Note that by Agmon's inequality we also get pointwise boundedness of all the signals.

13.11.2 Regulation to Zero

Using Lemma D.2, we get that $\|w\|$, $\|w_x\|$, $\|\Phi\|$, and $\|\Phi_x\|$ go to zero as $t \to \infty$. Using the relation, directly derived from (13.190),

$$h_x(x) = w_x(x) + \int_0^x \hat{l}_x(x, y) w(y) dy + \hat{l}(x, x) w(x),$$

and the bound (13.198), we get $\|h_x\| \to 0$ when $t \to \infty$. From the definition (13.189) we obtain the expression for $h_x(x)$,

$$h_x(x) = \psi_x(x) + \int_0^1 \hat{\theta}(\xi) \Phi_x(x, \xi)\, d\xi,$$

which implies that $\|\psi_x\|$ also goes to zero. From the expression of the spatial derivative of the observer error

$$e_x(x) = B_x(x) - \psi_x(x) - \int_0^1 \theta(\xi) \Phi_x(x, \xi)\, d\xi,$$

we obtain that $\|B_x\| \to 0$ as $t \to \infty$. Since $\|B\|$ is bounded, by Agmon's inequality this implies that $B(x, t) \to 0$ for all $x \in [0, 1]$ as $t \to \infty$. Since the transformation (13.149) is invertible, we get (13.187). □

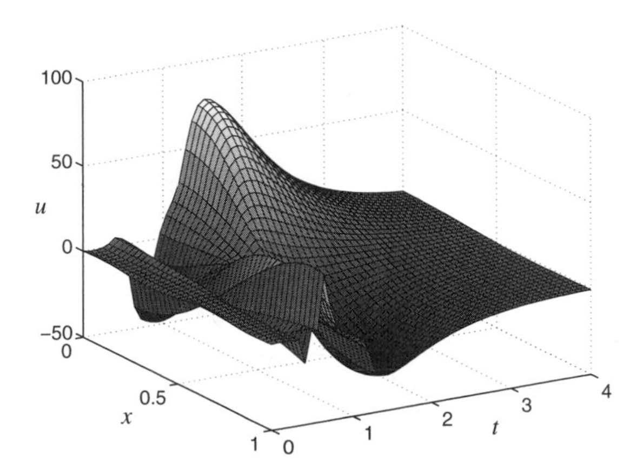

Figure 13.1 The closed-loop state $u(x, t)$ for the plant (13.1)–(13.3) with an adaptive output-feedback controller.

13.12 SIMULATIONS

First we present the results of numerical simulations for the plant (13.1)–(13.3). The parameters are $b(x) = 3 - 2x^2$, $\lambda(x) = 16 + 3\sin(2\pi x)$, $\varepsilon(x) \equiv 1$, $f(x, y) \equiv 0$, $g(x) \equiv 0$, and $q = 0$, so the plant is unstable. The evolution of the closed-loop state is shown in Figure 13.1. We can see that the regulation is achieved. The parameter estimates, shown in Figures 13.2, converge to some stabilizing values. The second simulation for the plant (13.1)–(13.3) shows tracking of the reference trajectory $u^r(0, t) = 3\sin(6t)$. The control effort and output evolution are shown in Figure 13.3. We can see that after the initial transient, the output follows the desired trajectory.

The simulation results for the Ginzburg-Landau equation are shown in Figures 13.4–13.9. In Figure 13.4 $a(x)$ is shown (solid line), while $a_0 = 0.0234(1 - j)$. The open-loop and closed-loop responses are shown in Figures 13.5 and 13.6, respectively. The control signal is shown in Figure 13.7. Even though the objective of the control design is stabilization and not necessarily identification of the parameters of the system, the final profile of the parameter estimate $\hat{\theta}$ turns out to be close to the true parameter $\theta(x)$, as can be seen in Figure 13.8. This suggests that a rather accurate knowledge of the parameter $a(x)$ is needed for the plant to be regulated to zero. To verify this intuitive statement, we approximate the unknown function $a(x)$ by its mean value $\bar{a}$ (plotted as a thin line in Fig. 13.4). The nonadaptive controller designed on the basis of this approximation fails to stabilize the plant, as illustrated in Figure 13.9.

13.13 NOTES AND REFERENCES

The adaptive schemes presented in this chapter admit several straightforward extensions.

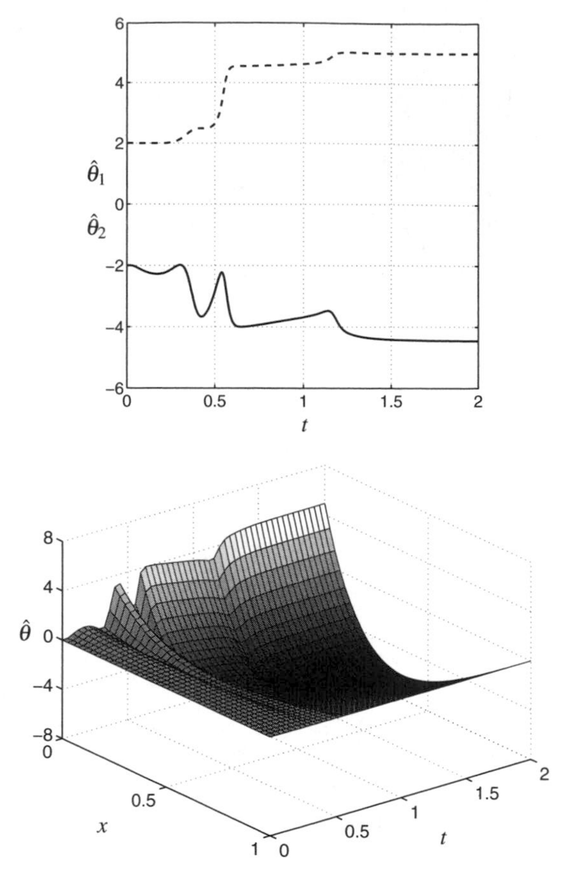

Figure 13.2 The parameter estimates $\hat{\theta}_1(t)$ (top, solid line), $\hat{\theta}_2(t)$ (top, dashed line), and $\hat{\theta}(x,t)$ (bottom).

First, it is possible to extend the designs to the case of Neumann actuation with either Dirichlet or Neumann sensing. Second, there is no conceptual difficulty in employing least squares update laws instead of gradient ones. Third, since the adaptive scheme involves online computation of the control gain, its robustness with respect to the numerical error can be analyzed in the same way as is done in Chapter 11 for the state-feedback problem.

In addition to conceptually easy extensions, there exist several challenging open problems, which we discuss here in order of increasing difficulty.

One such problem is adaptive identification. One cannot hope to exactly identify spatially varying unknown parameters with only boundary input. One practical approach to this problem is to approximate the unknown function with a high-order polynomial and then identify the coefficients of this polynomial. Under the assumption that the boundary input is sufficiently rich and the polynomial is of a sufficiently high order, one could identify the unknown parameter with a desired accuracy. Another issue here is to obtain the original parameters of the plant from

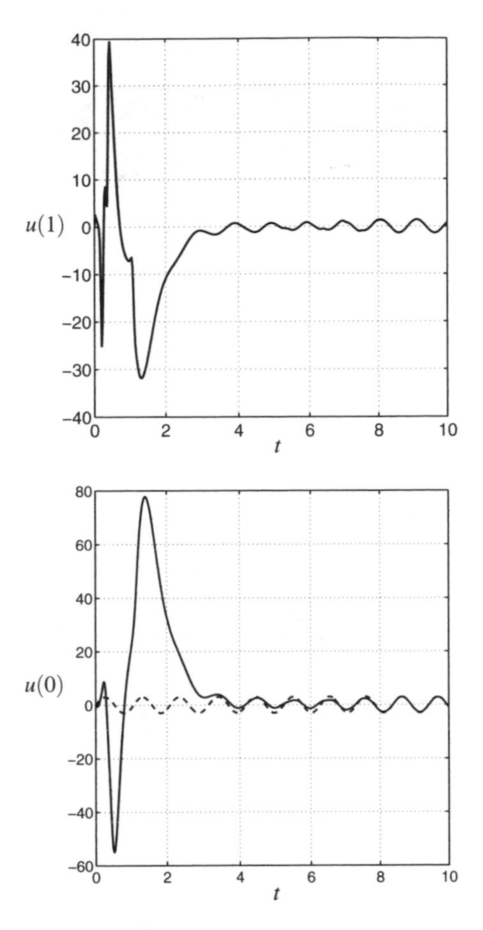

Figure 13.3 The adaptive trajectory tracking. Top: the control effort. Bottom: reference signal at the boundary (dashed line) and actual evolution of the output (solid line).

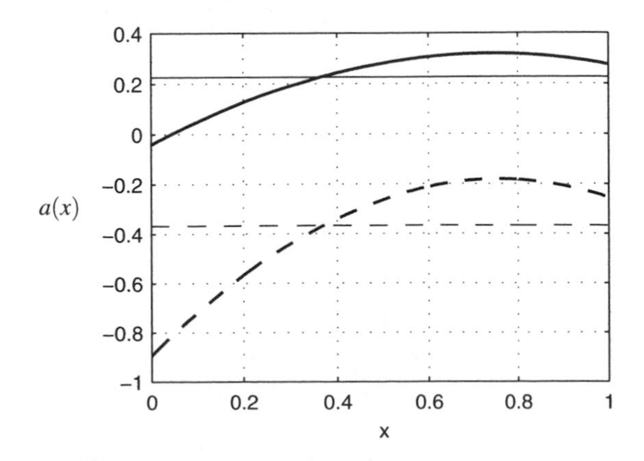

Figure 13.4 Real part (solid line) and imaginary part (dashed line) of the parameter $a(x)$ (bold line) together with the approximate constant $\bar{a}$ (thin line).

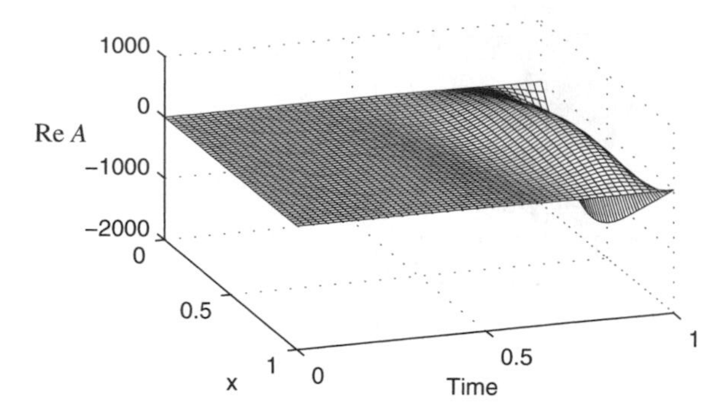

Figure 13.5 Real part of the state of the open-loop Ginzburg-Landau equation. The imaginary part is qualitatively the same.

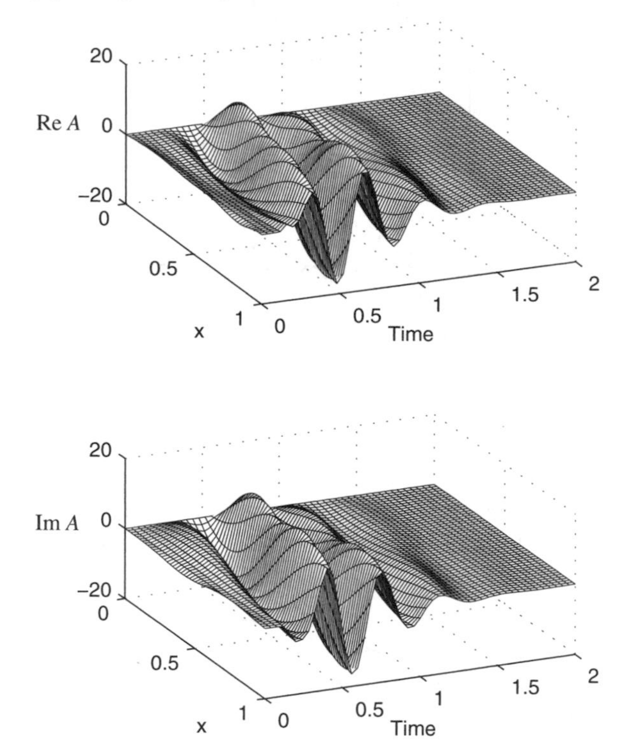

Figure 13.6 The closed-loop response of the Ginzburg-Landau equation.

the parameters of the observer canonical form (e.g., to obtain $\lambda(x)$ from $\theta(x)$ and θ_1). This so-called inverse problem is ill-posed in a sense that the smallest changes in $\theta(x)$ or θ_1 result in huge changes in $\lambda(x)$. Therefore, some regularization techniques would have to be used.

An important open problem is output-feedback adaptive control for plants with collocated sensing and actuation. The difficulty here is that the Volterra-type transformation into the observer canonical form results in the input appearing both at

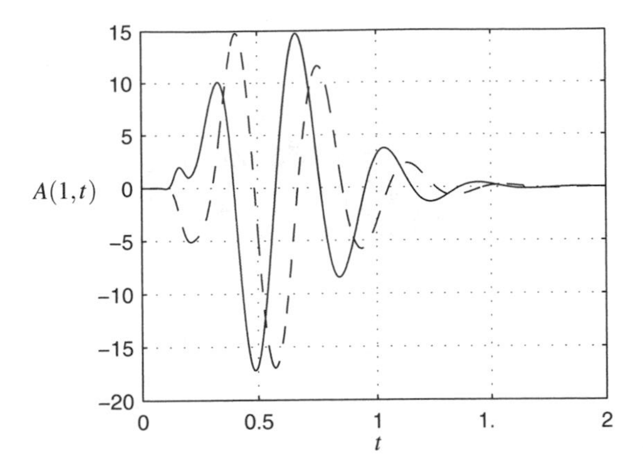

Figure 13.7 Real part (solid line) and imaginary part (dashed line) of the control signal for the Ginzburg-Landau equation.

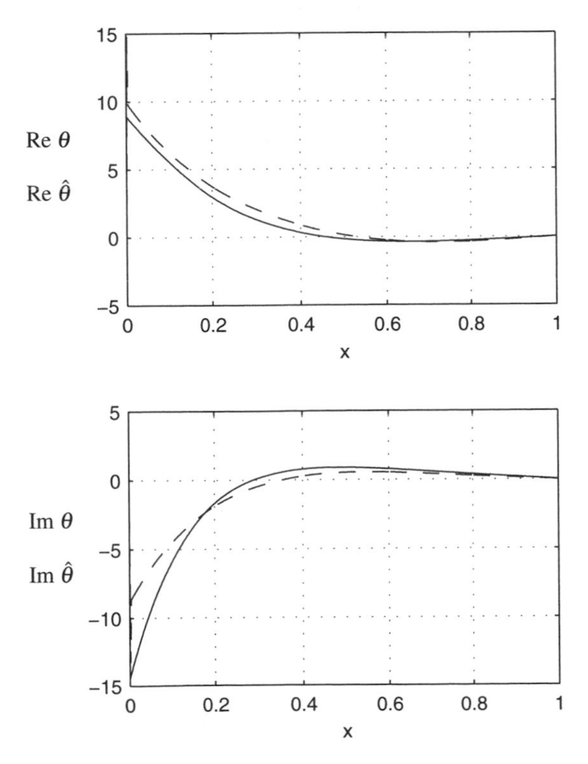

Figure 13.8 The parameter θ (solid line) with the final profile of its estimate $\hat{\theta}$ (dashed line) for the Ginzburg-Landau equation.

the boundary and inside the domain. Thus, the main obstacle here is the absence of parametrized families of controllers, since the transformed system is no longer in the class of plants considered in this book.

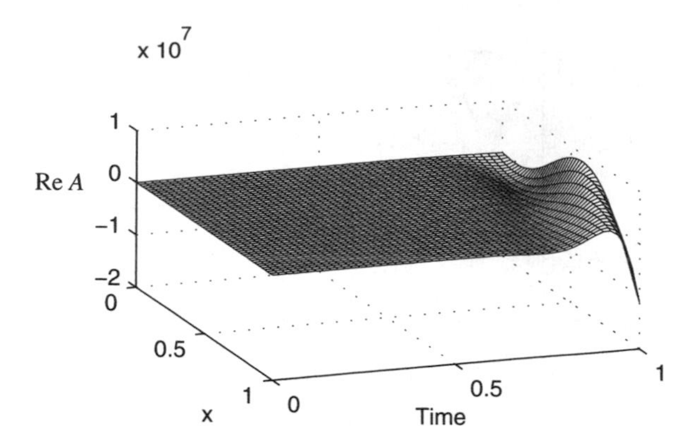

Figure 13.9 Real part of the state of the Ginzburg-Landau equation with nonadaptive controller designed on the basis of $\bar{a}$. The imaginary part is qualitatively the same.

The hardest open problem is output-feedback stabilization of plants with unknown diffusivity (or, in general, plants of unknown parameter multiplying the highest spatial derivative). This problem is very important in many applications, such as flow control (when the Reynolds number is unknown). As explained in Section 13.2, the fundamental difficulty is that one cannot separate the unknown parameter and the second derivative of the state.

Chapter Fourteen

Inverse Optimal Control

In this chapter we present backstepping designs that not only achieve stability but at the same time *minimize* some meaningful, positive definite, cost functionals. This property of a controller is referred to as *inverse optimality*. The controllers presented in the previous chapters do not possess this property. In this chapter we present redesigns that achieve inverse optimality. The idea of inverse optimality has a long history in control, starting from Kalman's paper [53] for linear systems and a paper by Moylan and Anderson [89] that extends the idea to nonlinear systems. More recent nonlinear efforts inspired by control Lyapunov functions are presented in the books by Freeman and Kokotovic [40] and Sepulchre, Jankovic, and Kokotovic [107] for deterministic systems, and in the book by Krstic and Deng [66] for Itô systems.

We pursue two ideas for combining inverse optimality and adaptivity, for which we use rather uninspired (but correct) names

- adaptive inverse optimal control, and
- inverse optimal adaptive control.

With these two ideas being differentiated only through the ordering of the terms "adaptive" and "inverse optimal," there is a significant danger of confusion. For this reason we carefully explain the difference between the two concepts.

By *adaptive inverse optimal* control we refer to a controller designed with known parameter values but which has a multiplicative gain that is tuned adaptively to a sufficient value that makes the controller stabilizing. The reason for this is that inverse optimal controllers have a very particular structure, or, to be more specific, a very particular dependence on a Lyapunov function (which is also a solution to a Hamilton-Jacobi equation), and because of this particular structure they require the use of sufficiently high gain. Analytical estimates of such a gain are usually very conservative. We employ adaptation in such a way that the designer can start with a zero gain on an inverse optimal feedback structure and the adaptation will tune the gain to a stabilizing value that is just sufficiently high.

By *inverse optimal adaptive* control we refer to an entirely different idea. In this case the plant parameters are considered to be unknown. The controller is designed not only as a stabilizing adaptive controller but also as an adaptive controller that is inverse optimal with respect to a meaningful cost functional. In the nonadaptive control setting this is now a standard concept, and the cost minimized by an inverse optimal controller includes a penalty on control and on the state. In the adaptive case, the situation is much more complex, and the cost needs to involve a penalty on the parameter estimate as well. The key challenge comes from the fact that inverse optimality has to be achieved, by design, for a system that has unknown parameters. Thankfully, this problem is solvable, as we show in this chapter. One can incorporate a particular positive definite penalty on the parameter estimation error in the cost, in addition to the penalties on the state and control.

In Sections 14.1–14.4 we deal with *adaptive inverse optimal* control: we design inverse optimal controllers for plants with *known* parameters, where these controllers minimize cost functionals that penalize both the control and the state. As mentioned above, to reduce the control effort and derive explicit bounds on the state and control, we also propose an adaptive control law that tunes a control gain to a sufficient (but nonconservative) value. These designs are compared with linear quadratic regulator (LQR) designs in Section 14.5, both conceptually and numerically.

In Sections 14.6–14.7 we deal with *inverse optimal adaptive* control, where we address an *unknown* parameter case and design an adaptive controller that is made inverse optimal. We show that the entire parameter-adaptive *nonlinear* system is optimal relative to a cost that simultaneously penalizes state transients, control transients, and infinite-time transients of the exponential of the square of the parameter estimation error.

14.1 NONADAPTIVE INVERSE OPTIMAL CONTROL

We consider the plant

$$u_t(x,t) = \varepsilon u_{xx}(x,t) + \lambda(x)u(x,t) + g(x)u(0,t) + \int_0^x f(x,y)u(y,t)\,dy, \tag{14.1}$$

$$u_x(0,t) = -qu(0,t), \tag{14.2}$$

$$u_x(1,t) = U(t), \tag{14.3}$$

and use the transformation

$$w(x,t) = u(x,t) - \int_0^x k(x,y)u(y,t)\,dy, \tag{14.4}$$

$$u(x,t) = w(x,t) + \int_0^x l(x,y)w(y,t)\,dy, \tag{14.5}$$

where $k(x,y)$ is the solution of (2.20), (2.21), (2.24) and $l(x,y)$ is the solution of (2.47)–(2.49). Then $w(x,t)$ satisfies the following PDE:

$$w_t(x,t) = \varepsilon w_{xx}(x,t) - cw(x,t), \tag{14.6}$$

$$w_x(0,t) = 0, \tag{14.7}$$

$$w(1,t) = u(1,t) - \int_0^1 k(1,y)u(y,t)\,dy. \tag{14.8}$$

Note that even though the kernels $k(x,y)$ and $l(x,y)$ are the same as in Chapter 2, here w satisfies the heat equation with a much more complicated boundary condition at $x = 1$. Therefore, one can think of $w(x,t)$ as a short way of writing the expression $u(x,t) - \int_0^x k(x,y)u(y,t)\,dy$.

The main result on inverse optimal stabilization is given by the following theorem.

THEOREM 14.1. *For any $c > 0$ and $\beta \geq 2$, the control law*

$$U^*(t) = -\frac{\beta}{R}\left(u(1,t) - \int_0^1 k(1,y)u(y,t)\,dy\right), \tag{14.9}$$

where $k(x,y)$ is the solution of (2.20), (2.21), (2.24), *achieves closed-loop stability of the system* (14.1)–(14.3),

$$\|u\| \leq Me^{-ct}\|u_0\|, \qquad M > 0, \tag{14.10}$$

while minimizing the cost functional

$$J = \int_0^\infty \left(Q[u](t) + R\,U(t)^2\right) dt \tag{14.11}$$

with

$$0 < R \leq \left(|l(1,1)| + L + \frac{1}{2}L^2\right)^{-1}, \qquad L = \int_0^1 |l_x(1,y)|\,dy, \tag{14.12}$$

and

$$\begin{aligned} Q[u](t) = \frac{1}{R}\beta^2 w^2(1,t) + 2\beta\left\{w(1,t)\int_0^1 l_x(1,y)w(y,t)\,dy + l(1,1)w^2(1,t)\right. \\ \left. + \frac{c}{\varepsilon}\int_0^1 w^2(x,t)\,dx + \int_0^1 w_x^2(x,t)\,dx\right\}, \end{aligned} \tag{14.13}$$

where $l(x,y)$ is the solution of (2.47)–(2.49), *and*

$$Q[u](t) \geq \frac{1}{R}\beta(\beta-2)w^2(1,t) + \frac{2\beta c}{\varepsilon}\int_0^1 w^2(x,t)\,dx + \beta\int_0^1 w_x^2(x,t)\,dx. \tag{14.14}$$

Proof. We start by proving stability. For the Lyapunov function

$$V(t) = \frac{1}{2}\int_0^1 w^2(x,t)\,dx, \tag{14.15}$$

we have (explicit dependence on time is dropped for clarity):

$$\begin{aligned}\dot{V} &= \varepsilon \int_0^1 w(x)w_{xx}(x)\,dx - c\int_0^1 w^2(x)\,dx \\ &= -2cV + \varepsilon w(1)w_x(1) - \varepsilon \int_0^1 w_x^2(x)\,dx \\ &= -2cV + \varepsilon w(1)U - \varepsilon w(1)\int_0^1 l_x(1,y)w(y)\,dy - \varepsilon l(1,1)w^2(1) - \varepsilon\|w_x\|^2 \\ &\le -2cV - \varepsilon\frac{\beta}{R}w^2(1) - \varepsilon\|w_x\|^2 + \varepsilon|l(1,1)|w^2(1) \\ &\quad + \varepsilon|w(1)|\left|w(1)\int_0^1 l_x(1,y)\,dy + \int_0^1 w_x(x)\int_0^x l_x(1,y)\,dy\,dx\right| \\ &\le -2cV - \varepsilon\frac{\beta}{R}w^2(1) - \frac{\varepsilon}{2}\|w_x\|^2 + \varepsilon|l(1,1)|w^2(1) + \varepsilon L w^2(1) + \frac{\varepsilon}{2}L^2w^2(1) \\ &\le -2cV - \frac{\varepsilon(\beta-1)}{R}w^2(1) - \frac{\varepsilon}{2}\|w_x\|^2. \end{aligned} \tag{14.16}$$

Since $\beta \ge 2$, we get $V(t) \le e^{-2ct}V(0)$. This, together with transformations (14.4) and (14.5), leads to (14.10).

Next, we show that the controller (14.9) minimizes the cost functional (14.11). Using the third line of (14.16) we can write Q in the form

$$Q[u](t) = \frac{1}{R}\beta^2 w^2(1,t) + 2\beta\left(w(1,t)U(t) - \frac{1}{\varepsilon}\dot{V}(t)\right). \tag{14.17}$$

Substituting (14.17) into (14.11), we obtain

$$\begin{aligned} J &= \int_0^\infty \left[\frac{\beta^2}{R}w^2(1,t) + 2\beta\left(w(1,t)U(t) - \frac{\dot{V}(t)}{\varepsilon}\right) + RU^2(t)\right]dt \\ &= \frac{2\beta}{\varepsilon}\left(V(0) - V(\infty)\right) + R\int_0^\infty \left[U(t) + \frac{\beta}{R}w(1,t)\right]^2 dt \\ &= \frac{2\beta}{\varepsilon}V(0) + R\int_0^\infty [U(t) - U^*(t)]^2\,dt. \end{aligned} \tag{14.18}$$

Therefore, the controller (14.9) delivers the minimum value $J = (\beta/\varepsilon)\|w_0\|^2$ for the cost functional (14.11).

To complete the proof, we show that Q is positive definite. Using (14.17) and (14.16), we get the estimate

$$\begin{aligned} Q[u] &\ge \frac{1}{R}\beta^2 w^2(1) - \frac{2}{R}\beta^2 w^2(1) + \frac{2\beta(\beta-1)}{R}w^2(1) + \beta\|w_x\|^2 + \frac{2\beta c}{\varepsilon}\|w\|^2 \\ &\ge \frac{\beta(\beta-2)}{R}w^2(1) + \beta\|w_x\|^2 + \frac{2\beta c}{\varepsilon}\|w\|^2. \end{aligned} \tag{14.19}$$

So, Q is a positive definite functional that bounds the H^1-norm of the state. □

Remark 14.1. If the Robin boundary condition (14.2) is replaced by the Dirichlet one, $u(0,t) = 0$, then the statement of Theorem 14.1 holds, with $k(x, y)$, $l(x, y)$ given by (2.71)–(2.73), (2.75)–(2.77) and $R^{-1} = |l(1, 1)| + (1/2)L^2$.

14.2 REDUCING CONTROL EFFORT THROUGH ADAPTATION

The control gain in (14.9) contains R that is rather conservative. We show how the control effort can be reduced by using adaptation to start with a zero gain and raise it online to a stabilizing value. The result is given by the following theorem.

THEOREM 14.2. *The controller*

$$\begin{aligned} U(t) &= -\hat{\theta}(t) w(1,t) \\ &= -\hat{\theta}(t)\left(u(1,t) - \int_0^1 k(1,y)u(y,t)\,dy\right) \end{aligned} \tag{14.20}$$

with an update law

$$\dot{\hat{\theta}}(t) = \gamma\, w^2(1,t), \qquad \hat{\theta}(0) = 0, \qquad \gamma > 0 \tag{14.21}$$

applied to system (14.1)–(14.3) *guarantees the following bounds for the state* $u(x,t)$*:*

$$\sup_{t\ge 0} \int_0^1 u^2(x,t)\,dx \le (1+\bar{l})^2 \left\{ \frac{2\varepsilon}{\gamma R^2} + (1+\bar{k})^2 \int_0^1 u^2(x,0)\,dx \right\}, \tag{14.22}$$

$$\int_0^\infty \int_0^1 u^2(x,t)\,dx\,dt \le \frac{(1+\bar{l})^2}{2c} \left\{ \frac{2\varepsilon}{\gamma R^2} + (1+\bar{k})^2 \int_0^1 u^2(x,0)\,dx \right\}, \tag{14.23}$$

and the following bound for control $U(t)$*:*

$$\int_0^\infty U(t)^2\,dt \le \frac{2\gamma R}{\varepsilon^2}\left((1+\bar{k})^2 \int_0^1 u^2(x,0)\,dx + \frac{4\varepsilon}{\gamma R^2}\right)^2, \tag{14.24}$$

where $\bar{l} = \sup_{(x,y)\in\mathcal{T}} l(x,y)$ *and* $\bar{k} = \sup_{(x,y)\in\mathcal{T}} k(x,y)$.

Proof. Denote $\tilde{\theta} = \theta - \hat{\theta}$,

$$\theta = \frac{\beta}{2R} = \frac{\beta}{2}\left(|l(1,1)| + L + \frac{1}{2}L^2\right). \tag{14.25}$$

Taking the Lyapunov function

$$V(t) = \frac{1}{2}\int_0^1 w^2(x,t)\,dx + \frac{\varepsilon}{2\gamma}\tilde{\theta}^2(t), \tag{14.26}$$

we get

$$
\begin{aligned}
\dot{V} = & \varepsilon w(1)\left(U - l(1,1)w(1) - \int_0^1 l_x(1,y)w(y)\,dy\right) \\
& -\varepsilon \int_0^1 w_x^2\,dx - c\int_0^1 w^2\,dx - \varepsilon\tilde{\theta}w^2(1) \\
\leq & -\varepsilon(\theta - R^{-1})w^2(1) - \frac{\varepsilon}{2}\int_0^1 w_x^2\,dx - c\int_0^1 w^2\,dx \\
\leq & -\frac{\varepsilon}{2R}(\beta - 2)w^2(1) - \frac{\varepsilon}{2}\int_0^1 w_x^2\,dx - c\int_0^1 w^2\,dx. \qquad (14.27)
\end{aligned}
$$

From (14.26) and (14.27) the following estimates easily follow:

$$
\sup_{t\geq 0}\int_0^1 w^2(x,t)\,dx \leq \int_0^1 w^2(x,0)\,dx + \frac{\varepsilon}{\gamma}\tilde{\theta}^2(0), \qquad (14.28)
$$

$$
\int_0^\infty \int_0^1 w^2(x,t)\,dx\,dt \leq \frac{\int_0^1 w^2(x,0)\,dx + \frac{\varepsilon}{\gamma}\tilde{\theta}^2(0)}{2c}, \qquad (14.29)
$$

and

$$
\begin{aligned}
\int_0^\infty U(t)^2\,dt \leq & \int_0^\infty \hat{\theta}^2(t)w^2(1,t)\,dt \\
\leq & 2\left(\theta^2 + \sup_{t\geq 0}\tilde{\theta}^2(t)\right)\int_0^\infty w^2(1,t)\,dt \\
\leq & 2\left(2\theta^2 + \frac{\gamma}{\varepsilon}\int_0^1 w^2(x,0)\,dx\right)\frac{R}{\varepsilon(\beta-2)}\left(\int_0^1 w^2(x,0)\,dx + \frac{\varepsilon}{\gamma}\theta^2\right) \\
\leq & \frac{\gamma R}{\varepsilon^2(\beta-2)}\left(\int_0^1 w^2(x,0)\,dx + \frac{\varepsilon\beta^2}{2\gamma R^2}\right)^2. \qquad (14.30)
\end{aligned}
$$

Using (14.4) and (14.5), we get the relationship between the norms of u and w:

$$
\|u\|^2 \leq (1+\bar{l})^2\|w\|^2, \qquad \|w\|^2 \leq (1+\bar{k})^2\|u\|^2. \qquad (14.31)
$$

Now (14.28)–(14.30) with (14.31) and $\beta = 2\sqrt{2}$ give estimates (14.22)–(14.24), respectively. □

The controller (14.20) is not inverse optimal, but it performs better than the stabilizing controller (2.82) and uses less gain than the nonadaptive inverse optimal one. As the numerical simulations in Section 14.4 show, performance improvement can be quite substantial: for the considered settings, the adaptive controller uses several dozen times less gain than the nonadaptive one would.

14.3 DIRICHLET ACTUATION

There are two ways to address the case of Dirichlet actuation. One way is to use (14.20) to express $u(1,t)$:

$$u(1,t) = -\frac{1}{\hat{\theta}} u_x(1,t) + \int_0^1 k(1,y)u(y,t)\,dy, \tag{14.32}$$

with the dynamics of $\hat{\theta}$ given by (14.21). The only restriction is that $\hat{\theta}(0) > 0$, which ensures $\hat{\theta}(t) > 0$ for $t \geq 0$. The downside of this approach is that the resulting controller puts a penalty on $u_x(1,t)$ instead of $u(1,t)$, so it cannot be called optimal as a Dirichlet controller.

To get a true inverse optimal Dirichlet controller, we write, similarly to (14.20)–(14.21):

$$u(1,t) = -\hat{\theta} w_x(1,t), \qquad \dot{\hat{\theta}} = \gamma w_x^2(1,t), \tag{14.33}$$

or, using the expression for $w_x(1,t)$,

$$w_x(1,t) = u_x(1,t) - k(1,1)u(1,t) - \int_0^1 k_x(1,y)u(y,t)\,dy, \tag{14.34}$$

the controller can be written as

$$u(1,t) = -\frac{\hat{\theta}}{1 + \frac{1}{2}\hat{\theta}\int_0^1 (\lambda(y) + c)\,dy}\left(u_x(1,t) - \int_0^1 k_x(1,y)u(y,t)\,dy\right). \tag{14.35}$$

The initial value of the adaptive gain $\hat{\theta}(0)$ can be taken to be zero in meaningful stabilization problems where $\lambda(x) \geq 0$ (if this condition is not satisfied, then one can increase c to make sure that $\lambda(x) + c \geq 0$). Note the structural similarity of the controllers (14.32) and (14.35). Both employ an integral operator of u measured for all $x \in [0, 1)$ and a measurement of u_x at $x = 1$. The optimality of the controller (14.35) (for the case of constant $\hat{\theta}$) can be proved along the same lines as in the proof of Theorem 14.2.

14.4 DESIGN EXAMPLE

Let us illustrate the designs of the previous two sections by constructing explicit controllers for the reaction-diffusion plant

$$u_t(x,t) = u_{xx}(x,t) + \lambda u(x,t), \tag{14.36}$$

$$u(0,t) = 0, \tag{14.37}$$

$$u_x(1,t) = U(t). \tag{14.38}$$

In Chapter 3 we derived explicit formulas for $k(x,y)$ and $l(x,y)$:

$$k(x,y) = -\lambda y \frac{I_1\left(\sqrt{\lambda(x^2 - y^2)}\right)}{\sqrt{\lambda(x^2 - y^2)}}, \qquad l(x,y) = -\lambda y \frac{J_1\left(\sqrt{\lambda(x^2 - y^2)}\right)}{\sqrt{\lambda(x^2 - y^2)}}. \tag{14.39}$$

The stabilizing controller is

$$
\begin{aligned}
U(t) &= k(1,1)u(1,t) + \int_0^1 k(1,y)u(y,t)\,dy \\
&= -\frac{\lambda}{2}u(1,t) - \int_0^1 \lambda y \frac{I_2\left(\sqrt{\lambda(1-y^2)}\right)}{1-y^2} u(y,t)\,dy. \qquad (14.40)
\end{aligned}
$$

To construct the inverse optimal controller, we compute

$$
\begin{aligned}
L &= \int_0^1 |l_x(1,y)|\,dy = \int_0^1 \frac{\lambda y}{1-y^2}\left|J_2\left(\sqrt{\lambda(1-y^2)}\right)\right| dy \\
&= \lambda \int_0^{\sqrt{\lambda}} \frac{|J_2(z)|}{z}\,dz \le \lambda + 1. \qquad (14.41)
\end{aligned}
$$

From Remark 14.1. we have $R^{-1} = (\lambda+(\lambda+1)^2)/2$. The controller (14.9) becomes

$$
U(t) = -\beta\frac{(\lambda+1)^2+\lambda}{2}\left(u(1,t) + \int_0^1 \lambda y \frac{I_1\left(\sqrt{\lambda(1-y^2)}\right)}{\sqrt{\lambda(1-y^2)}} u(y,t)\,dy\right), \qquad (14.42)
$$

where $\beta \ge 2$.

In Figure 14.1 we compare the performances of the adaptive inverse optimal controller and the stabilizing controller. The parameters of the plant are $\lambda = 10$, $c = 0$, and the initial condition is $u_0(x) = \sin(\pi x)$. The adaptation gain is set to $\gamma = 2$. We can see that, compared to the stabilizing controller, the adaptive inverse optimal controller achieves better performance with less control effort. We can also estimate how much gain is actually saved compared to using the nonadaptive controller. As (14.42) shows, the gain of the nonadaptive controller should be greater than or equal to $(\lambda+1)^2+\lambda = 131$, while the adaptive gain is less than or equal to 2.85 at all times (Fig. 14.1, bottom), which is approximately 46 times less than the nominal gain.

14.5 COMPARISON WITH THE LQR APPROACH

The inverse optimal controllers presented above should be compared to some other optimal controllers, such as the traditional infinite-dimensional linear quadratic regulator (LQR).

The LQR approach results in stability and good performance but requires the solution of an infinite-dimensional Riccati equation. The inverse optimal design does not require the solution of a Riccati equation. Instead, one needs to solve a *linear* hyperbolic PDE, an object that is conceptually simpler than an operator Riccati equation (which is *quadratic*). Moreover, efficient numerical methods exist that exploit the structure of this hyperbolic PDE and give large computational savings compared to the numerical solution of the operator Riccati equation.

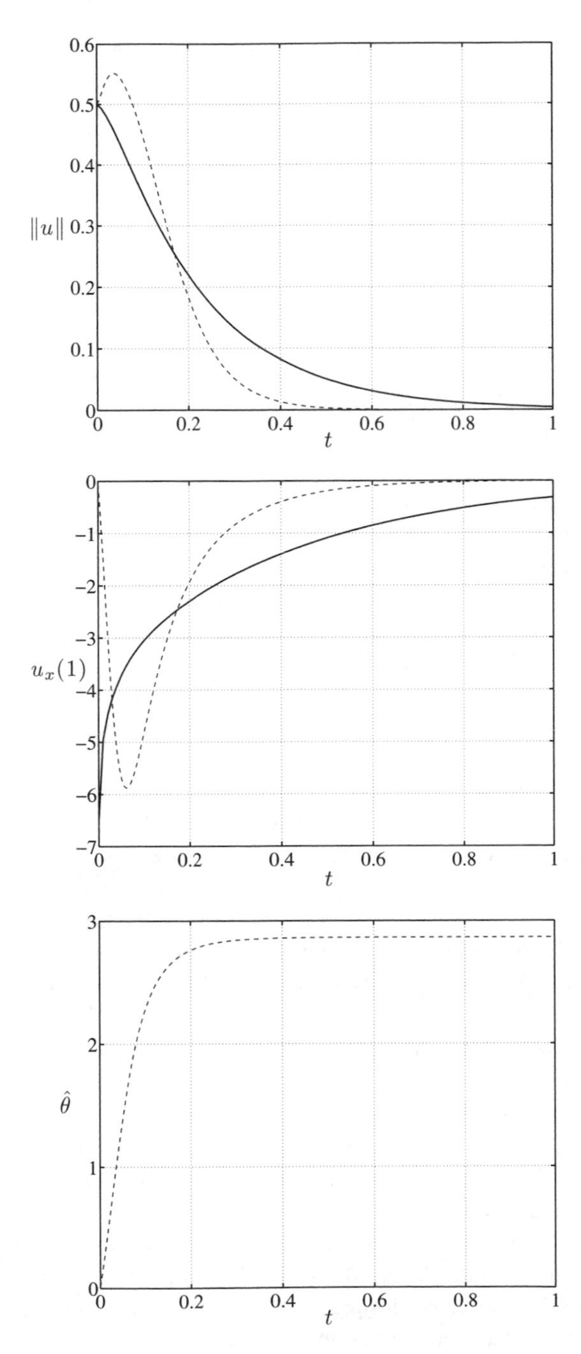

Figure 14.1 Comparison between stabilizing (solid line) and inverse optimal (dashed line) controllers. Top: L^2-norm of the state. Middle: control effort. Bottom: adaptive gain.

In fact, a Riccati equation is explicitly solved in the inverse optimal backstepping approach. The solution is

$$\mathcal{P} = (\mathrm{Id} - \mathcal{K})^*(\mathrm{Id} - \mathcal{K}), \tag{14.43}$$

where Id is the identity operator, $\mathcal{K}$ is the operator corresponding to the transformation $u(x) \mapsto \int_0^x k(x,\xi)u(\xi)d\xi$, and $*$ denotes the adjoint of the operator.

The inverse optimal approach may appear somewhat limited in that it does not solve the optimal control problem for an *arbitrary* cost functional but instead solves it for a very specialized cost functional (14.11)–(14.13). However, this cost functional is positive definite and bounded from below by the H^1-norm, which constrains both the (spatial) energy of the transients and the (spatial) peaks of the transients. The penalty on the control R seems to become smaller as the plant becomes more unstable (see (14.12)), but in fact only the ratio between Q and R matters, and it cannot be estimated in our design. This is another difference from the LQR approach: in the LQR approach the penalties on the state and the control are constant, whereas in our design these penalties are free to change. In both approaches the optimal value of the functional changes with parameters of the plant and grows unbounded as the plant's level of instability tends to infinity.

We should also mention that inverse optimal controllers have an infinite gain margin (not possessed by ordinary stabilizing controllers), which allows the gain to be of any size, provided it is above a minimal stabilizing value (note that β in Theorem 14.1 is an arbitrary number greater than 2). They also have the 60° phase margin (the ability to tolerate a certain class of passive but possibly large actuator unmodeled dynamics), which, although not stated here, is provable.

To conclude the comparison, we show the results of numerical simulation for both designs. The parameters of the open-loop plant are set to $\varepsilon = 1$, $\lambda(x) = 14-16(x-1/2)^2$, and $q = 2$, and all other parameters are zero. The initial condition is $u_0(x) = 1 + \sin(3\pi x/2)$. With these parameters the uncontrolled plant has one unstable eigenvalue at 7.8. A static LQR controller with unity penalties on the state and control is implemented using the most popular Galerkin approximation with $N = 100$ steps, although more advanced techniques exist [20]. Computation of the gain kernel for the inverse optimal controller using the scheme (2.85)–(2.86) took as much as 20 times less time than computation of the Riccati kernel. This suggests that our method may be of even more interest in 2D and 3D problems, where the cost of solving operator Riccati equations becomes nearly prohibitive [20]. Symbolic computation using the series (2.43) is also possible; the first four terms in the series are sufficient for practical implementation, giving a relative error of 0.6% with respect to the true solution. The results of the simulations of the closed-loop system are presented in Figure 14.2. The system was discretized using a BTCS finite-difference method with 100 steps. We used the Dirichlet adaptive controller with $c = 0$, $\gamma = 1$, $\hat{\theta}(0) = 0.5$. The control effort and the L^2-norm of the closed-loop state are shown in Figure 14.2. We see that the LQR controller shows just slightly better performance.

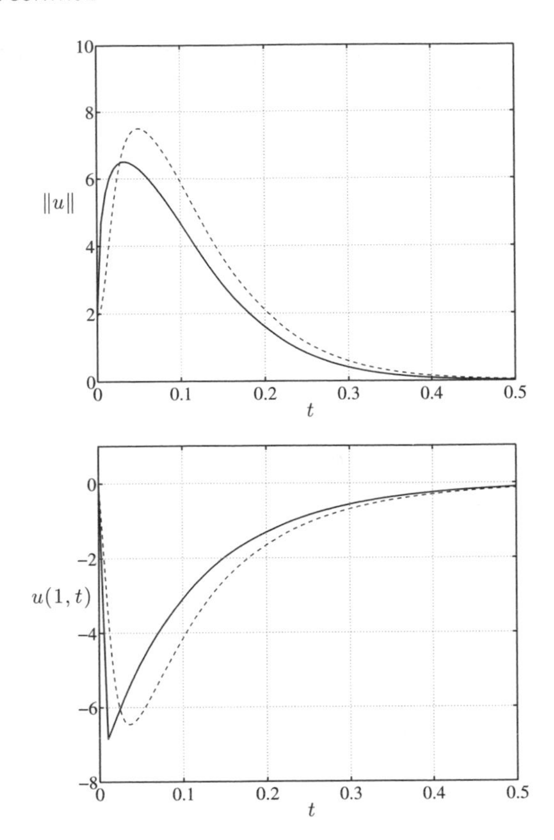

Figure 14.2 The comparison between LQR (solid line) and inverse optimal (dashed line) controllers. Top: L^2-norm of the state. Bottom: control effort.

14.6 INVERSE OPTIMAL ADAPTIVE CONTROL

Starting with this section we present the *inverse optimal adaptive control* design for a plant with a parametric uncertainty. Note that this design is completely different from the adaptive inverse optimal design presented in Section 14.2. There, we used adaptation only to reduce the conservativeness of the nominal inverse optimal design; the plant was known. Here, we design a truly optimal adaptive controller for an unknown plant. This controller is optimal with respect to a cost functional that puts positive definite penalties on state and control and at the same time penalizes the parameter estimation through an exponential of its square.

We should mention that a comparison of this adaptive scheme with LQR design (similar to the comparison made in Section 14.5) is not meaningful. Unlike our controllers, optimal adaptive controllers obtained as a certainty equivalence combination of linear quadratic optimal control and standard parameter estimation do not actually possess any optimality property, not to mention that it is not feasible to compute a solution to a Riccati equation at each time step.

To make the design ideas clear, we confine ourselves to the plant with just one scalar unknown parameter, although it is possible to combine the ideas presented here with the adaptive controllers designed in Chapter 11 for general plants with functional parametric uncertainties.

Consider the reaction-diffusion plant

$$u_t(x,t) = u_{xx}(x,t) + \lambda u(x,t), \tag{14.44}$$

$$u(0,t) = 0, \tag{14.45}$$

$$u_x(1,t) = U(t), \tag{14.46}$$

where λ is an unknown constant parameter. With a backstepping transformation,

$$w(x,t,\hat{\lambda}) = u(x,t) - \int_0^x k(x,y,\hat{\lambda})u(y,t)\,dy, \tag{14.47}$$

$$k(x,y,\hat{\lambda}) = -\hat{\lambda}y\frac{I_1\left(\sqrt{\hat{\lambda}(x^2-y^2)}\right)}{\sqrt{\hat{\lambda}(x^2-y^2)}}, \tag{14.48}$$

and its inverse,

$$u(x,t) = w(x,t,\hat{\lambda}) + \int_0^x l(x,y,\hat{\lambda})w(y,t,\hat{\lambda})\,dy, \tag{14.49}$$

$$l(x,y,\hat{\lambda}) = -\hat{\lambda}y\frac{J_1\left(\sqrt{\hat{\lambda}(x^2-y^2)}\right)}{\sqrt{\hat{\lambda}(x^2-y^2)}}, \tag{14.50}$$

we map (14.44)–(14.46) into

$$\frac{\mathrm{d}}{\mathrm{d}t}w(x,t,\hat{\lambda}(t)) = w_{xx}(x,t,\hat{\lambda}(t)) + \dot{\hat{\lambda}}(t)\int_0^x \frac{y}{2}w(y,t,\hat{\lambda}(t))\,dy$$
$$+(\lambda-\hat{\lambda}(t))w(x,t,\hat{\lambda}(t)), \tag{14.51}$$

$$w(0,t,\hat{\lambda}(t)) = 0, \tag{14.52}$$

$$w_x(1,t,\hat{\lambda}(t)) = U(t) - k(1,1,\hat{\lambda}(t))u(1,t)$$
$$-\int_0^1 k_x(1,y,\hat{\lambda}(t))u(y,t)\,dy. \tag{14.53}$$

In Chapter 8 we showed that the adaptive controller

$$U^\circ(t) = k(1,1,\hat{\lambda}(t))u(1,t) + \int_0^1 k_x(1,y,\hat{\lambda}(t))u(y,t)\,dy \tag{14.54}$$

with the parameter update law

$$\dot{\hat{\lambda}}(t) = \frac{\int_0^1 w(x,t,\hat{\lambda}(t))^2\,dx}{1+\int_0^1 w(x,t,\hat{\lambda}(t))^2\,dx} \tag{14.55}$$

is globally stabilizing and achieves regulation of u to zero.

Here we propose a very different controller and show its inverse optimality. This controller is given by

$$U^*(t) = -\left(1 + 3|\hat{\lambda}(t)| + |\hat{\lambda}(t)|^2\right)\left(u(1,t) - \int_0^1 k(1,y,\hat{\lambda}(t))u(y,t)\,dy\right). \tag{14.56}$$

14.7 STABILITY AND INVERSE OPTIMALITY OF THE ADAPTIVE SCHEME

Before we state the stability result, we introduce the following functional:

$$\begin{aligned}\mathcal{L}(u,\hat{\lambda})(t) = {} & 2\int_0^1 w_x(x,t,\hat{\lambda}(t))^2\,dx - \frac{\int_0^1 w(x,t,\hat{\lambda}(t))^2\,dx}{1+\int_0^1 w(x,t,\hat{\lambda}(t))^2\,dx} \\ & \times \int_0^1 w(x,t,\hat{\lambda}(t)) \int_0^x yw(y,t,\hat{\lambda}(t))\,dy\,dx \\ & + 2l(1,1,\hat{\lambda}(t))w(1,t,\hat{\lambda}(t))^2 \\ & + 2w(1,t,\hat{\lambda}(t)) \int_0^1 l_x(1,y,\hat{\lambda}(t))w(y,t,\hat{\lambda}(t))\,dy \\ & + (1+3|\hat{\lambda}(t)| + |\hat{\lambda}(t)|^2)w(1,t,\hat{\lambda}(t))^2. \end{aligned} \tag{14.57}$$

This functional is positive definite, as shown by the following lemma.

Lemma 14.1. *For the functional* (14.57), *the following holds:*

$$\mathcal{L}(u,\hat{\lambda})(t) \geq \left(1 - \frac{2}{\pi^2\sqrt{3}}\right)\int_0^1 w_x(x,t,\hat{\lambda}(t))^2\,dx \geq \frac{2\int_0^1 u(x,t)^2\,dx}{\left(1+\frac{1}{2}|\hat{\lambda}(t)|\right)^2}. \tag{14.58}$$

Proof. We start with the inequality (1.20),

$$\left|\int_0^1 w(x)\left(\int_0^x yw(y)\,dy\right)dx\right| \leq \frac{1}{2\sqrt{3}}\|w\|^2 \leq \frac{2}{\pi^2\sqrt{3}}\|w_x\|^2. \tag{14.59}$$

Using (14.41) and Agmon's and Young's inequalities, we get

$$2w(1)\int_0^1 l_x(1,y,\hat{\lambda})w(y)\,dy \geq -\|w_x\|^2 - (1+|\hat{\lambda}|)^2 w(1)^2. \tag{14.60}$$

Estimates (14.59) and (14.60), along with the fact that $2l(1,1,\hat{\lambda}) = -\hat{\lambda}$, give the first line of (14.58). The second line of (14.58) follows from (14.49), inequality (B.1) from Appendix B, and estimates $\sup_{0\leq y\leq x\leq 1}|l(x,y,\hat{\lambda})| \leq |\hat{\lambda}/2|$ and $\frac{\pi^2}{4}(1 - \frac{2}{\pi^2\sqrt{3}}) = 2.18 > 2$. □

THEOREM 14.3 (Stabilization). *The closed-loop system consisting of the plant* (14.44)–(14.46), *parameter update law* (14.55), *and the controller* (14.56) *is globally Lyapunov stable in the sense of the norm*

$$\int_0^1 u(x,t)^2\,dx + (\hat{\lambda}(t)-\lambda)^2, \tag{14.61}$$

and furthermore,

$$\int_0^1 u(x,t)^2\,dx \to 0, \qquad \hat{\lambda}(t) \to \lambda_\infty \tag{14.62}$$

as $t \to \infty$, where λ_∞ is some constant.

Proof. We use the Lyapunov functional

$$V(t) = \left(1 + \int_0^1 w(x,t,\hat{\lambda}(t))^2\,dx\right) e^{(\hat{\lambda}(t)-\lambda)^2} - 1. \tag{14.63}$$

An easy calculation yields

$$\begin{aligned}\dot{V}(t) = & -e^{(\hat{\lambda}(t)-\lambda)^2}\mathcal{L}(u,\hat{\lambda})(t)\\ & -e^{(\hat{\lambda}(t)-\lambda)^2}\left(1+3|\hat{\lambda}(t)|+|\hat{\lambda}(t)|^2\right) w(1,t,\hat{\lambda}(t))^2.\end{aligned} \tag{14.64}$$

By deriving an upper and lower bound on (14.63), from $\dot{V} \le 0$ we get

$$\Upsilon(t) \le e^{\Upsilon(0)} - 1, \tag{14.65}$$

where

$$\Upsilon(t) = \int_0^1 w(x,t,\hat{\lambda}(t))^2\,dx + (\hat{\lambda}(t)-\lambda)^2. \tag{14.66}$$

The functions $k(x,y,\lambda)$ and $l(x,y,\lambda)$ have the following bounds:

$$\sup_{0\le y\le x\le 1} |k(x,y,\theta)| \le |\theta| e^{\sqrt{|\theta|}}, \tag{14.67}$$

$$\sup_{0\le y\le x\le 1} |l(x,y,\theta)| \le |\theta|. \tag{14.68}$$

Then a lengthy but routine calculation, starting from (14.65), (14.67), (14.68), yields

$$\Omega(t) \le \Sigma(1+|\lambda|+\Sigma), \tag{14.69}$$

where

$$\Omega(t) = \left(\int_0^1 u(x,t)^2\,dx + (\hat{\lambda}(t)-\lambda)^2\right)^{1/2}, \tag{14.70}$$

$$\Sigma(\Omega_0) = \left(\exp\left\{\left(1+(|\lambda|+|\Omega_0|)\,e^{\sqrt{|\lambda|+|\Omega_0|}}\right)^2 \Omega_0^2\right\} - 1\right)^{1/2}, \tag{14.71}$$

$$\Omega_0 = \Omega(0). \tag{14.72}$$

Since (14.69) is a class $\mathcal{K}$ function of Σ, and Σ is a class $\mathcal{K}$ function of Ω_0, this proves global stability in the norm $\Omega(t)$. The regulation result is argued in a similar way as in Chapter 8 (despite the fact that the Lyapunov function is different), using (14.64) and (14.58). □

Optimality of the controller (14.56) is established by the following theorem.

Theorem 14.4 (Inverse optimality). *Consider the system consisting of the plant* (14.44)–(14.46) *with the parameter update law* (14.55). *The controller* (14.56) *minimizes the cost functional*

$$J = \lim_{\sigma\to\infty}\Bigg\{e^{(\hat\lambda(\sigma)-\lambda)^2} - 1 + \int_0^\sigma e^{(\hat\lambda(t)-\lambda)^2}\left[\mathcal{L}(u,\hat\lambda)(t) + \frac{U(t)^2}{1+3|\hat\lambda(t)|+|\hat\lambda(t)|^2}\right]dt\Bigg\}. \quad (14.73)$$

The minimum of the cost functional is

$$J^* = \left(1 + \int_0^1 w(x,0,\hat\lambda(0))^2\,dx\right)e^{(\hat\lambda(0)-\lambda)^2} - 1. \quad (14.74)$$

Proof. A straightforward calculation yields

$$J = V(0) + \int_0^\infty e^{(\hat\lambda(t)-\lambda)^2} \times \frac{\left[U(t) + \left(1+3|\hat\lambda(t)|+|\hat\lambda(t)|^2\right)w(1,t,\hat\lambda(t))\right]^2}{1+3|\hat\lambda(t)|+|\hat\lambda(t)|^2}\,dt, \quad (14.75)$$

from which the results of the theorem follow. □

14.8 NOTES AND REFERENCES

The approach presented in Sections 14.1–14.4 is the extension of inverse optimal designs for finite-dimensional systems [66] and was first presented in [108].

The results in Sections 14.6–14.7 are based on [64]. We should point out that the form of optimality in these sections is different from the one presented in [79]. The result in [79] is an inverse optimality result, in the spirit of the classical inverse optimality theory for nonlinear systems [89, 107], stated in the context of a parameter-adaptive tracking problem for globally stabilizable nonlinear finite-dimensional systems with uncertain parameters. The interesting aspect of this result is that the adaptive controller is truly a minimizer of a meaningful cost functional, which includes, besides the positive definite penalties on the state and control (with complicated nonlinear scaling in terms of the parameter estimate), a simple *terminal penalty on the parameter estimation error.*

The cost functional (14.73), which is optimized by the controller (14.56), provides a clue as to what form of cost penalties in state, control, and parameter error are meaningful to pursue in possible future developments of direct (rather than inverse) optimal adaptive control. Parameter estimation transients are penalized (through an exponential-of-square penalty), but they are not penalized in a way that would demand convergence of the parameter estimate to the true parameter. This is consistent with the fact that parameter convergence requires persistence of excitation, which is normally not present in problems where the state is being regulated to zero.

Appendix A

Adaptive Backstepping for Nonlinear ODEs—The Basics

The adaptive designs for parabolic PDEs presented in this book are motivated by the adaptive backstepping designs for ODEs [68]. Many of these designs serve as an inspiration, but they are not exactly copied, since most of them cannot be extended exactly from finite dimension to infinite dimension. For example, the *tuning function* design for ODEs serves as an inspiration for the Lyapunov-based design for PDEs, but it is not exactly reproducible because the nonlinear growth of the adaptive feedback law increases with each step of the finite-dimensional backstepping design. Similarly, the *modular* design for ODEs serves as an inspiration for the certainty equivalence design for PDEs, but it is not exactly copied since the nonlinear damping tool cannot be applied to PDEs (at least not in the same manner), as its nonlinearity grows with each step of backstepping.

In this appendix we present a tutorial introduction to adaptive backstepping designs for ODEs [68], with a focus on nonlinear ODEs in the *parametric strict feedback* form. To help the reader without any prior exposure to finite-dimensional backstepping, we start in Section A.1 with a presentation of a nonadaptive version of backstepping when the plant parameters are known. The presentation is based on a tutorial example, and the exposition is detailed and pedagogical. Then, in Section A.2 we present the tuning function design, in Section A.3 we present various modular designs, and in Section A.4 we present extensions from the full-state-feedback case to the output-feedback case.

A.1 NONADAPTIVE BACKSTEPPING—THE KNOWN PARAMETER CASE

We introduce the idea of backstepping by carrying out a *nonadaptive* design for the example system

$$\dot{x}_1 = x_2 + \varphi(x_1)^{\mathrm{T}}\theta\,, \qquad \varphi(0) = 0, \tag{A.1}$$

$$\dot{x}_2 = u, \tag{A.2}$$

where θ is a *known* parameter vector and $\varphi(x_1)$ is a smooth nonlinear function with the property $\varphi(0) = 0$. Our goal is to stabilize the equilibrium

$$x_1 = 0, \tag{A.3}$$

$$x_2 = -\varphi(0)^{\mathrm{T}}\theta = 0\,. \tag{A.4}$$

Backstepping design is recursive. First, the state x_2 is treated as a *virtual control* for the x_1-equation (A.1), and a *stabilizing function*

$$\alpha_1(x_1) = -c_1 x_1 - \varphi(x_1)^{\mathrm{T}}\theta, \qquad c_1 > 0 \tag{A.5}$$

is designed to stabilize (A.1), assuming that $x_2 = \alpha_1(x_1)$ can be implemented. Since this is not the case, we define

$$z_1 = x_1, \tag{A.6}$$

$$z_2 = x_2 - \alpha_1(x_1), \tag{A.7}$$

where z_2 is an error variable expressing the fact that x_2 is not the true control. Differentiating z_1 and z_2 with respect to time, the complete system (A.1), (A.2) is expressed in the error coordinates (A.6), (A.7):

$$\begin{aligned}
\dot{z}_1 &= \dot{x}_1 \\
&= x_2 + \varphi^{\mathrm{T}}\theta \\
&= z_2 + \alpha_1 + \varphi^{\mathrm{T}}\theta \\
&= -c_1 z_1 + z_2, &\text{(A.8)} \\
\dot{z}_2 &= \dot{x}_2 - \dot{\alpha}_1 \\
&= u - \frac{\partial \alpha_1}{\partial x_1}\dot{x}_1 \\
&= u - \frac{\partial \alpha_1}{\partial x_1}\left(x_2 + \varphi^{\mathrm{T}}\theta\right). &\text{(A.9)}
\end{aligned}$$

It is important to observe that the time derivative $\dot{\alpha}_1$ is implemented analytically, without a differentiator.

For the system (A.8), (A.9) we now design a control law $u = \alpha_2(x_1, x_2)$ to render the time derivative of a Lyapunov function negative definite. The design can be completed with the simplest Lyapunov function,

$$V(x_1, x_2) = \frac{1}{2}z_1^2 + \frac{1}{2}z_2^2. \tag{A.10}$$

Its derivative for (A.8), (A.9) is

$$\begin{aligned}
\dot{V} &= z_1(-c_1 z_1 + z_2) + z_2\left[u - \frac{\partial \alpha_1}{\partial x_1}\left(x_2 + \varphi^{\mathrm{T}}\theta\right)\right] \\
&= -c_1 z_1^2 + z_2\left[u + z_1 - \frac{\partial \alpha_1}{\partial x_1}\left(x_2 + \varphi^{\mathrm{T}}\theta\right)\right]. &\text{(A.11)}
\end{aligned}$$

An obvious way to achieve negativity of $\dot{V}$ is to employ u to make the bracketed expression equal to $-c_2 z_2$ with $c_2 > 0$, namely,

$$u = \alpha_2(x_1, x_2) = -c_2 z_2 - z_1 + \frac{\partial \alpha_1}{\partial x_1}\left(x_2 + \varphi^{\mathrm{T}}\theta\right). \tag{A.12}$$

This control may not be the best choice because it cancels some terms that may contribute to the negativity of $\dot{V}$. Backstepping design offers enough flexibility to avoid cancelation. However, for the sake of clarity, we will assume that none of the nonlinearities are useful, so that they all need to be canceled, as in the control law (A.12). This control law yields

$$\dot{V} = -c_1 z_1^2 - c_2 z_2^2, \tag{A.13}$$

which means that the equilibrium $z = 0$ is globally asymptotically stable. In view of (A.6), (A.7), the same is true about $x = 0$. The resulting closed-loop system in the z-coordinates is linear:

$$\begin{bmatrix} \dot{z}_1 \\ \dot{z}_2 \end{bmatrix} = \begin{bmatrix} -c_1 & 1 \\ -1 & -c_2 \end{bmatrix} \begin{bmatrix} z_1 \\ z_2 \end{bmatrix}. \tag{A.14}$$

In the next four sections we present adaptive nonlinear designs through examples. Summaries of general design procedures are also provided but without technical details, for which the reader is referred to the text on nonlinear and adaptive control design by Krstic, Kanallakopoulos, and Kokotovic [68].

The two main methodologies for adaptive backstepping design are the *tuning functions design*, discussed in Section A.2, and the *modular design*, discussed in Section A.3. These sections assume that the full state is available for feedback. Section A.4 presents designs in which only the output is measured. Section A.5 discusses various extensions to more general classes of systems.

A.2 TUNING FUNCTIONS DESIGN

In the tuning functions design both the controller and the parameter update law are designed recursively. At each consecutive step a *tuning function* is designed as a potential update law. The tuning functions are not implemented as update laws. Instead, the stabilizing functions use them to compensate for the effects of parameter estimation transients. Only the final tuning function is used as the parameter update law.

A.2.1 Introductory Examples

The tuning functions design will be introduced through examples of increasing complexity:

A	**B**	**C**
$\dot{x}_1 = u + \varphi(x_1)^{\mathrm{T}}\theta$	$\dot{x}_1 = x_2 + \varphi(x_1)^{\mathrm{T}}\theta$	$\dot{x}_1 = x_2 + \varphi(x_1)^{\mathrm{T}}\theta$
	$\dot{x}_2 = u$	$\dot{x}_2 = x_3$
		$\dot{x}_3 = u$

The adaptive problem arises because the parameter vector θ is *unknown*. The nonlinearity $\varphi(x_1)$ is known.

The systems A, B, and C differ structurally: the number of integrators between the control u and the unknown parameter θ increases from zero at A to two at C. Design A is the simplest to develop because the control u and the uncertainty $\varphi(x_1)^{\mathrm{T}}\theta$ are "matched"; that is, the control does not have to overcome integrator transients in order to counteract the effects of the uncertainty. Design C is the hardest because the control must act through two integrators before it reaches the uncertainty.

A.2.1.1 Design A

Let $\hat{\theta}$ be an estimate of the unknown parameter θ in the system

$$\dot{x}_1 = u + \varphi^{\mathrm{T}}\theta\,. \tag{A.15}$$

If this estimate were correct, $\hat{\theta} = \theta$, then the control law

$$u = -c_1 x_1 - \varphi(x_1)^{\mathrm{T}}\hat{\theta} \tag{A.16}$$

would achieve global asymptotic stability of $x = 0$. Because the parameter estimation error

$$\tilde{\theta} = \theta - \hat{\theta} \tag{A.17}$$

is non-zero, we have

$$\dot{x}_1 = -c_1 x_1 + \varphi(x_1)^{\mathrm{T}}\tilde{\theta}\,, \tag{A.18}$$

that is, the parameter estimation error $\tilde{\theta}$ continues to act as a disturbance that may destabilize the system.

Our task is to find an update law for $\hat{\theta}(t)$ that preserves the boundedness of $x(t)$ and achieves its regulation to zero. To this end, we consider the Lyapunov function

$$V_1(x, \hat{\theta}) \;=\; \frac{1}{2}x_1^2 + \frac{1}{2}\tilde{\theta}^{\mathrm{T}}\Gamma^{-1}\tilde{\theta}, \tag{A.19}$$

where Γ is a positive definite symmetric matrix. The derivative of V_1 is

$$\begin{aligned}\dot{V}_1 &= -c_1 x_1^2 + x_1\varphi(x_1)^{\mathrm{T}}\tilde{\theta} - \tilde{\theta}^{\mathrm{T}}\Gamma^{-1}\dot{\hat{\theta}} \\ &= -c_1 x_1^2 + \tilde{\theta}^{\mathrm{T}}\Gamma^{-1}\left(\Gamma\varphi(x_1)x_1 - \dot{\hat{\theta}}\right).\end{aligned} \tag{A.20}$$

Our goal is to select an update law for $\dot{\hat{\theta}}$ to guarantee

$$\dot{V}_1 \leq 0. \tag{A.21}$$

The only way this can be achieved for any unknown $\tilde{\theta}$ is to choose

$$\dot{\hat{\theta}} = \Gamma\varphi(x_1)x_1. \tag{A.22}$$

This choice yields

$$\dot{V}_1 = -c_1 x_1^2\,, \tag{A.23}$$

which guarantees global stability of the equilibrium $x_1 = 0$, $\hat{\theta} = \theta$, and hence the boundedness of $x_1(t)$ and $\hat{\theta}(t)$. By LaSalle's invariance theorem, all the trajectories of the closed-loop adaptive system converge to the set where $\dot{V}_1 = 0$, that is, to the set where $c_1 x_1^2 = 0$, which implies that

$$\lim_{t\to\infty} x_1(t) = 0\,. \tag{A.24}$$

The update law (A.22) is driven by the vector $\varphi(x_1)$, called the *regressor*, and the state x_1. This is a typical form of an update law in the tuning functions design: the speed of adaptation is dictated by the nonlinearity $\varphi(x_1)$ and the state x_1.

A.2.1.2 Design B

For the system

$$\begin{aligned} \dot{x}_1 &= x_2 + \varphi(x_1)^{\mathrm{T}}\theta \\ \dot{x}_2 &= u, \end{aligned} \tag{A.25}$$

we have already designed a nonadaptive controller in Section A.1. To design an adaptive controller, we replace the unknown θ by its estimate $\hat{\theta}$ both in the stabilizing function (A.5) and in the change of coordinate (A.7):

$$z_2 = x_2 - \alpha_1(x_1, \hat{\theta}) \tag{A.26}$$

$$\alpha_1(x_1, \hat{\theta}) = -c_1 z_1 - \varphi^{\mathrm{T}}\hat{\theta}\,. \tag{A.27}$$

Because in the system (A.25) the control input is separated from the unknown parameter by an integrator, the control law (A.12) will be strengthened by a term $\nu_2(x_1, x_2, \hat{\theta})$ that will compensate for the parameter estimation transients:

$$\begin{aligned} u &= \alpha_2(x_1, x_2, \hat{\theta}) \\ &= -c_2 z_2 - z_1 + \frac{\partial \alpha_1}{\partial x_1}\left(x_2 + \varphi^{\mathrm{T}}\hat{\theta}\right) + \nu_2(x_1, x_2, \hat{\theta}). \end{aligned} \tag{A.28}$$

The resulting system in the z-coordinates is

$$\begin{aligned} \dot{z}_1 &= z_2 + \alpha_1 + \varphi^{\mathrm{T}}\theta \\ &= -c_1 z_1 + z_2 + \varphi^{\mathrm{T}}\tilde{\theta}, \end{aligned} \tag{A.29}$$

$$\begin{aligned} \dot{z}_2 &= \dot{x}_2 - \dot{\alpha}_1 \\ &= u - \frac{\partial \alpha_1}{\partial x_1}\left(x_2 + \varphi^{\mathrm{T}}\theta\right) - \frac{\partial \alpha_1}{\partial \hat{\theta}}\dot{\hat{\theta}} \\ &= -z_1 - c_2 z_2 - \frac{\partial \alpha_1}{\partial x_1}\varphi^{\mathrm{T}}\tilde{\theta} - \frac{\partial \alpha_1}{\partial \hat{\theta}}\dot{\hat{\theta}} + \nu_2(x_1, x_2, \hat{\theta}), \end{aligned} \tag{A.30}$$

or, in vector form,

$$\begin{bmatrix} \dot{z}_1 \\ \dot{z}_2 \end{bmatrix} = \begin{bmatrix} -c_1 & 1 \\ -1 & -c_2 \end{bmatrix} \begin{bmatrix} z_1 \\ z_2 \end{bmatrix} + \begin{bmatrix} \varphi^{\mathrm{T}} \\ -\frac{\partial \alpha_1}{\partial x_1}\varphi^{\mathrm{T}} \end{bmatrix} \tilde{\theta} + \begin{bmatrix} 0 \\ -\frac{\partial \alpha_1}{\partial \hat{\theta}}\dot{\hat{\theta}} + \nu_2(x_1, x_2, \hat{\theta}) \end{bmatrix}. \tag{A.31}$$

The ν_2-term can now be chosen to eliminate the last brackets:

$$\nu_2(x_1, x_2, \hat{\theta}) = \frac{\partial \alpha_1}{\partial \hat{\theta}}\dot{\hat{\theta}}. \tag{A.32}$$

This expression is implementable because $\dot{\hat{\theta}}$ will be available from the update law. Thus we obtain the *error system*

$$\begin{bmatrix} \dot{z}_1 \\ \dot{z}_2 \end{bmatrix} = \begin{bmatrix} -c_1 & 1 \\ -1 & -c_2 \end{bmatrix} \begin{bmatrix} z_1 \\ z_2 \end{bmatrix} + \begin{bmatrix} \varphi^{\mathrm{T}} \\ -\frac{\partial \alpha_1}{\partial x_1}\varphi^{\mathrm{T}} \end{bmatrix} \tilde{\theta}. \tag{A.33}$$

When the parameter error $\tilde{\theta}$ is zero, this system becomes the linear asymptotically stable system (A.14).

Our remaining task is to select the update law $\dot{\hat{\theta}} = \Gamma\tau_2(x, \hat{\theta})$. Consider the Lyapunov function

$$
\begin{aligned}
V_2(x_1, x_2, \hat{\theta}) &= V_1 + \frac{1}{2}z_2^2 \\
&= \frac{1}{2}z_1^2 + \frac{1}{2}z_2^2 + \frac{1}{2}\tilde{\theta}^{\mathrm{T}}\Gamma^{-1}\tilde{\theta}.
\end{aligned} \tag{A.34}
$$

Because $\dot{\tilde{\theta}} = -\dot{\hat{\theta}}$, the derivative of V_2 is

$$
\begin{aligned}
\dot{V}_2 &= -c_1 z_1^2 - c_2 z_2^2 + [z_1, \ z_2] \begin{bmatrix} \varphi^{\mathrm{T}} \\ -\dfrac{\partial \alpha_1}{\partial x_1}\varphi^{\mathrm{T}} \end{bmatrix} \tilde{\theta} - \tilde{\theta}^{\mathrm{T}}\Gamma^{-1}\dot{\hat{\theta}} \\
&= -c_1 z_1^2 - c_2 z_2^2 + \tilde{\theta}^{\mathrm{T}}\Gamma^{-1}\left(\Gamma\left[\varphi, -\frac{\partial \alpha_1}{\partial x_1}\varphi\right]\begin{bmatrix} z_1 \\ z_2 \end{bmatrix} - \dot{\hat{\theta}}\right).
\end{aligned} \tag{A.35}
$$

The only way to eliminate the unknown parameter error $\tilde{\theta}$ is to select the update law

$$
\begin{aligned}
\dot{\hat{\theta}} &= \Gamma\tau_2(x, \hat{\theta}) \\
&= \Gamma\left[\varphi, -\frac{\partial \alpha_1}{\partial x_1}\varphi\right]\begin{bmatrix} z_1 \\ z_2 \end{bmatrix} \\
&= \Gamma\left(\varphi z_1 - \frac{\partial \alpha_1}{\partial x_1}\varphi z_2\right).
\end{aligned} \tag{A.36}
$$

Then $\dot{V}_2$ is nonpositive:

$$
\dot{V}_2 = -c_1 z_1^2 - c_2 z_2^2, \tag{A.37}
$$

which means that the global stability of $z = 0, \ \tilde{\theta} = 0$ is achieved. Moreover, by applying the LaSalle argument mentioned in Design A, we prove that $z(t) \to 0$ as $t \to \infty$. Finally, from (A.27), it follows that the equilibrium $x = 0, \ \hat{\theta} = \theta$ is globally stable and $x(t) \to 0$ as $t \to \infty$.

The crucial property of the control law in Design B is that it incorporates the ν_2-term (A.32), which is proportional to $\dot{\hat{\theta}}$ and compensates for the effect of parameter estimation transients on the coordinate change (A.26). It is this departure from the certainty equivalence principle that makes the adaptive stabilization possible for systems with nonlinearities of arbitrary growth.

By comparing (A.36) with (A.22), we note that the first term, φz_1, is the potential update law for the z_1-system. The functions

$$
\tau_1(x_1) = \varphi z_1 \tag{A.38}
$$

$$
\tau_2(x_1, x_2, \hat{\theta}) = \tau_1(x_1) - \tfrac{\partial \alpha_1}{\partial x_1}\varphi z_2 \tag{A.39}
$$

are referred to as *tuning functions* because of their role as potential update laws for intermediate systems in the backstepping procedure.

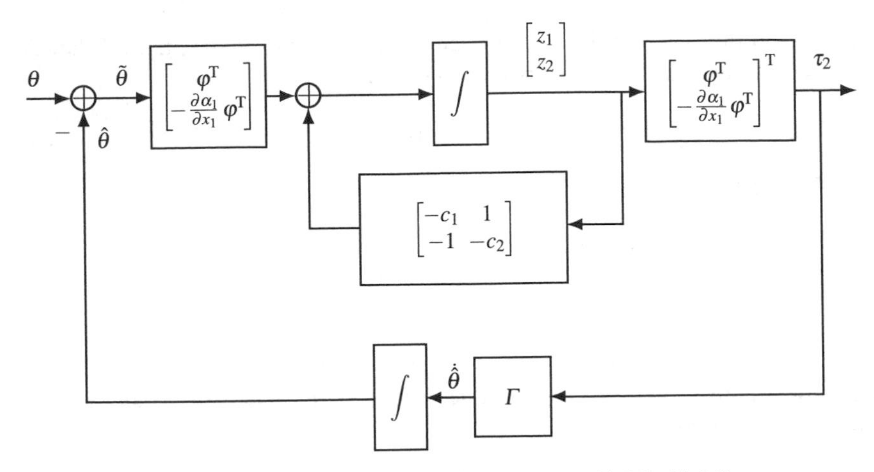

Figure A.1 The closed-loop adaptive system (A.33), (A.36).

A.2.1.3 Design C

The system

$$\begin{aligned}\dot{x}_1 &= x_2 + \varphi(x_1)^{\mathrm{T}}\theta \\ \dot{x}_2 &= x_3 \\ \dot{x}_3 &= u\end{aligned} \tag{A.40}$$

is obtained by augmenting system (A.25) with an integrator. The control law $\alpha_2(x_1, x_2, \hat{\theta})$ designed in (A.28) can no longer be directly applied because x_3 is a state and not a control input. We "step back" through the integrator $\dot{x}_3 = u$ and design the control law for the actual input u. However, we keep the stabilizing function α_2 and use it to define the third error coordinate

$$z_3 = x_3 - \alpha_2(x_1, x_2, \hat{\theta})\,. \tag{A.41}$$

The parameter update law (A.36) will have to be modified with an additional z_3-term. Instead of $\dot{\hat{\theta}}$ in (A.32), the compensating term ν_2 will now use the potential update law (A.39) for the system (A.33):

$$\nu_2(x_1, x_2, \hat{\theta}) = \frac{\partial \alpha_1}{\partial \hat{\theta}}\Gamma\tau_2(x_1, x_2, \hat{\theta}). \tag{A.42}$$

Hence, the role of the tuning function τ_2 is to substitute for the actual update law in compensating for the effects of parameter estimation transients.

With (A.26), (A.27), (A.41), (A.39), (A.42) and the stabilizing function α_2 in (A.28), we have

$$\begin{bmatrix}\dot{z}_1\\ \dot{z}_2\end{bmatrix} = \begin{bmatrix}-c_1 & 1\\ -1 & -c_2\end{bmatrix}\begin{bmatrix}z_1\\ z_2\end{bmatrix} + \begin{bmatrix}\varphi^{\mathrm{T}}\\ -\dfrac{\partial \alpha_1}{\partial x_1}\varphi^{\mathrm{T}}\end{bmatrix}\tilde{\theta} + \begin{bmatrix}0\\ z_3 + \dfrac{\partial \alpha_1}{\partial \hat{\theta}}(\Gamma\tau_2 - \dot{\hat{\theta}})\end{bmatrix}. \tag{A.43}$$

This system differs from the error system (A.33) only in its last term. Likewise, instead of the Lyapunov inequality (A.37) we have

$$\dot{V}_2 = -c_1 z_1^2 - c_2 z_2^2 + z_2 z_3 + z_2 \frac{\partial \alpha_1}{\partial \hat{\theta}}(\Gamma \tau_2 - \dot{\hat{\theta}}) + \tilde{\theta}^{\mathrm{T}}(\tau_2 - \Gamma^{-1}\dot{\hat{\theta}}). \qquad \text{(A.44)}$$

Differentiating (A.41), we get

$$\begin{aligned}\dot{z}_3 &= u - \frac{\partial \alpha_2}{\partial x_1}\left(x_2 + \varphi^{\mathrm{T}}\theta\right) - \frac{\partial \alpha_2}{\partial x_2}x_3 - \frac{\partial \alpha_2}{\partial \hat{\theta}}\dot{\hat{\theta}} \\ &= u - \frac{\partial \alpha_2}{\partial x_1}\left(x_2 + \varphi^{\mathrm{T}}\hat{\theta}\right) - \frac{\partial \alpha_2}{\partial x_2}x_3 - \frac{\partial \alpha_2}{\partial \hat{\theta}}\dot{\hat{\theta}} - \frac{\partial \alpha_2}{\partial x_1}\varphi^{\mathrm{T}}\tilde{\theta}. \qquad \text{(A.45)}\end{aligned}$$

We now stabilize the (z_1, z_2, z_3)-system (A.43), (A.45) with respect to the Lyapunov function

$$\begin{aligned}V_3(x, \hat{\theta}) &= V_2 + \frac{1}{2}z_3^2 \\ &= \frac{1}{2}z_1^2 + \frac{1}{2}z_2^2 + \frac{1}{2}z_3^2 + \frac{1}{2}\tilde{\theta}^{\mathrm{T}}\Gamma^{-1}\tilde{\theta}. \qquad \text{(A.46)}\end{aligned}$$

Its derivative along (A.43) and (A.45) is

$$\begin{aligned}\dot{V}_3 = &-c_1 z_1^2 - c_2 z_2^2 + z_2 \frac{\partial \alpha_1}{\partial \hat{\theta}}(\Gamma \tau_2 - \dot{\hat{\theta}}) \\ &+ z_3\left[z_2 + u - \frac{\partial \alpha_2}{\partial x_1}\left(x_2 + \varphi^{\mathrm{T}}\hat{\theta}\right) - \frac{\partial \alpha_2}{\partial x_2}x_3 - \frac{\partial \alpha_2}{\partial \hat{\theta}}\dot{\hat{\theta}}\right] \\ &+ \tilde{\theta}^{\mathrm{T}}\left(\tau_2 - \frac{\partial \alpha_2}{\partial x_1}\varphi z_3 - \Gamma^{-1}\dot{\hat{\theta}}\right). \qquad \text{(A.47)}\end{aligned}$$

Again we must eliminate the unknown parameter error $\tilde{\theta}$ from $\dot{V}_3$. For this we must choose the update law as

$$\begin{aligned}\dot{\hat{\theta}} &= \Gamma \tau_3(x_1, x_2, x_3, \hat{\theta}) \\ &= \Gamma\left(\tau_2 - \frac{\partial \alpha_2}{\partial x_1}\varphi z_3\right) \\ &= \Gamma\left[\varphi,\ -\frac{\partial \alpha_1}{\partial x_1}\varphi,\ \frac{\partial \alpha_2}{\partial x_1}\varphi\right]\begin{bmatrix} z_1 \\ z_2 \\ z_3 \end{bmatrix}. \qquad \text{(A.48)}\end{aligned}$$

Upon inspection of the bracketed terms in $\dot{V}_3$, we pick the control law:

$$\begin{aligned}u = &= \alpha_3(x_1, x_2, x_3, \hat{\theta}) \\ &= -z_2 - c_3 z_3 + \frac{\partial \alpha_2}{\partial x_1}\left(x_2 + \varphi^{\mathrm{T}}\hat{\theta}\right) + \frac{\partial \alpha_2}{\partial x_2}x_3 + \nu_3. \qquad \text{(A.49)}\end{aligned}$$

The compensation term ν_3 is yet to be chosen. Substituting (A.49) into (A.47), we get

$$\dot{V}_3 = -c_1 z_1^2 - c_2 z_2^2 - c_3 z_3^2 + z_2 \frac{\partial \alpha_1}{\partial \hat{\theta}}(\Gamma \tau_2 - \dot{\hat{\theta}}) + z_3\left(\nu_3 - \frac{\partial \alpha_2}{\partial \hat{\theta}}\dot{\hat{\theta}}\right). \qquad \text{(A.50)}$$

From this expression it is clear that ν_3 should cancel $\frac{\partial\alpha_2}{\partial\hat{\theta}}\dot{\hat{\theta}}$. In order to cancel the cross-term $z_2\frac{\partial\alpha_1}{\partial\hat{\theta}}(\Gamma\tau_2-\dot{\hat{\theta}})$ with ν_3, it appears that we would need to divide by z_3. However, the variable z_3 might take a zero value during the transient, and should be regulated to zero to accomplish the control objective. We resolve this difficulty by noting that

$$\begin{aligned}\dot{\hat{\theta}}-\Gamma\tau_2 &= \dot{\hat{\theta}}-\Gamma\tau_3+\Gamma\tau_3-\Gamma\tau_2\\ &= \dot{\hat{\theta}}-\Gamma\tau_3-\Gamma\frac{\partial\alpha_2}{\partial x_1}\varphi z_3\,,\end{aligned} \tag{A.51}$$

so that $\dot{V}_3$ in (A.50) is rewritten as

$$\dot{V}_3=-c_1z_1^2-c_2z_2^2-c_3z_3^2+z_3\left(\nu_3-\frac{\partial\alpha_2}{\partial\hat{\theta}}\Gamma\tau_3+\frac{\partial\alpha_1}{\partial\hat{\theta}}\Gamma\frac{\partial\alpha_2}{\partial x_1}\varphi z_2\right). \tag{A.52}$$

From (A.52) the choice of ν_3 is immediate:

$$\nu_3(x_1,x_2,x_3,\hat{\theta})=\frac{\partial\alpha_2}{\partial\hat{\theta}}\Gamma\tau_3-\frac{\partial\alpha_1}{\partial\hat{\theta}}\Gamma\frac{\partial\alpha_2}{\partial x_1}\varphi z_2\,. \tag{A.53}$$

The resulting $\dot{V}_3$ is

$$\dot{V}_3=-c_1z_1^2-c_2z_2^2-c_3z_3^2\,, \tag{A.54}$$

which guarantees that the equilibrium $x=0,\ \hat{\theta}=\theta$ is globally stable, and $x(t)\to 0$ as $t\to\infty$.

The Lyapunov design leading to (A.52) is effective but does not reveal the stabilization mechanism. To provide further insight we write the (z_1,z_2,z_3)-system (A.43), (A.45) with u given in (A.49) but with ν_3 yet to be selected:

$$\begin{bmatrix}\dot{z}_1\\ \dot{z}_2\\ \dot{z}_3\end{bmatrix}=\begin{bmatrix}-c_1 & 1 & 0\\ -1 & -c_2 & 1\\ 0 & -1 & -c_3\end{bmatrix}\begin{bmatrix}z_1\\ z_2\\ z_3\end{bmatrix}+\begin{bmatrix}\varphi^{\mathrm{T}}\\ -\frac{\partial\alpha_1}{\partial x_1}\varphi^{\mathrm{T}}\\ -\frac{\partial\alpha_2}{\partial x_1}\varphi^{\mathrm{T}}\end{bmatrix}\tilde{\theta}
+\begin{bmatrix}0\\ \frac{\partial\alpha_1}{\partial\hat{\theta}}(\Gamma\tau_2-\dot{\hat{\theta}})\\ \nu_3-\frac{\partial\alpha_2}{\partial\hat{\theta}}\Gamma\tau_3\end{bmatrix}. \tag{A.55}$$

While ν_3 can cancel the matched term $\frac{\partial\alpha_2}{\partial\hat{\theta}}\dot{\hat{\theta}}$, it cannot cancel the term $\frac{\partial\alpha_1}{\partial\hat{\theta}}(\Gamma\tau_2-\dot{\hat{\theta}})$ in the second equation. By substituting (A.51), we note that $\frac{\partial\alpha_1}{\partial\hat{\theta}}(\Gamma\tau_2-\dot{\hat{\theta}})$ has z_3 as a factor, and absorb it into the "system matrix":

$$\begin{bmatrix} \dot{z}_1 \\ \dot{z}_2 \\ \dot{z}_3 \end{bmatrix} = \begin{bmatrix} -c_1 & 1 & 0 \\ -1 & -c_2 & 1 + \frac{\partial \alpha_1}{\partial \hat{\theta}} \Gamma \frac{\partial \alpha_2}{\partial x_1} \varphi \\ 0 & -1 & -c_3 \end{bmatrix} \begin{bmatrix} z_1 \\ z_2 \\ z_3 \end{bmatrix} + \begin{bmatrix} \varphi^{\mathrm{T}} \\ -\frac{\partial \alpha_1}{\partial x_1} \varphi^{\mathrm{T}} \\ -\frac{\partial \alpha_2}{\partial x_1} \varphi^{\mathrm{T}} \end{bmatrix} \tilde{\theta} + \begin{bmatrix} 0 \\ 0 \\ \nu_3 - \frac{\partial \alpha_2}{\partial \hat{\theta}} \Gamma \tau_3 \end{bmatrix}. \tag{A.56}$$

Now ν_3 in (A.53) yields

$$\begin{bmatrix} \dot{z}_1 \\ \dot{z}_2 \\ \dot{z}_3 \end{bmatrix} = \begin{bmatrix} -c_1 & 1 & 0 \\ -1 & -c_2 & 1 + \frac{\partial \alpha_1}{\partial \hat{\theta}} \Gamma \frac{\partial \alpha_2}{\partial x_1} \varphi \\ 0 & -1 - \frac{\partial \alpha_1}{\partial \hat{\theta}} \Gamma \frac{\partial \alpha_2}{\partial x_1} \varphi & -c_3 \end{bmatrix} \begin{bmatrix} z_1 \\ z_2 \\ z_3 \end{bmatrix} + \begin{bmatrix} \varphi^{\mathrm{T}} \\ -\frac{\partial \alpha_1}{\partial x_1} \varphi^{\mathrm{T}} \\ -\frac{\partial \alpha_2}{\partial x_1} \varphi^{\mathrm{T}} \end{bmatrix} \tilde{\theta}. \tag{A.57}$$

This choice, which places the term $-\frac{\partial \alpha_1}{\partial \hat{\theta}} \Gamma \frac{\partial \alpha_2}{\partial x_1} \varphi$ at the (2,3) position in the system matrix, achieves skew symmetry with its positive image above the diagonal. What could not be achieved by pursuing a linear-like form was achieved by designing a nonlinear system where the nonlinearities are "balanced" rather than canceled.

A.2.2 General Recursive Design Procedure

A systematic backstepping design with tuning functions has been developed for the class of nonlinear systems transformable into the *parametric strict feedback form*:

$$\begin{aligned} \dot{x}_1 &= x_2 + \varphi_1(x_1)^{\mathrm{T}} \theta \\ \dot{x}_2 &= x_3 + \varphi_2(x_1, x_2)^{\mathrm{T}} \theta \\ &\vdots \\ \dot{x}_{n-1} &= x_n + \varphi_{n-1}(x_1, \ldots, x_{n-1})^{\mathrm{T}} \theta \\ \dot{x}_n &= \beta(x) u + \varphi_n(x)^{\mathrm{T}} \theta \\ y &= x_1, \end{aligned} \tag{A.58}$$

where β and

$$F(x) = [\varphi_1(x_1),\ \varphi_2(x_1, x_2),\ \cdots,\ \varphi_n(x)] \tag{A.59}$$

Table A.1 Summary of the tuning functions design for tracking. (For notational convenience we define $z_0 \triangleq 0$, $\alpha_0 \triangleq 0$, $\tau_0 \triangleq 0$.)

$$z_i = x_i - y_r^{(i-1)} - \alpha_{i-1} \tag{A.60}$$

$$\alpha_i(\bar{x}_i, \hat{\theta}, \bar{y}_r^{(i-1)}) = -z_{i-1} - c_i z_i - w_i^{\mathrm{T}}\hat{\theta} + \sum_{k=1}^{i-1}\left(\frac{\partial \alpha_{i-1}}{\partial x_k} x_{k+1} + \frac{\partial \alpha_{i-1}}{\partial y_r^{(k-1)}} y_r^{(k)}\right) + \nu_i \tag{A.61}$$

$$\nu_i(\bar{x}_i, \hat{\theta}, \bar{y}_r^{(i-1)}) = +\frac{\partial \alpha_{i-1}}{\partial \hat{\theta}}\Gamma\tau_i + \sum_{k=2}^{i-1}\frac{\partial \alpha_{k-1}}{\partial \hat{\theta}}\Gamma w_i z_k \tag{A.62}$$

$$\tau_i(\bar{x}_i, \hat{\theta}, \bar{y}_r^{(i-1)}) = \tau_{i-1} + w_i z_i \tag{A.63}$$

$$w_i(\bar{x}_i, \hat{\theta}, \bar{y}_r^{(i-2)}) = \varphi_i - \sum_{k=1}^{i-1}\frac{\partial \alpha_{i-1}}{\partial x_k}\varphi_k \tag{A.64}$$

$$i = 1, \ldots, n$$

$$\bar{x}_i = (x_1, \ldots, x_i), \qquad \bar{y}_r^{(i)} = (y_r, \dot{y}_r, \ldots, y_r^{(i)})$$

Adaptive control law:

$$u = \frac{1}{\beta(x)}\left[\alpha_n(x, \hat{\theta}, \bar{y}_r^{(n-1)}) + y_r^{(n)}\right] \tag{A.65}$$

Parameter update law:

$$\dot{\hat{\theta}} = \Gamma\tau_n(x, \hat{\theta}, \bar{y}_r^{(n-1)}) = \Gamma W z \tag{A.66}$$

are smooth nonlinear functions, and $\beta(x) \neq 0$, $\forall x \in \mathbb{R}^n$. (Broader classes of systems that can be controlled using adaptive backstepping are listed in Section A.5.)

The general design summarized in Table A.1 achieves asymptotic tracking, that is, the output $y = x_1$ of the system (A.58) is forced to asymptotically track the reference output $y_r(t)$ whose first n derivatives are assumed to be known, bounded, and piecewise continuous.

The closed-loop system has the form

$$\dot{z} = A_z(z, \hat{\theta}, t)z + W(z, \hat{\theta}, t)^{\mathrm{T}}\tilde{\theta} \tag{A.67}$$

$$\dot{\hat{\theta}} = \Gamma W(z, \hat{\theta}, t)z\,, \tag{A.68}$$

where

$$A_z(z,\hat{\theta},t) = \begin{bmatrix} -c_1 & 1 & 0 & \cdots & 0 \\ -1 & -c_2 & 1+\sigma_{23} & \cdots & \sigma_{2n} \\ 0 & -1-\sigma_{23} & \ddots & \ddots & \vdots \\ \vdots & \vdots & \ddots & \ddots & 1+\sigma_{n-1,n} \\ 0 & -\sigma_{2n} & \cdots & -1-\sigma_{n-1,n} & -c_n \end{bmatrix}. \tag{A.69}$$

and

$$\sigma_{jk}(x,\hat{\theta}) = -\frac{\partial \alpha_{j-1}}{\partial \hat{\theta}} \Gamma w_k . \tag{A.70}$$

Because of the skew symmetry of the off-diagonal part of the matrix A_z, it is easy to see that the Lyapunov function

$$V_n = \frac{1}{2} z^{\mathrm{T}} z + \frac{1}{2} \tilde{\theta}^{\mathrm{T}} \Gamma^{-1} \tilde{\theta} \tag{A.71}$$

has the derivative

$$\dot{V}_n = -\sum_{k=1}^{n} c_k z_k^2 , \tag{A.72}$$

which guarantees that the equilibrium $z = 0$, $\hat{\theta} = \theta$ is globally stable and $z(t) \to 0$ as $t \to \infty$. This means, in particular, that the system state and the control input are bounded and asymptotic tracking is achieved: $\lim_{t\to\infty} \left[y(t) - y_{\mathrm{r}}(t) \right] = 0$.

To help understand how the control design of Table A.1 leads to the closed-loop system (A.67)–(A.70), we provide an interpretation of the matrix A_z for $n = 5$:

$$A_z = \begin{bmatrix} -c_1 & 1 & & & \\ -1 & -c_2 & 1 & & \\ & -1 & -c_3 & 1 & \\ & & -1 & -c_4 & 1 \\ & & & -1 & -c_5 \end{bmatrix} + \begin{bmatrix} & 0 & 0 & 0 & 0 \\ 0 & & \sigma_{23} & \sigma_{24} & \sigma_{25} \\ 0 & -\sigma_{23} & & \sigma_{34} & \sigma_{35} \\ 0 & -\sigma_{24} & -\sigma_{34} & & \sigma_{45} \\ 0 & -\sigma_{25} & -\sigma_{35} & -\sigma_{45} & \end{bmatrix}. \tag{A.73}$$

If the parameters were known, $\hat{\theta} \equiv \theta$, in which case we would not use adaptation, $\Gamma = 0$, the stabilizing functions (A.61) would be implemented with $\nu_i \equiv 0$, and hence $\sigma_{i,j} = 0$. Then A_z would be just the above constant tridiagonal asymptotically stable matrix. When the parameters are unknown, we use $\Gamma > 0$ and, owing to the change of variable $z_i = x_i - y_{\mathrm{r}}^{(i-1)} - \alpha_{i-1}$, in each of the $\dot{z}_i$-equations a term $-\frac{\partial \alpha_{i-1}}{\partial \hat{\theta}} \dot{\hat{\theta}} = \sum_{k=1}^{n} \sigma_{ik} z_k$ appears. The term $\nu_i = -\sum_{k=1}^{i} \sigma_{ik} z_k - \sum_{k=2}^{i-1} \sigma_{ki} z_k$ in the stabilizing function (A.61) is crucial in compensating the effect of $\dot{\hat{\theta}}$. The σ_{ik}-terms above the diagonal in (A.73) come from $\dot{\hat{\theta}}$. Their skew-symmetric negative images come from feedback ν_i.

It can be shown that the resulting closed-loop system (A.67), (A.68), as well as each intermediate system, has a *strict passivity* property from $\tilde{\theta}$ as the input to τ_i as the output. The loop around this operator is closed (see Fig. A.1) with the vector integrator with gain Γ, which is a passive block. It follows from passivity theory that this feedback connection of one strictly passive and one passive block is globally stable.

A.3 MODULAR DESIGN

In the tuning functions design the controller and the identifier were derived in an interlaced fashion. This interlacing led to considerable controller complexity and inflexibility in the choice of the update law.

It is not hard to extend various standard identifiers for linear systems to nonlinear systems. It is therefore desirable to have adaptive designs where the controller can be combined with different identifiers (gradient, least squares, passivity based, etc.). We refer to such adaptive designs as *modular*.

In nonlinear systems it is important to be careful when using the "certainty equivalence" approach, that is, when connecting an off-the-shelf identifier with a stabilizing (in the nonadaptive sense) controller. To illustrate this, let us consider the error system

$$\dot{x} = -x + \varphi(x)\tilde{\theta} \tag{A.74}$$

obtained by applying the controller $u = -x - \varphi(x)\hat{\theta}$ to the scalar system $\dot{x} = u + \varphi(x)\theta$. The parameter estimators commonly used in adaptive linear control generate bounded estimates $\hat{\theta}(t)$ with convergence rates not faster than exponential. Suppose that $\tilde{\theta}(t) = e^{-t}$ and $\varphi(x) = x^3$, which, upon substitution in (A.74), gives

$$\dot{x} = -x + x^3 e^{-t} . \tag{A.75}$$

For initial conditions $|x_0| > \sqrt{\frac{3}{2}}$, the system (A.75) is not only unstable but its solution escapes to infinity in finite time:

$$x(t) \to \infty \quad \text{as} \quad t \to \frac{1}{3} \ln \frac{x_0^2}{x_0^2 - 3/2} . \tag{A.76}$$

From this example we conclude that for nonlinear systems we need stronger controllers that prevent the unbounded behavior caused by $\tilde{\theta}$.

A.3.1 Controller Design

We strengthen the controller for the preceding example, $u = -x - \varphi(x)\hat{\theta}$, with a *nonlinear damping* term $-\varphi(x)^2 x$, that is, $u = -x - \varphi(x)\hat{\theta} - \varphi(x)^2 x$. With this stronger controller, the closed-loop system is

$$\dot{x} = -x - \varphi(x)^2 x + \varphi(x)\tilde{\theta}. \tag{A.77}$$

To see that x is bounded whenever $\tilde{\theta}$ is, we consider the Lyapunov function $V = \frac{1}{2}x^2$. Its derivative along the solutions of (A.77) is

$$\begin{aligned} \dot{V} &= -x^2 - \varphi(x)^2 x^2 + x\varphi(x)\tilde{\theta} \\ &= -x^2 - \left[\varphi(x)x - \frac{1}{2}\tilde{\theta}\right]^2 + \frac{1}{4}\tilde{\theta}^2 \\ &\leq -x^2 + \frac{1}{4}\tilde{\theta}^2. \end{aligned} \tag{A.78}$$

From this inequality it is clear that $|x(t)|$ will not grow larger than $\frac{1}{2}|\tilde{\theta}(t)|$, because then $\dot{V}$ becomes negative and $V = \frac{1}{2}x^2$ decreases. Thanks to the nonlinear damping, the boundedness of $\tilde{\theta}(t)$ guarantees that $x(t)$ is bounded.

To show how nonlinear damping is incorporated into a higher-order backstepping design, we consider the system

$$\begin{aligned} \dot{x}_1 &= x_2 + \varphi(x_1)^{\mathrm{T}}\theta \\ \dot{x}_2 &= u. \end{aligned} \tag{A.79}$$

Viewing x_2 as a control input, we first design a control law $\alpha_1(x_1, \hat{\theta})$ to guarantee that the state x_1 in $\dot{x}_1 = x_2 + \varphi(x_1)^{\mathrm{T}}\theta$ is bounded whenever $\tilde{\theta}$ is bounded. In the first stabilizing function we include a nonlinear damping term[1] $-\kappa_1|\varphi(x_1)|^2 x_1$:

$$\alpha_1(x_1, \hat{\theta}) = -c_1 x_1 - \varphi(x_1)^{\mathrm{T}}\hat{\theta} - \kappa_1|\varphi(x_1)|^2 x_1, \quad c_1, \kappa_1 > 0. \tag{A.80}$$

Then we define the error variable $z_2 = x_2 - \alpha_1(x_1, \hat{\theta})$, and for uniformity denote $z_1 = x_1$. The first equation is now

$$\dot{z}_1 = -c_1 z_1 - \kappa_1|\varphi|^2 z_1 + \varphi^{\mathrm{T}}\tilde{\theta} + z_2. \tag{A.81}$$

If z_2 were zero, the Lyapunov function $V_1 = \frac{1}{2}z_1^2$ would have the derivative

$$\begin{aligned} \dot{V}_1 &= -c_1 z_1^2 - \kappa_1|\varphi|^2 z_1^2 + z_1\varphi^{\mathrm{T}}\tilde{\theta} \\ &= -c_1 z_1^2 - \kappa_1\left|\varphi z_1 - \frac{1}{2\kappa_1}\tilde{\theta}\right|^2 + \frac{1}{4\kappa_1}|\tilde{\theta}|^2 \\ &\leq -c_1 z_1^2 + \frac{1}{4\kappa_1}|\tilde{\theta}|^2, \end{aligned} \tag{A.82}$$

so that z_1 would be bounded whenever $\tilde{\theta}$ was bounded. With $z_2 \neq 0$ we have

$$\dot{V}_1 \leq -c_1 z_1^2 + \frac{1}{4\kappa_1}|\tilde{\theta}|^2 + z_1 z_2. \tag{A.83}$$

Differentiating $z_2 = x_2 - \alpha_1(x_1, \hat{\theta})$, the second equation in (A.79) yields

$$\begin{aligned} \dot{z}_2 &= \dot{x}_2 - \dot{\alpha}_1 \\ &= u - \frac{\partial \alpha_1}{\partial x_1}\left(x_2 + \varphi^{\mathrm{T}}\theta\right) - \frac{\partial \alpha_1}{\partial \hat{\theta}}\dot{\hat{\theta}}. \end{aligned} \tag{A.84}$$

[1]The Euclidian norm of a vector v is denoted as $|v| = \sqrt{v^{\mathrm{T}}v}$.

The derivative of the Lyapunov function

$$V_2 = V_1 + \frac{1}{2}z_2^2 = \frac{1}{2}|z|^2 \tag{A.85}$$

along the solutions of (A.81) and (A.84) is

$$\begin{aligned}\dot{V}_2 \leq & -c_1 z_1^2 + \frac{1}{4\kappa_1}|\tilde{\theta}|^2 + z_1 z_2 + z_2\left[u - \frac{\partial \alpha_1}{\partial x_1}\left(x_2 + \varphi^{\mathrm{T}}\theta\right) - \frac{\partial \alpha_1}{\partial \hat{\theta}}\dot{\hat{\theta}}\right] \\ \leq & -c_1 z_1^2 + \frac{1}{4\kappa_1}|\tilde{\theta}|^2 \\ & + z_2\left[u + z_1 - \frac{\partial \alpha_1}{\partial x_1}\left(x_2 + \varphi^{\mathrm{T}}\hat{\theta}\right) - \left(\frac{\partial \alpha_1}{\partial x_1}\varphi^{\mathrm{T}}\tilde{\theta} + \frac{\partial \alpha_1}{\partial \hat{\theta}}\dot{\hat{\theta}}\right)\right].\end{aligned} \tag{A.86}$$

We note that now, in addition to the $\tilde{\theta}$-dependent disturbance term $\frac{\partial \alpha_1}{\partial x_1}\varphi^{\mathrm{T}}\tilde{\theta}$, we also have a $\dot{\hat{\theta}}$-dependent disturbance $\frac{\partial \alpha_1}{\partial \hat{\theta}}\dot{\hat{\theta}}$. No such term appeared in the scalar system (A.77). We now use the nonlinear damping terms $-\kappa_2\left|\frac{\partial \alpha_1}{\partial x_1}\varphi\right|^2 z_2$ and $-g_2\left|\frac{\partial \alpha_1}{\partial \hat{\theta}}^{\mathrm{T}}\right|^2 z_2$ to counteract the effects of both $\tilde{\theta}$ and $\dot{\hat{\theta}}$:

$$u = -z_1 - c_2 z_2 - \kappa_2\left|\frac{\partial \alpha_1}{\partial x_1}\varphi\right|^2 z_2 - g_2\left|\frac{\partial \alpha_1}{\partial \hat{\theta}}^{\mathrm{T}}\right|^2 z_2 + \frac{\partial \alpha_1}{\partial x_1}\left(x_2 + \varphi^{\mathrm{T}}\hat{\theta}\right), \tag{A.87}$$

where $c_2, \kappa_2, g_2 > 0$. Upon completing the squares as in (A.82), we get

$$\dot{V}_2 \leq -c_1 z_1^2 - c_2 z_2^2 + \left(\frac{1}{4\kappa_1} + \frac{1}{4\kappa_2}\right)|\tilde{\theta}|^2 + \frac{1}{4g_2}|\dot{\hat{\theta}}|^2, \tag{A.88}$$

which means that the state of the error system

$$\begin{aligned}\dot{z} = & \begin{bmatrix} -c_1 - \kappa_2|\varphi|^2 & 1 \\ -1 & -c_2 - \kappa_2\left|\frac{\partial \alpha_1}{\partial x_1}\varphi\right|^2 - g_2\left|\frac{\partial \alpha_1}{\partial \hat{\theta}}^{\mathrm{T}}\right|^2 \end{bmatrix} z \\ & + \begin{bmatrix} \varphi^{\mathrm{T}} \\ -\frac{\partial \alpha_1}{\partial x_1}\varphi^{\mathrm{T}} \end{bmatrix}\tilde{\theta} + \begin{bmatrix} 0 \\ -\frac{\partial \alpha_1}{\partial \hat{\theta}} \end{bmatrix}\dot{\hat{\theta}}\end{aligned} \tag{A.89}$$

is bounded whenever the disturbance inputs $\tilde{\theta}$ and $\dot{\hat{\theta}}$ are bounded. Moreover, since V_2 is quadratic in z (see (A.85)), we can use (A.88) to show that the boundedness of z is guaranteed also when $\dot{\hat{\theta}}$ is square integrable but not bounded. This observation is crucial for the modular design with passive identifiers where $\dot{\hat{\theta}}$ cannot be a priori guaranteed to be bounded.

The recursive controller design for the parametric strict feedback system (A.58) is summarized in Table A.2.

Table A.2 Summary of the controller design in the modular approach. (For notational convenience we define $z_0 \triangleq 0$, $\alpha_0 \triangleq 0$.)

$$z_i = x_i - y_r^{(i-1)} - \alpha_{i-1} \tag{A.90}$$

$$\alpha_i(\bar{x}_i, \hat{\theta}, \bar{y}_r^{(i-1)}) = -z_{i-1} - c_i z_i - w_i^{\mathrm{T}}\hat{\theta} + \sum_{k=1}^{i-1}\left(\frac{\partial \alpha_{i-1}}{\partial x_k}x_{k+1} + \frac{\partial \alpha_{i-1}}{\partial y_r^{(k-1)}}y_r^{(k)}\right) - s_i z_i \tag{A.91}$$

$$w_i(\bar{x}_i, \hat{\theta}, \bar{y}_r^{(i-2)}) = \varphi_i - \sum_{k=1}^{i-1}\frac{\partial \alpha_{i-1}}{\partial x_k}\varphi_k \tag{A.92}$$

$$s_i(\bar{x}_i, \hat{\theta}, \bar{y}_r^{(i-2)}) = \kappa_i |w_i|^2 + g_i \left|\frac{\partial \alpha_{i-1}}{\partial \hat{\theta}}^{\mathrm{T}}\right|^2 \tag{A.93}$$

$$i = 1, \ldots, n$$

$$\bar{x}_i = (x_1, \ldots, x_i), \qquad \bar{y}_r^{(i)} = (y_r, \dot{y}_r, \ldots, y_r^{(i)})$$

Adaptive control law:

$$u = \frac{1}{\beta(x)}\left[\alpha_n(x, \hat{\theta}, \bar{y}_r^{(n-1)}) + y_r^{(n)}\right] \tag{A.94}$$

Controller module guarantees:

If $\tilde{\theta} \in \mathcal{L}_\infty$ and $\dot{\hat{\theta}} \in \mathcal{L}_2$ or $\mathcal{L}_\infty$ then $x \in \mathcal{L}_\infty$

Comparing the expression for the stabilizing function (A.91) in the modular design with the expression (A.61) for the tuning functions design we see that, while the stabilization in the tuning functions design is achieved with the terms ν_i, in the modular design this is accomplished with the nonlinear damping term $-s_i z_i$, where

$$s_i(\bar{x}_i, \hat{\theta}, \bar{y}_r^{(i-2)}) = \kappa_i |w_i|^2 + g_i \left|\frac{\partial \alpha_{i-1}}{\partial \hat{\theta}}^{\mathrm{T}}\right|^2 . \tag{A.95}$$

The resulting error system is

$$\dot{z} = A_z(z, \hat{\theta}, t)z + W(z, \hat{\theta}, t)^{\mathrm{T}}\tilde{\theta} + Q(z, \hat{\theta}, t)^{\mathrm{T}}\dot{\hat{\theta}} \tag{A.96}$$

where A_z, W, Q are:

$$A_z(z,\hat{\theta},t) = \begin{bmatrix} -c_1 - s_1 & 1 & 0 & \cdots & 0 \\ -1 & -c_2 - s_2 & 1 & \ddots & \vdots \\ 0 & -1 & \ddots & \ddots & 0 \\ \vdots & \ddots & \ddots & \ddots & 1 \\ 0 & \cdots & 0 & -1 & -c_n - s_n \end{bmatrix}$$

$$W(z,\hat{\theta},t)^{\mathrm{T}} = \begin{bmatrix} w_1^{\mathrm{T}} \\ w_2^{\mathrm{T}} \\ \vdots \\ w_n^{\mathrm{T}} \end{bmatrix}, \qquad Q(z,\hat{\theta},t)^{\mathrm{T}} = \begin{bmatrix} 0 \\ -\dfrac{\partial \alpha_1}{\partial \hat{\theta}} \\ \vdots \\ -\dfrac{\partial \alpha_{n-1}}{\partial \hat{\theta}} \end{bmatrix}. \tag{A.97}$$

Since the controller module guarantees that x is bounded whenever $\tilde{\theta}$ is bounded and $\dot{\hat{\theta}}$ is either bounded or square integrable, then we need identifiers that independently guarantee these properties. Both the boundedness and the square integrability requirements for $\dot{\hat{\theta}}$ are essentially conditions that limit the speed of adaptation, and only one of them needs to be satisfied to achieve boundedness. The modular design needs *slow adaptation* because the controller does not cancel the effect of $\dot{\hat{\theta}}$, as was the case in the tuning functions design.

In addition to boundedness of $x(t)$, our goal is to achieve asymptotic tracking, that is, to regulate $z(t)$ to zero. With z and $\hat{\theta}$ bounded, it is not hard to prove that $z(t) \to 0$ provided

$$W(z(t),\hat{\theta}(t),t)^{\mathrm{T}}\tilde{\theta}(t) \to 0 \text{ and } \dot{\hat{\theta}}(t) \to 0.$$

Let us factor the regressor matrix W, using (A.97), (A.92) and (A.59) as

$$W(z,\hat{\theta},t)^{\mathrm{T}} = \begin{bmatrix} 1 & 0 & \cdots & 0 \\ -\dfrac{\partial \alpha_1}{\partial x_1} & 1 & \ddots & \vdots \\ \vdots & \ddots & \ddots & 0 \\ -\dfrac{\partial \alpha_{n-1}}{\partial x_1} & \cdots & -\dfrac{\partial \alpha_{n-1}}{\partial x_{n-1}} & 1 \end{bmatrix} F(x)^{\mathrm{T}} \triangleq N(z,\hat{\theta},t)F(x)^{\mathrm{T}}. \tag{A.98}$$

Since the matrix $N(z,\hat{\theta},t)$ is invertible, the tracking condition $W(z(t), \hat{\theta}(t),t)^{\mathrm{T}}\tilde{\theta}(t) \to 0$ becomes $F(x(t))^{\mathrm{T}}\tilde{\theta}(t) \to 0$.

In the next two subsections we develop identifiers for the general parametric model

$$\dot{x} = f(x,u) + F(x,u)^{\mathrm{T}}\theta. \tag{A.99}$$

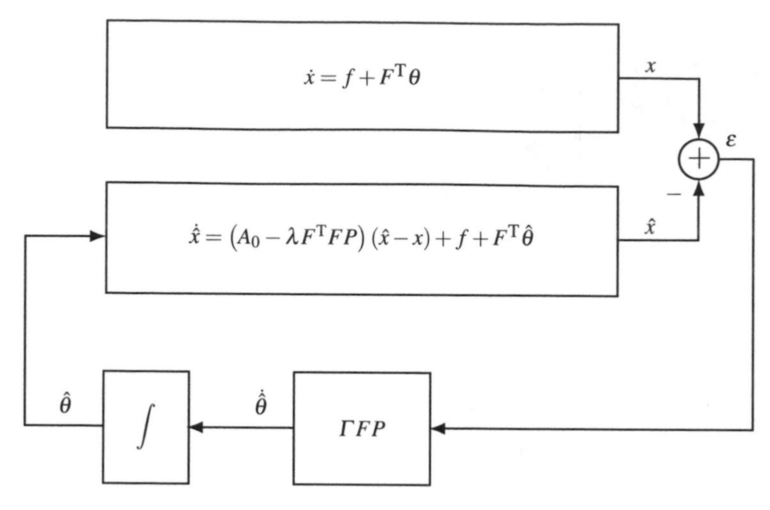

Figure A.2 The passive identifier.

The parametric strict-feedback system (A.58) is a special case of this model with $F(x, u)$ given by (A.59) and $f(x, u) = [x_2, \ \ldots, \ x_n, \ \beta_0(x)u]^{\mathrm{T}}$.

Before we present the design of identifiers, we summarize the properties required from the identifier module:

$$\text{(i)} \quad \tilde{\theta} \in \mathcal{L}_\infty \text{ and } \dot{\hat{\theta}} \in \mathcal{L}_2 \text{ or } \mathcal{L}_\infty,$$

$$\text{(ii)} \quad \text{if } x \in \mathcal{L}_\infty \text{ then } F(x(t))^{\mathrm{T}}\tilde{\theta}(t) \to 0 \text{ and } \dot{\hat{\theta}}(t) \to 0.$$

We present two types of identifiers: the *passive* identifier and the *swapping* identifier.

A.3.2 Passive Identifier

For the parametric model (A.99) we implement the "observer"

$$\dot{\hat{x}} = \left[A_0 - \lambda F(x, u)^{\mathrm{T}} F(x, u) P\right](\hat{x} - x) + f(x, u) + F(x, u)^{\mathrm{T}}\hat{\theta}, \qquad \text{(A.100)}$$

where $\lambda > 0$ and A_0 is an arbitrary constant matrix such that

$$P A_0 + A_0^{\mathrm{T}} P = -I, \quad P = P^{\mathrm{T}} > 0. \qquad \text{(A.101)}$$

The block diagram of the passive identifier is shown in Figure A.2.
By direct substitution it can be seen that the observer error

$$\varepsilon = x - \hat{x} \qquad \text{(A.102)}$$

is governed by

$$\dot{\varepsilon} = \left[A_0 - \lambda F(x, u)^{\mathrm{T}} F(x, u) P\right]\varepsilon + F(x, u)^{\mathrm{T}}\tilde{\theta}. \qquad \text{(A.103)}$$

The observer error system (A.103) has a *strict passivity* property from the input $\tilde{\theta}$ to the output $F(x, u)P\varepsilon$.

A standard result of passivity theory is that the equilibrium $\tilde{\theta} = 0, \ \varepsilon = 0$ of the negative feedback connection of one strictly passive and one passive system

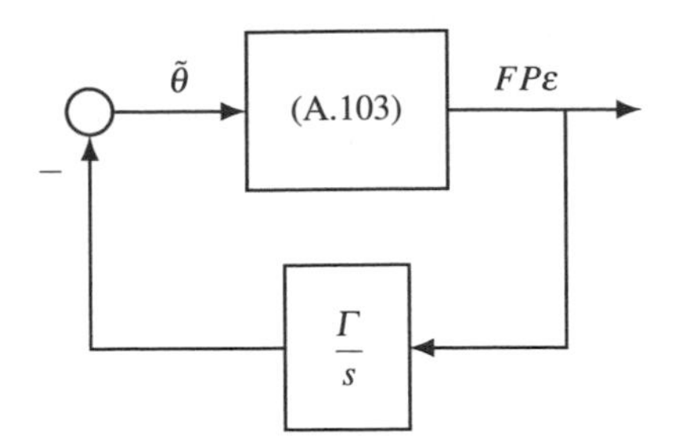

Figure A.3 Negative feedback connection of the strictly passive system (A.103) with the passive system $\frac{\Gamma}{s}$.

is globally stable. Using integral feedback, such a connection can be formed as in Figure A.3. This suggests the use of the following update law:

$$\dot{\hat{\theta}} = \Gamma F(x,u)P\varepsilon\,, \qquad \Gamma = \Gamma^{\mathrm{T}} > 0\,. \tag{A.104}$$

To analyze the stability properties of the passive identifier we use the Lyapunov function

$$V = \tilde{\theta}^{\mathrm{T}}\Gamma^{-1}\tilde{\theta} + \varepsilon^{\mathrm{T}}P\varepsilon\,. \tag{A.105}$$

After uncomplicated calculations, its derivative can be shown to satisfy

$$\dot{V} \le -\varepsilon^{\mathrm{T}}\varepsilon - \frac{\lambda}{\bar{\lambda}(\Gamma)^2}|\dot{\hat{\theta}}|^2\,. \tag{A.106}$$

This guarantees the boundedness of $\tilde{\theta}$ and ε, even when $\lambda = 0$. However, $\dot{\hat{\theta}}$ cannot be shown to be bounded (unless x and u are known to be bounded). Instead, for the passive identifier one can show that $\dot{\hat{\theta}}$ is square integrable. For this we must use $\lambda > 0$, that is, we rely on the nonlinear damping term $-\lambda F(x,u)^{\mathrm{T}}F(x,u)P$ in the observer. The boundedness of $\tilde{\theta}$ and the square integrability of $\dot{\hat{\theta}}$ imply (see Table A.2) that x is bounded.

To prove the tracking, we need to show that the identifier guarantees that, whenever x is bounded, $F(x(t))^{\mathrm{T}}\tilde{\theta}(t) \to 0$ and $\dot{\hat{\theta}}(t) \to 0$. Both properties are established by Barbalat's lemma. The latter property can easily be shown to follow from the square integrability of $\dot{\hat{\theta}}$. The regulation of $F(x)^{\mathrm{T}}\tilde{\theta}$ to zero follows upon showing that both $\varepsilon(t)$ and $\dot{\varepsilon}(t)$ converge to zero. While the convergence of $\varepsilon(t)$ follows by deducing its square integrability from (A.106), the convergence of $\dot{\varepsilon}(t)$ follows from the fact that its integral, $\int_0^\infty \dot{\varepsilon}(\tau)d\tau = \varepsilon(\infty) - \varepsilon(0) = -\varepsilon(0)$, exists.

A.3.3 Swapping Identifier

For the parametric model (A.99) we implement two filters,

$$\dot{\Omega}^{\mathrm{T}} = \left[A_0 - \lambda F(x,u)^{\mathrm{T}}F(x,u)P\right]\Omega^{\mathrm{T}} + F(x,u)^{\mathrm{T}} \tag{A.107}$$

$$\dot{\Omega}_0 = \left[A_0 - \lambda F(x,u)^{\mathrm{T}}F(x,u)P\right](\Omega_0 - x) - f(x,u)\,, \tag{A.108}$$

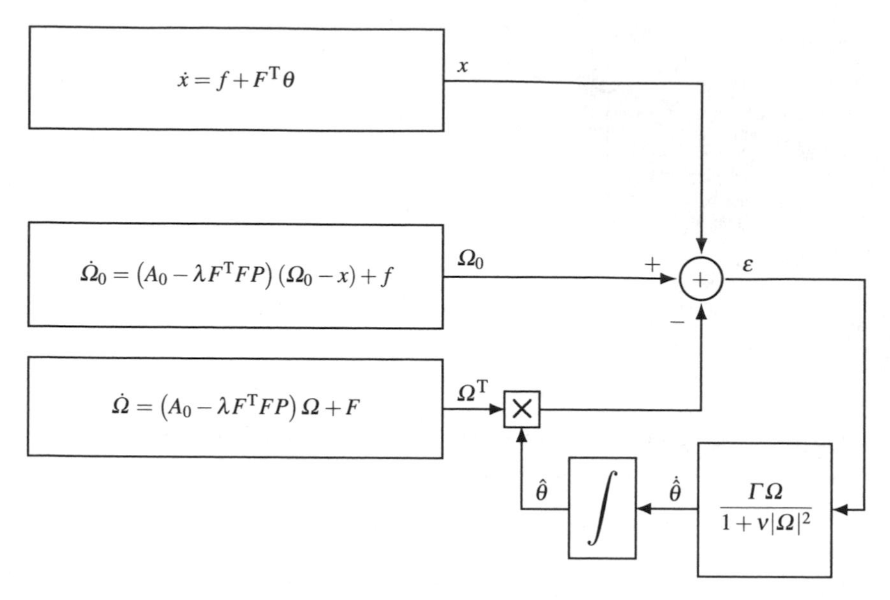

Figure A.4 The swapping identifier.

where $\lambda \geq 0$ and A_0 is as defined in (A.101). The *estimation error*,

$$\varepsilon = x + \Omega_0 - \Omega^{\mathrm{T}}\hat{\theta}, \tag{A.109}$$

can be written in the form

$$\varepsilon = \Omega^{\mathrm{T}}\tilde{\theta} + \tilde{\varepsilon}, \tag{A.110}$$

where $\tilde{\varepsilon} \stackrel{\Delta}{=} x + \Omega_0 - \Omega^{\mathrm{T}}\theta$ decays exponentially because it is governed by

$$\dot{\tilde{\varepsilon}} = \left[A_0 - \lambda F(x,u)^{\mathrm{T}} F(x,u) P\right]\tilde{\varepsilon}. \tag{A.111}$$

The filters (A.107) and (A.108) have converted the dynamic model (A.99) into the linear static parametric model (A.110), to which we can apply standard estimation algorithms. As our update law we will employ either the gradient

$$\dot{\hat{\theta}} = \Gamma \frac{\Omega\varepsilon}{1 + \nu \mathrm{tr}\{\Omega^{\mathrm{T}}\Omega\}}, \qquad \begin{array}{l} \Gamma = \Gamma^{\mathrm{T}} > 0 \\ \nu \geq 0 \end{array} \tag{A.112}$$

or the least squares algorithm

$$\begin{aligned} \dot{\hat{\theta}} &= \Gamma \frac{\Omega\varepsilon}{1 + \nu \mathrm{tr}\{\Omega^{\mathrm{T}}\Omega\}} \\ \dot{\Gamma} &= -\Gamma \frac{\Omega\Omega^{\mathrm{T}}}{1 + \nu \mathrm{tr}\{\Omega^{\mathrm{T}}\Omega\}}\Gamma, \qquad \begin{array}{l} \Gamma(0) = \Gamma(0)^{\mathrm{T}} > 0 \\ \nu \geq 0. \end{array} \end{aligned} \tag{A.113}$$

By allowing $\nu = 0$, we encompass unnormalized gradient and least squares update laws. The complete swapping identifier is shown in Figure A.4.

The update law normalization, $\nu > 0$, and the nonlinear damping, $\lambda > 0$, are two different means for slowing down the identifier in order to guarantee the boundedness and square integrability of $\dot{\hat{\theta}}$.

For the gradient update law (A.112), the identifier properties (boundedness of $\tilde{\theta}$ and $\dot{\hat{\theta}}$ and regulation of $F(x)\tilde{\theta}$ and $\dot{\hat{\theta}}$) are established via the Lyapunov function

$$V = \frac{1}{2}\tilde{\theta}^{\mathrm{T}}\Gamma^{-1}\tilde{\theta} + \tilde{\varepsilon}P\tilde{\varepsilon} \tag{A.114}$$

whose derivative is

$$\dot{V} \leq -\frac{3}{4}\frac{\varepsilon^{\mathrm{T}}\varepsilon}{1+\nu\, tr\{\Omega^{\mathrm{T}}\Omega\}}. \tag{A.115}$$

The Lyapunov function for the least squares update law (A.113) is $V = \tilde{\theta}^{\mathrm{T}}\Gamma(t)^{-1}\tilde{\theta} + \tilde{\varepsilon}P\tilde{\varepsilon}$.

A.4 OUTPUT FEEDBACK DESIGNS

For linear systems, a common solution to the output-feedback problem is a stabilizing state-feedback controller employing the state estimates from an exponentially converging observer. Unfortunately, this approach is not applicable to nonlinear systems. Additional difficulties arise when the nonlinear plant has unknown parameters because *adaptive* observers, in general, are not exponentially convergent.

These obstacles have been overcome for systems transformable into the *output-feedback form*:

$$\begin{aligned} \dot{x} &= Ax + \phi(y) + \Phi(y)a + \begin{bmatrix} 0 \\ b \end{bmatrix}\sigma(y)u\,, \qquad x \in \mathbb{R}^n \\ y &= e_1^{\mathrm{T}}x\,, \end{aligned} \tag{A.116}$$

where only the output y is available for measurement,

$$A = \begin{bmatrix} 0 & \\ \vdots & I_{n-1} \\ 0 & \cdots\ 0 \end{bmatrix}, \tag{A.117}$$

$$\phi(y) = \begin{bmatrix} \varphi_{0,1}(y) \\ \vdots \\ \varphi_{0,n}(y) \end{bmatrix}, \quad \Phi(y) = \begin{bmatrix} \varphi_{1,1}(y) & \cdots & \varphi_{q,1}(y) \\ \vdots & & \vdots \\ \varphi_{1,n}(y) & \cdots & \varphi_{q,n}(y) \end{bmatrix}, \tag{A.118}$$

and the vectors of unknown constant parameters are:

$$a = [a_1, \ldots, a_q]^{\mathrm{T}}, \quad b = [b_m, \ldots, b_0]^{\mathrm{T}}. \tag{A.119}$$

We make the following assumptions: the sign of b_m is known; the polynomial $B(s) = b_m s^m + \cdots + b_1 s + b_0$ is known to be Hurwitz; and $\sigma(y) \neq 0\ \forall y \in \mathbb{R}$. An important restriction is that the nonlinearities $\phi(y)$ and $\Phi(y)$ are allowed to depend only on the output y. Even when θ is known this restriction is needed to achieve global stability.

Table A.3 State estimation filters.

Filters:

$$\dot{\xi} = A_0\xi + ky + \phi(y) \tag{A.120}$$
$$\dot{\Xi} = A_0\Xi + \Phi(y) \tag{A.121}$$
$$\dot{\lambda} = A_0\lambda + e_n\sigma(y)u \tag{A.122}$$

$$v_j = A_0^j\lambda, \qquad j = 0, \ldots, m \tag{A.123}$$
$$\Omega^{\mathrm{T}} = [v_m, \ldots, v_1, v_0, \ \Xi] \tag{A.124}$$

We define the parameter-dependent state estimate

$$\hat{x} = \xi + \Omega^{\mathrm{T}}\theta\,, \tag{A.125}$$

which employs the filters given in Table A.3, with the vector $k = [k_1, \ldots, k_n]^{\mathrm{T}}$ chosen so that the matrix

$$A_0 = A - ke_1^{\mathrm{T}} \tag{A.126}$$

is Hurwitz, that is,

$$PA_0 + A_0^{\mathrm{T}}P = -I, \quad P = P^{\mathrm{T}} > 0\,. \tag{A.127}$$

The state estimation error

$$\varepsilon = x - \hat{x} \tag{A.128}$$

is readily shown to satisfy

$$\dot{\varepsilon} = A_0\varepsilon\,. \tag{A.129}$$

The following two expressions for $\dot{y}$ are instrumental in the backstepping design:

$$\dot{y} = \omega_0 + \omega^{\mathrm{T}}\theta + \varepsilon_2 \tag{A.130}$$
$$= b_m v_{m,2} + \omega_0 + \bar{\omega}^{\mathrm{T}}\theta + \varepsilon_2\,, \tag{A.131}$$

where

$$\omega_0 = \varphi_{0,1} + \xi_2 \tag{A.132}$$
$$\omega = [v_{m,2}, v_{m-1,2}, \ldots, v_{0,2}, \ \Phi_{(1)} + \Xi_{(2)}]^{\mathrm{T}} \tag{A.133}$$
$$\bar{\omega} = [0, v_{m-1,2}, \ldots, v_{0,2}, \ \Phi_{(1)} + \Xi_{(2)}]^{\mathrm{T}}\,. \tag{A.134}$$

Since the states $x_2, \ldots, x_n$ are not measured, the backstepping design is applied to the system

$$\dot{y} = b_m v_{m,2} + \omega_0 + \bar{\omega}^{\mathrm{T}}\theta + \varepsilon_2 \tag{A.135}$$

$$\dot{v}_{m,i} = v_{m,i+1} - k_i v_{m,1}\,, \qquad i = 2, \ldots, \rho - 1 \tag{A.136}$$

$$\dot{v}_{m,\rho} = \sigma(y)u + v_{m,\rho+1} - k_\rho v_{m,1}\,. \tag{A.137}$$

The order of this system is equal to the relative degree of the plant (A.116).

A.4.1 Output-Feedback Design with Tuning Functions

The output-feedback design with tuning functions is summarized in Table A.4. The resulting error system is

$$\dot{z} = A_z(z,t)z + W_\varepsilon(z,t)\varepsilon_2 + W_\theta(z,t)^{\mathrm{T}}\tilde{\theta} - b_m\left(\dot{y}_{\mathrm{r}} + \bar{\alpha}_1\right)e_1\tilde{\rho} \tag{A.138}$$

where

$$A_z = \begin{bmatrix} -c_1 - d_1 & \hat{b}_m & 0 & \cdots & \cdots & 0 \\ -\hat{b}_m & -c_2 - d_2\left(\frac{\partial\alpha_1}{\partial y}\right)^2 & 1+\sigma_{23} & \sigma_{24} & \cdots & \sigma_{2,\rho} \\ 0 & -1-\sigma_{23} & \ddots & \ddots & \ddots & \vdots \\ \vdots & -\sigma_{24} & \ddots & \ddots & \ddots & \sigma_{\rho-2,\rho} \\ \vdots & \vdots & \ddots & \ddots & \ddots & 1+\sigma_{\rho-1,\rho} \\ 0 & -\sigma_{2,\rho} & \cdots & -\sigma_{\rho-2,\rho} & -1-\sigma_{\rho-1,\rho} & -c_\rho - d_\rho\left(\frac{\partial\alpha_{\rho-1}}{\partial y}\right)^2 \end{bmatrix} \tag{A.139}$$

and

$$W_\varepsilon(z,t) = \begin{bmatrix} 1 \\ -\dfrac{\partial\alpha_1}{\partial y} \\ \vdots \\ -\dfrac{\partial\alpha_{\rho-1}}{\partial y} \end{bmatrix}, \qquad W_\theta(z,t)^{\mathrm{T}} = W_\varepsilon(z,t)\omega^{\mathrm{T}} - \hat{\rho}\left(\dot{y}_{\mathrm{r}} + \bar{\alpha}_1\right)e_1 e_1^{\mathrm{T}}\,. \tag{A.140}$$

The nonlinear damping terms $-d_i\left(\frac{\partial\alpha_{i-1}}{\partial y}\right)^2$ in (A.139) are included to counteract the exponentially decaying state estimation error ε_2. The variable $\hat{\rho}$ is an estimate of $\rho = 1/b_m$.

A.4.2 Output-Feedback Modular Design

In addition to $\mathrm{sgn}(b_m)$, in the modular design we assume that a positive constant ς_m is known such that $|b_m| \geq \varsigma_m$.

Table A.4 Output-feedback tuning functions design.

$$z_1 = y - y_r \tag{A.141}$$

$$z_i = v_{m,i} - \hat{\rho} y_r^{(i-1)} - \alpha_{i-1}, \qquad i = 2, \ldots, \rho \tag{A.142}$$

$$\alpha_1 = \hat{\rho}\bar{\alpha}_1, \qquad \bar{\alpha}_1 = -(c_1 + d_1) z_1 - \omega_0 - \bar{\omega}^T \hat{\theta} \tag{A.143}$$

$$\alpha_2 = -\hat{b}_m z_1 - \left[c_2 + d_2 \left(\frac{\partial \alpha_1}{\partial y} \right)^2 \right] z_2 + \left(\dot{y}_r + \frac{\partial \alpha_1}{\partial \hat{\rho}} \right) \dot{\hat{\rho}} + \frac{\partial \alpha_1}{\partial \hat{\theta}} \Gamma \tau_2 + \beta_2 \tag{A.144}$$

$$\alpha_i = -z_{i-1} - \left[c_i + d_i \left(\frac{\partial \alpha_{i-1}}{\partial y} \right)^2 \right] z_i + \left(y_r^{(i-1)} + \frac{\partial \alpha_{i-1}}{\partial \hat{\rho}} \right) \dot{\hat{\rho}}$$

$$+ \frac{\partial \alpha_{i-1}}{\partial \hat{\theta}} \Gamma \tau_i - \sum_{j=2}^{i-1} \frac{\partial \alpha_{j-1}}{\partial \hat{\theta}} \Gamma \frac{\partial \alpha_{i-1}}{\partial y} z_j + \beta_i, \qquad i = 3, \ldots, \rho \tag{A.145}$$

$$\beta_i = \frac{\partial \alpha_{i-1}}{\partial y} \left(\omega_0 + \omega^T \hat{\theta} \right) + \frac{\partial \alpha_{i-1}}{\partial \xi} (A_0 \xi + k y + \phi) + \frac{\partial \alpha_{i-1}}{\partial \Xi} (A_0 \Xi + \Phi)$$

$$+ \sum_{j=1}^{i-1} \frac{\partial \alpha_{i-1}}{\partial y_r^{(j-1)}} y_r^{(j)} + k_i v_{m,1} + \sum_{j=1}^{m+i-1} \frac{\partial \alpha_{i-1}}{\partial \lambda_j} (-k_j \lambda_1 + \lambda_{j+1}) \tag{A.146}$$

$$\tau_1 = \left(\omega - \hat{\rho} (\dot{y}_r + \bar{\alpha}_1) e_1 \right) z_1 \tag{A.147}$$

$$\tau_i = \tau_{i-1} - \frac{\partial \alpha_{i-1}}{\partial y} \omega z_i, \qquad i = 2, \ldots, \rho \tag{A.148}$$

Adaptive control law:

$$u = \frac{1}{\sigma(y)} \left(\alpha_\rho - v_{m,\rho+1} + \hat{\rho} y_r^{(\rho)} \right) \tag{A.149}$$

Parameter update laws:

$$\dot{\hat{\theta}} = \Gamma \tau_\rho \tag{A.150}$$

$$\dot{\hat{\rho}} = -\gamma \, \mathrm{sgn}(b_m) (\dot{y}_r + \bar{\alpha}_1) z_1 \tag{A.151}$$

Table A.5 Output-feedback controller in the modular design.

$$z_1 = y - y_r \tag{A.152}$$

$$z_i = v_{m,i} - \frac{1}{\hat{b}_m} y_r^{(i-1)} - \alpha_{i-1}, \qquad i = 2, \ldots, \rho \tag{A.153}$$

$$\alpha_1 = -\frac{\operatorname{sgn}(b_m)}{\varsigma_m}(c_1 + s_1) z_1 + \frac{1}{\hat{b}_m}\bar{\alpha}_1, \qquad \bar{\alpha}_1 = -\omega_0 - \bar{\omega}^{\mathrm{T}}\hat{\theta} \tag{A.154}$$

$$\alpha_2 = -\hat{b}_m z_1 - (c_2 + s_2) z_2 + \beta_2 \tag{A.155}$$

$$\alpha_i = -z_{i-1} - (c_i + s_i) z_i + \beta_i, \qquad i = 3, \ldots, \rho \tag{A.156}$$

$$\beta_i = \frac{\partial \alpha_{i-1}}{\partial y}\left(\omega_0 + \omega^{\mathrm{T}}\hat{\theta}\right) + \frac{\partial \alpha_{i-1}}{\partial \xi}(A_0\xi + ky + \phi) + \frac{\partial \alpha_{i-1}}{\partial \Xi}(A_0\Xi + \Phi)$$
$$+ \sum_{j=1}^{i-1} \frac{\partial \alpha_{i-1}}{\partial y_r^{(j-1)}} y_r^{(j)} + k_i v_{m,1} + \sum_{j=1}^{m+i-1} \frac{\partial \alpha_{i-1}}{\partial \lambda_j}(-k_j\lambda_1 + \lambda_{j+1}) \tag{A.157}$$

$$s_1 = d_1 + \kappa_1 \left|\bar{\omega} + \frac{1}{\hat{b}_m}(\dot{y}_r + \bar{\alpha}_1) e_1\right|^2 \tag{A.158}$$

$$s_2 = d_2\left(\frac{\partial \alpha_1}{\partial y}\right)^2 + \kappa_2\left|\frac{\partial \alpha_1}{\partial y}\omega - z_1 e_1\right|^2 + g_2\left|\frac{\partial \alpha_1}{\partial \hat{\theta}}^{\mathrm{T}} - \frac{1}{\hat{b}_m^2}\dot{y}_r e_1\right|^2 \tag{A.159}$$

$$s_i = d_i\left(\frac{\partial \alpha_{i-1}}{\partial y}\right)^2 + \kappa_i\left|\frac{\partial \alpha_{i-1}}{\partial y}\omega\right|^2 + g_i\left|\frac{\partial \alpha_{i-1}}{\partial \hat{\theta}}^{\mathrm{T}} - \frac{1}{\hat{b}_m^2} y_r^{(i-1)} e_1\right|^2, \qquad i = 3, \ldots, \rho \tag{A.160}$$

Adaptive control law:

$$u = \frac{1}{\sigma(y)}\left(\alpha_\rho - v_{m,\rho+1} + \frac{1}{\hat{b}_m} y_r^{(\rho)}\right) \tag{A.161}$$

The complete design of the control law is summarized in Table A.5. The resulting error system is

$$\dot{z} = A_z^*(z,t)z + W_\varepsilon(z,t)\varepsilon_2 + W_\theta^*(z,t)^{\mathrm{T}}\tilde{\theta} + Q(z,t)^{\mathrm{T}}\dot{\hat{\theta}}, \tag{A.162}$$

where

$$A_z^*(z,t) = \begin{bmatrix} -\frac{|b_m|}{\varsigma_m}(c_1+s_1) & b_m & 0 & \cdots & 0 \\ -b_m & -(c_2+s_2) & 1 & \ddots & \vdots \\ 0 & -1 & \ddots & \ddots & 0 \\ \vdots & \ddots & \ddots & \ddots & 1 \\ 0 & \cdots & 0 & -1 & -(c_\rho+s_\rho) \end{bmatrix}, \tag{A.163}$$

$$W_\varepsilon(z,t)=\begin{bmatrix}1\\-\frac{\partial\alpha_1}{\partial y}\\\vdots\\-\frac{\partial\alpha_{\rho-1}}{\partial y}\end{bmatrix}\qquad W_\theta^*(z,t)^{\mathrm T}=\begin{bmatrix}\bar\omega^{\mathrm T}+\frac{1}{\hat b_m}\left(\dot y_{\mathrm r}+\bar\alpha_1\right)e_1^{\mathrm T}\\-\frac{\partial\alpha_1}{\partial y}\omega^{\mathrm T}+z_1e_1^{\mathrm T}\\-\frac{\partial\alpha_2}{\partial y}\omega^{\mathrm T}\\\vdots\\-\frac{\partial\alpha_{\rho-1}}{\partial y}\omega^{\mathrm T}\end{bmatrix}\tag{A.164}$$

$$Q(z,t)^{\mathrm T}=\begin{bmatrix}0\\-\frac{\partial\alpha_1}{\partial\hat\theta}+\frac{1}{\hat b_m^2}\dot y_{\mathrm r}e_1^{\mathrm T}\\\vdots\\-\frac{\partial\alpha_{\rho-1}}{\partial\hat\theta}+\frac{1}{\hat b_m^2}y_{\mathrm r}^{(\rho-1)}e_1^{\mathrm T}\end{bmatrix}.\tag{A.165}$$

A.4.2.1 Passive Identifier

For the parametric model (A.130), we introduce the scalar observer

$$\dot{\hat y}=-\left(c_0+\kappa_0|\omega|^2\right)\left(\hat y-y\right)+\omega_0+\omega^{\mathrm T}\hat\theta\,.\tag{A.166}$$

The observer error

$$\varepsilon=y-\hat y\tag{A.167}$$

is governed by

$$\dot\varepsilon=-\left(c_0+\kappa_0|\omega|^2\right)\varepsilon+\omega^{\mathrm T}\tilde\theta+\varepsilon_2.\tag{A.168}$$

The parameter update law is

$$\dot{\hat\theta}=\operatorname*{Proj}_{\hat b_m}\{\Gamma\omega\varepsilon\}\,,\qquad \begin{array}{l}\hat b_m(0)\mathrm{sgn}b_m>\varsigma_m\\ \Gamma=\Gamma^{\mathrm T}>0\end{array}\,,\tag{A.169}$$

where the projection operator is employed to guarantee that $|\hat b_m(t)|\geq\varsigma_m>0$, $\forall t\geq 0$.

A.4.2.2 Swapping Identifier

The estimation error

$$\varepsilon=y-\xi_1-\Omega_1^{\mathrm T}\hat\theta\tag{A.170}$$

satisfies the following equation linear in the parameter error:

$$\varepsilon=\Omega_1^{\mathrm T}\tilde\theta+\varepsilon_1\,.\tag{A.171}$$

The update law for $\hat{\theta}$ is either the gradient:

$$\dot{\hat{\theta}} = \operatorname*{Proj}_{\hat{b}_m}\left\{\Gamma\frac{\Omega_1\varepsilon}{1+\nu|\Omega_1|^2}\right\}, \qquad \begin{array}{l}\hat{b}_m(0)\mathrm{sgn}b_m > \varsigma_m\\ \Gamma = \Gamma^{\mathrm{T}} > 0\\ \nu > 0,\end{array} \qquad (A.172)$$

or the least squares:

$$\begin{aligned}\dot{\hat{\theta}} &= \operatorname*{Proj}_{\hat{b}_m}\left\{\Gamma\frac{\Omega_1\varepsilon}{1+\nu|\Omega_1|^2}\right\}, & &\hat{b}_m(0)\mathrm{sgn}b_m > \varsigma_m\\ \dot{\Gamma} &= -\Gamma\frac{\Omega_1\Omega_1^{\mathrm{T}}}{1+\nu|\Omega_1|^2}\Gamma, & &\Gamma(0) = \Gamma(0)^{\mathrm{T}} > 0\\ & & &\nu > 0.\end{aligned} \qquad (A.173)$$

A.5 EXTENSIONS

Adaptive nonlinear control designs presented in the preceding sections are applicable to classes of nonlinear systems broader than the parametric strict feedback systems (A.58).

Pure feedback systems

$$\begin{aligned}\dot{x}_i &= x_{i+1} + \varphi_i(x_1,\ldots,x_{i+1})^{\mathrm{T}}\theta, \qquad i = 1,\ldots,n-1\\ \dot{x}_n &= \left(\beta_0(x) + \beta(x)^{\mathrm{T}}\theta\right)u + \varphi_0(x) + \varphi_n(x)^{\mathrm{T}}\theta,\end{aligned} \qquad (A.174)$$

where $\varphi_0(0) = 0$, $\varphi_1(0) = \cdots = \varphi_n(0) = 0$, $\beta_0(0) \neq 0$. Because of the dependence of φ_i on x_{i+1}, the regulation or tracking for pure feedback systems is, in general, not global, even when θ is known.

Unknown virtual control coefficients

$$\begin{aligned}\dot{x}_i &= b_i x_{i+1} + \varphi_i(x_1,\ldots,x_i)^{\mathrm{T}}\theta, \qquad i = 1,\ldots,n-1\\ \dot{x}_n &= b_n\beta(x)u + \varphi_n(x_1,\ldots,x_n)^{\mathrm{T}}\theta,\end{aligned} \qquad (A.175)$$

where, in addition to the unknown vector θ, the constant coefficients b_i are also unknown. The unknown b_i-coefficients are frequent in applications ranging from electric motors to flight dynamics. The signs of b_i, $i = 1,\ldots,n$, are assumed to be known. In the tuning functions design, in addition to estimating b_i, we also estimate its inverse, $\rho_i = 1/b_i$. In the modular design we assume that in addition to $\mathrm{sgn}b_i$, a positive constant ς_i is known such that $|b_i| \geq \varsigma_i$. Then, instead of estimating $\rho_i = 1/b_i$, we use the inverse of the estimate $\hat{b}_i$, that is $1/\hat{b}_i$, where $\hat{b}_i(t)$ is kept away from zero by using parameter projection.

Multi-input systems

$$\begin{aligned}\dot{X}_i &= B_i(\bar{X}_i)X_{i+1} + \Phi_i(\bar{X}_i)^{\mathrm{T}}\theta, \quad i = 1,\ldots,n-1\\ \dot{X}_n &= B_n(X)u + \Phi_n(X)^{\mathrm{T}}\theta,\end{aligned} \qquad (A.176)$$

where X_i is a ν_i-vector, $\nu_1 \leq \nu_2 \leq \cdots \leq \nu_n$, $\bar{X}_i = \left[X_1^{\mathrm{T}}, \ldots, X_i^{\mathrm{T}}\right]^{\mathrm{T}}$, $X = \bar{X}_n$. and the matrices $B_i(\bar{X}_i)$ have full rank for all $\bar{X}_i \in \mathbb{R}^{\sum_{j=1}^{i} \nu_j}$. The input u is a ν_n-vector. The matrices B_i can be allowed to be unknown provided they are constant and positive definite.

Block strict feedback systems

$$
\begin{aligned}
\dot{x}_i &= x_{i+1} + \varphi_i(x_1, \ldots, x_i, \zeta_1, \ldots, \zeta_i)^{\mathrm{T}}\theta, \qquad i = 1, \ldots, \rho - 1 \\
\dot{x}_\rho &= \beta(x, \zeta)u + \varphi_\rho(x, \zeta)^{\mathrm{T}}\theta \qquad\qquad\qquad\qquad (\mathrm{A}.177) \\
\dot{\zeta}_i &= \Phi_{i,0}(\bar{x}_i, \bar{\zeta}_i) + \Phi_i(\bar{x}_i, \bar{\zeta}_i)^{\mathrm{T}}\theta, \qquad i = 1, \ldots, \rho
\end{aligned}
$$

with the following notation: $\bar{x}_i = [x_1, \ldots, x_i]^{\mathrm{T}}$, $\bar{\zeta}_i = \left[\zeta_1^{\mathrm{T}}, \ldots, \zeta_i^{\mathrm{T}}\right]^{\mathrm{T}}$, $x = \bar{x}_\rho$, and $\zeta = \bar{\zeta}_\rho$. Each ζ_i-subsystem of (A.177) is assumed to be bounded-input bounded-state (BIBS) stable with respect to the input $(\bar{x}_i, \bar{\zeta}_{i-1})$. For this class of systems it is quite simple to modify the procedure in Tables A.1 and A.2. Because of the dependence of φ_i on $\bar{\zeta}_i$, the stabilizing function α_i is augmented by the term $+\sum_{k=1}^{i-1} \frac{\partial\alpha_{i-1}}{\partial\zeta_k}\Phi_{k,0}$, and the regressor w_i is augmented by $-\sum_{k=1}^{i-1} \Phi_i \left(\frac{\partial\alpha_{i-1}}{\partial\zeta_k}\right)^{\mathrm{T}}$.

Partial state feedback systems

In many physical systems there are unmeasured states, as in the output-feedback form (A.116), but there are also states other than the output $y = x_1$ that are measured. An example of such a system is

$$
\begin{aligned}
\dot{x}_1 &= x_2 + \varphi_1(x_1)^{\mathrm{T}}\theta \\
\dot{x}_2 &= x_3 + \varphi_2(x_1, x_2)^{\mathrm{T}}\theta \\
\dot{x}_3 &= x_4 + \varphi_3(x_1, x_2)^{\mathrm{T}}\theta \\
\dot{x}_4 &= x_5 + \varphi_4(x_1, x_2)^{\mathrm{T}}\theta \\
\dot{x}_5 &= u + \varphi_5(x_1, x_2, x_5)^{\mathrm{T}}\theta\,.
\end{aligned}
$$

The states x_3 and x_4 are assumed not to be measured. To apply the adaptive backstepping designs presented in this chapter, we combine the state-feedback techniques with the output-feedback techniques. The subsystem (x_2, x_3, x_4) is in the output-feedback form with x_2 as a measured output, so we employ a state estimator for (x_2, x_3, x_4) using the filters introduced in Section A.4.

Appendix B

Poincaré and Agmon Inequalities

Lemma B.1 (Poincaré inequality). *For any $w \in H^1(0,1)$, the following inequalities hold:*

$$\int_0^1 w^2(x)\,dx \le 2w^2(1) + 4\int_0^1 w_x^2(x)\,dx, \tag{B.1}$$

$$\int_0^1 w^2(x)\,dx \le 2w^2(0) + 4\int_0^1 w_x^2(x)\,dx. \tag{B.2}$$

Remark B.1. The tight versions of (B.1), (B.2) are

$$\int_0^1 (w(x) - w(1))^2\,dx \le \frac{4}{\pi^2}\int_0^1 w_x^2(x)\,dx\,, \tag{B.3}$$

$$\int_0^1 (w(x) - w(0))^2\,dx \le \frac{4}{\pi^2}\int_0^1 w_x^2(x)\,dx, \tag{B.4}$$

which are sometimes called "variations of Wirtinger's inequality." The proof of (B.3), (B.4) is far more complicated than the proof of (B.1), (B.2) and is given in [46].

Proof.

$$\begin{aligned}\int_0^1 w^2\,dx &= xw^2|_0^1 - 2\int_0^1 xww_x\,dx\\ &= w^2(1) - 2\int_0^1 xww_x\,dx\\ &\le w^2(1) + \frac{1}{2}\int_0^1 w^2dx + 2\int_0^1 x^2w_x^2\,dx\,.\end{aligned}$$

Subtracting the second term from both sides of the inequality we get the first inequality in (B.1):

$$\begin{aligned}\frac{1}{2}\int_0^1 w^2(x)\,dx &\le w^2(1) + 2\int_0^1 x^2w_x^2(x)\,dx\\ &\le w^2(1) + 2\int_0^1 w_x^2\,dx\,.\end{aligned} \tag{B.5}$$

The second inequality is obtained in a similar fashion. □

Lemma B.2 (Agmon's inequality). *For any $w \in H^1(0,1)$, the following inequalities hold:*

$$\max_{x\in[0,1]} |w(x)|^2 \leq w(0)^2 + 2\|w\|\|w_x\|, \tag{B.6}$$

$$\max_{x\in[0,1]} |w(x)|^2 \leq w(1)^2 + 2\|w\|\|w_x\|. \tag{B.7}$$

Proof.

$$\int_0^x w(x)w_x(x)\,dx = \int_0^x \partial_x \frac{1}{2}w^2(x)dx = \frac{1}{2}w(x)^2 - \frac{1}{2}w(0)^2. \tag{B.8}$$

Taking the absolute value on both sides and using the triangle inequality gives

$$\frac{1}{2}|w(x)^2| \leq \int_0^x |w(x)||w_x(x)|\,dx + \frac{1}{2}w(0)^2, \tag{B.9}$$

which can be rewritten as

$$|w(x)|^2 \leq w(0)^2 + 2\int_0^x |w(x)||w_x(x)|\,dx. \tag{B.10}$$

The right-hand side of this inequality does not depend on x, and therefore

$$\max_{x\in[0,1]} |w(x)|^2 \leq w(0)^2 + 2\int_0^1 |w(x)||w_x(x)|\,dx. \tag{B.11}$$

Using the Cauchy-Schwartz inequality we get the first inequality of (B.6). The second inequality is obtained in a similar fashion. □

Appendix C

Bessel Functions

C.1 BESSEL FUNCTION J_N

The function $y(x) = J_n(x)$ is a solution to the following ODE:

$$x^2 y''_{xx} + x y'_x + (x^2 - n^2) y = 0. \tag{C.1}$$

See Figure C.1.
Series representation:

$$J_n(x) = \sum_{m=0}^{\infty} \frac{(-1)^m (x/2)^{n+2m}}{m!(m+n)!} \tag{C.2}$$

Properties:

$$2n J_n(x) = x(J_{n-1}(x) + J_{n+1}(x)) \tag{C.3}$$

$$J_n(-x) = (-1)^n J_n(x) \tag{C.4}$$

Differentiation:

$$\frac{d}{dx} J_n(x) = \frac{1}{2}(J_{n-1}(x) - J_{n+1}(x)) = \frac{n}{x} J_n(x) - J_{n+1}(x) \tag{C.5}$$

$$\frac{d}{dx}(x^n J_n(x)) = x^n J_{n-1}, \qquad \frac{d}{dx}(x^{-n} J_n(x)) = -x^{-n} J_{n+1} \tag{C.6}$$

Asymptotic properties:

$$J_n(x) \approx \frac{1}{n!}\left(\frac{x}{2}\right)^n, \qquad x \to 0 \tag{C.7}$$

$$J_n(x) \approx \sqrt{\frac{2}{\pi x}} \cos\left(x - \frac{\pi n}{2} - \frac{\pi}{4}\right), \qquad x \to \infty \tag{C.8}$$

C.2 MODIFIED BESSEL FUNCTION I_N

The function $y(x) = I_n(x)$ is a solution to the following ODE:

$$x^2 y''_{xx} + x y'_x - (x^2 + n^2) y = 0. \tag{C.9}$$

See Figure C.2.
Series representation:

$$I_n(x) = \sum_{m=0}^{\infty} \frac{(x/2)^{n+2m}}{m!(m+n)!} \tag{C.10}$$

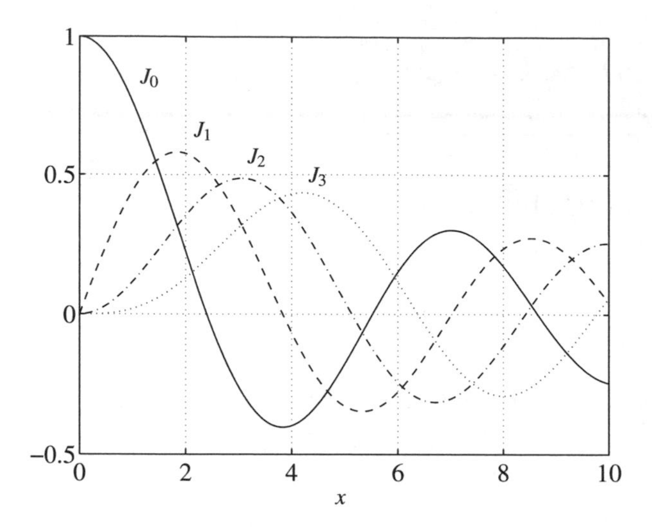

Figure C.1 Bessel functions J_n.

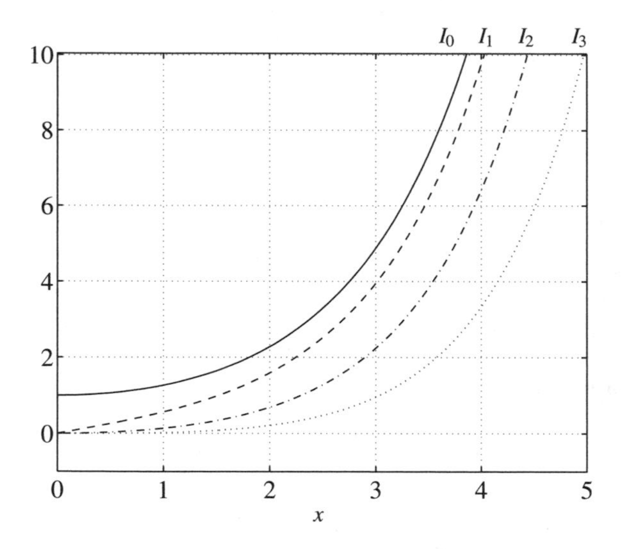

Figure C.2 Modified Bessel functions I_n.

Relationship to $J_n(x)$:

$$I_n(x) = i^{-n} J_n(ix), \qquad I_n(ix) = i^n J_n(x) \tag{C.11}$$

Properties:

$$2n I_n(x) = x(I_{n-1}(x) - I_{n+1}(x)) \tag{C.12}$$

$$I_n(-x) = (-1)^n I_n(x) \tag{C.13}$$

Differentiation:

$$\frac{d}{dx} I_n(x) = \frac{1}{2}(I_{n-1}(x) + I_{n+1}(x)) = \frac{n}{x} I_n(x) + I_{n+1}(x) \qquad \text{(C.14)}$$

$$\frac{d}{dx}(x^n I_n(x)) = x^n I_{n-1}, \qquad \frac{d}{dx}(x^{-n} I_n(x)) = x^{-n} I_{n+1} \qquad \text{(C.15)}$$

Asymptotic properties:

$$I_n(x) \approx \frac{1}{n!}\left(\frac{x}{2}\right)^n, \qquad x \to 0 \qquad \text{(C.16)}$$

$$I_n(x) \approx \frac{e^x}{\sqrt{2\pi x}}, \qquad x \to \infty \qquad \text{(C.17)}$$

Appendix D

Barbalat's and Other Lemmas for Proving Adaptive Regulation

LEMMA D.1 (Barbalat). *Consider the function $\phi : \mathbb{R}_+ \to \mathbb{R}$. If ϕ is uniformly continuous and $\lim_{t\to\infty} \int_0^\infty \phi(\tau)d\tau$ exists and is finite, then*

$$\lim_{t\to\infty} \phi(t) = 0\,. \tag{D.1}$$

Proof. Suppose that (D.1) does not hold; that is, either the limit does not exist or it is not equal to zero. Then there exists $\varepsilon > 0$ such that for every $T > 0$ one can find $t_1 \geq T$ with $|\phi(t_1)| > \varepsilon$. Since ϕ is uniformly continuous, there is a positive constant $\delta(\varepsilon)$ such that $|\phi(t) - \phi(t_1)| < \varepsilon/2$ for all $t_1 \geq 0$ and all t such that $|t - t_1| \leq \delta(\varepsilon)$. Hence, for all $t \in [t_1, t_1 + \delta(\varepsilon)]$, we have

$$\begin{aligned} |\phi(t)| &= |\phi(t) - \phi(t_1) + \phi(t_1)| \\ &\geq |\phi(t_1)| - |\phi(t) - \phi(t_1)| \\ &> \varepsilon - \frac{\varepsilon}{2} = \frac{\varepsilon}{2}\,, \end{aligned} \tag{D.2}$$

which implies that

$$\left|\int_{t_1}^{t_1+\delta(\varepsilon)} \phi(\tau)d\tau\right| = \int_{t_1}^{t_1+\delta(\varepsilon)} |\phi(\tau)|\,d\tau\, \frac{\varepsilon\delta(\varepsilon)}{2}\,, \tag{D.3}$$

where the first equality holds, since $\phi(t)$ does not change sign on $[t_1, t_1 + \delta(\varepsilon)]$. Noting that

$$\int_0^{t_1+\delta(\varepsilon)} \phi(\tau)d\tau = \int_0^{t_1} \phi(\tau)d\tau + \int_{t_1}^{t_1+\delta(\varepsilon)} \phi(\tau)d\tau\,,$$

we conclude that $\int_0^t \phi(\tau)d\tau$ cannot converge to a finite limit as $t \to \infty$, which contradicts the assumption of the lemma. Thus, $\lim_{t\to\infty} \phi(t) = 0$. □

COROLLARY D.1. *Consider the function $\phi : \mathbb{R}_+ \to \mathbb{R}$. If $\phi, \dot{\phi} \in \mathcal{L}_\infty$, and $\phi \in \mathcal{L}_p$ for some $p \in [1, \infty)$, then*

$$\lim_{t\to\infty} \phi(t) = 0\,. \tag{D.4}$$

LEMMA D.2 ([81]). *Suppose that the function $f(t)$ defined on $[0, \infty)$ satisfies the following conditions:*

(i) $f(t) \geq 0$ for all $t \in [0, \infty)$,

(ii) $f(t)$ is differentiable on $[0, \infty)$ and there exists a constant M such that $f'(t) \leq M,\ \forall t \geq 0$,

(iii) $\int_0^\infty f(t)\,dt < \infty$.
Then

$$\lim_{t\to\infty} f(t) = 0. \tag{D.5}$$

Proof. If $M \leq 0$, then $f(t)$ is nonincreasing. Hence conditions (*i*) and (*iii*) immediately imply (D.5).

Suppose that $M > 0$. We argue by contradication. If (D.5) is not true, then there exist a positive constant δ and a sequence $\{t_n\}$ ($n = 1, 2, \ldots$) with $t_n \to \infty$ as $n \to \infty$ such that

$$f(t_n) \geq \delta, \quad n = 1, 2, \ldots .$$

Let

$$F(t) = f(t) - M(t - t_n) - f(t_n).$$

By condition (*ii*), we have

$$F'(t) = f'(t) - M \leq 0.$$

Therefore, we deduce

$$F(t) \geq F(t_n) = 0, \quad \forall 0 \leq t \leq t_n,$$

that is,

$$f(t) \geq M(t - t_n) + f(t_n), \quad \forall 0 \leq t \leq t_n.$$

Thus we have

$$\begin{aligned} \int_{t_n-\delta/M}^{t_n} f(t)dt &\geq \int_{t_n-\delta/M}^{t_n} [M(t-t_n) + f(t_n)]dt \\ &= \frac{\delta f(t_n)}{M} - \frac{\delta^2}{2M} \\ &\geq \frac{\delta^2}{2M}, \qquad n = 1, 2, \ldots \end{aligned} \tag{D.6}$$

which contradicts condition *(iii)*. □

Lemma D.3. *Let v, l_1, and l_2 be real-valued functions defined on R_+, and let c be a positive constant. If l_1 and l_2 are nonnegative and integrable and satisfy the differential inequality*

$$\dot{v} \leq -cv + l_1(t)v + l_2(t), \quad v(0) \geq 0\,, \tag{D.7}$$

then v is bounded and integrable.

Proof. Using the fact that

$$v(t) \leq w(t)\,, \tag{D.8}$$

$$\dot{w} = -cw + l_1(t)w + l_2(t)\,, \tag{D.9}$$

$$w(0) = v(0)\,, \tag{D.10}$$

(the comparison principle), and applying the variation of constants formula, the differential inequality (D.7) is rewritten as

$$\begin{aligned} v(t) &\le v(0)e^{\int_0^t [-c+l_1(s)]ds} + \int_0^t e^{\int_\tau^t [-c+l_1(s)]\,ds} l_2(\tau)\,d\tau \\ &\le v(0)e^{-ct} e^{\int_0^\infty l_1(s)\,ds} + \int_0^t e^{-c(t-\tau)} l_2(\tau)\,d\tau\, e^{\int_0^\infty l_1(s)ds} \\ &\le \left[v(0)e^{-ct} + \int_0^t e^{-c(t-\tau)} l_2(\tau)d\tau \right] e^{\|l_1\|_1}. \end{aligned} \tag{D.11}$$

By taking a supremum of $e^{-c(t-\tau)}$ over $[0, \infty)$, we get boundedness of v. Integrating (D.11) over $[0, \infty)$, we get

$$\int_0^t v(\tau)\,d\tau \le \left(\frac{1}{c} v(0) + \int_0^t \left[\int_0^\tau e^{-c(\tau-s)} l_2(s)\,ds \right] d\tau \right) e^{\|l_1\|_1}. \tag{D.12}$$

Changing the order of integration in the double integral, we get the integrability of v. □

Appendix E

Basic Parabolic PDEs and Their Exact Solutions

To ease the reader with no prior exposure to PDEs into the subject of this book, we provide a few elementary constructions of explicit solutions to basic reaction-diffusion PDEs.

E.1 REACTION-DIFFUSION EQUATION WITH DIRICHLET BOUNDARY CONDITIONS

Consider the reaction-diffusion equation

$$w_t = \varepsilon w_{xx} - cw\,, \tag{E.1}$$

$$w(0,t) = w(1,t) = 0, \tag{E.2}$$

$$w(x,0) = w_0(x). \tag{E.3}$$

To find the solution of this problem, we use the method of separation of variables. Let us assume that the solution $w(x,t)$ can be written as a product of a function of space and a function of time,

$$w(x,t) = X(x)T(t)\,. \tag{E.4}$$

Substituting (E.4) into the PDE (E.1), we get

$$X(x)\dot{T}(t) = \varepsilon X''(x)T(t) - cX(x)T(t). \tag{E.5}$$

Gathering like terms on opposite sides yields

$$\frac{\dot{T}(t)}{T(t)} = \frac{\varepsilon X''(x) - cX(x)}{X(x)}. \tag{E.6}$$

Since the function on the left depends only on time and the function on the right depends only on the spatial variable, the equality can hold only if both functions are constant. Let us denote this constant by σ. We then get two ODEs:

$$\dot{T} = \sigma T \tag{E.7}$$

with initial condition $T(0) = T_0$, and

$$\varepsilon X'' - (c+\sigma)X = 0 \tag{E.8}$$

with boundary conditions $X(0) = X(1) = 0$ (they follow from the PDE boundary conditions). The solution to (E.7) is given by

$$T = T_0 e^{\sigma t}. \tag{E.9}$$

The solution to (E.8) has the form

$$X(x) = A\sinh(\sqrt{(c+\sigma)/\varepsilon x}) + B\cosh(\sqrt{(c+\sigma)/\varepsilon x}), \qquad \text{(E.10)}$$

where A and B are constants that should be determined from the boundary conditions. We have:

$$B = 0, \qquad A\sinh(\sqrt{(c+\sigma)/\varepsilon}) = 0. \qquad \text{(E.11)}$$

The last equality can hold true only if $\sqrt{(c+\sigma)/\varepsilon} = \pi nj$ for $n = 0, 1, 2, \ldots$, so that

$$\sigma = -c - \varepsilon\pi^2 n^2, \qquad n = 0, 1, 2, \ldots \qquad \text{(E.12)}$$

Substituting (E.9), (E.10) into (E.4) yields

$$w_n(x,t) = T_0 A_n e^{-(c+\varepsilon\pi^2 n^2)t}\sin(\pi n x), \quad n = 0, 1, 2, \ldots. \qquad \text{(E.13)}$$

For linear PDEs the sum of particular solutions is also a solution (the principle of superposition). Therefore the formal general solution of (E.1)–(E.3) is given by

$$w(x,t) = \sum_{n=1}^{\infty} C_n e^{-(c+\varepsilon\pi^2 n^2)t}\sin(\pi n x), \qquad \text{(E.14)}$$

where $C_n = A_n T_0$.

To determine the constants C_n, let us set $t = 0$ in (E.14) and multiply both sides of the resulting equality by $\sin(\pi m x)$:

$$w_0(x)\sin(\pi m x) = \sin(\pi m x)\sum_{n=1}^{\infty} C_n \sin(\pi n x). \qquad \text{(E.15)}$$

Then, using the identity

$$\int_0^1 \sin(\pi m x)\sin(\pi n x)dx = \begin{Bmatrix} 1/2, & n = m \\ 0, & n \neq m \end{Bmatrix}, \qquad \text{(E.16)}$$

we get

$$C_m = \frac{1}{2}\int_0^1 w_0(x)\sin(\pi m x)dx. \qquad \text{(E.17)}$$

Substituting this expression into (E.14), we get the solution:

$$w(x,t) = 2\sum_{n=1}^{\infty} e^{-(c+\varepsilon\pi^2 n^2)t}\sin(\pi n x)\int_0^1 \sin(\pi n y)w_0(y)\,dy. \qquad \text{(E.18)}$$

Even though this solution has been obtained formally, it can be proved that this is indeed a well-defined solution in the sense that it is unique, has continuous spatial derivatives up to a second order, and depends continuously on the initial data.

E.2 REACTION-DIFFUSION EQUATION WITH NEUMANN BOUNDARY CONDITIONS

Consider the reaction-diffusion equation

$$w_t = \varepsilon w_{xx} - cw\,, \tag{E.19}$$

$$w_x(0,t) = w_x(1,t) = 0\,, \tag{E.20}$$

$$w(x,0) = w_0(x)\,. \tag{E.21}$$

Using the method of separation of variables, one gets the solution

$$w(x,t) = 2\sum_{n=0}^{\infty} e^{-(c+\varepsilon\pi^2 n^2)t}\cos(\pi n x)\int_0^1 \cos(\pi n y) w_0(y)\,dy\,. \tag{E.22}$$

E.3 REACTION-DIFFUSION EQUATION WITH MIXED BOUNDARY CONDITIONS

Consider the reaction-diffusion equation

$$w_t = \varepsilon w_{xx} - cw\,, \tag{E.23}$$

$$w_x(0,t) = w(1,t) = 0\,, \tag{E.24}$$

$$w(x,0) = w_0(x)\,. \tag{E.25}$$

Using the method of separation of variables, one gets the solution

$$w(x,t) = 2\sum_{n=0}^{\infty} e^{-(c+\varepsilon\pi^2(n+\frac{1}{2})^2)t}\cos\left[\pi\left(n+\frac{1}{2}\right)x\right] \times \int_0^1 \cos\left[\pi\left(n+\frac{1}{2}\right)y\right] w_0(y)\,dy. \tag{E.26}$$

References

[1] Aamo, O.M., Krstic, M.: Global stabilization of a nonlinear Ginzburg-Landau model of vortex shedding. European Journal of Control **10**, 105–116 (2004)

[2] Aamo, O.M., Smyshlyaev, A., Krstic, M.: Boundary control of the linearized Ginzburg-Landau model of vortex shedding. SIAM Journal of Control and Optimization **43**, 1953–1971 (2005)

[3] Aamo, O.M., Smyshlyaev, A., Krstic, M., Foss, B.: Stabilization of a Ginzburg-Landau model of vortex shedding by output-feedback boundary control. IEEE Transactions on Automatic Control **52**, 742–748 (2007)

[4] Ablowitz, M.J., Kruskal, M.D., Ladik, J.F.: Solitary wave collisions. SIAM Journal of Applied Mathematics **36**(3), 428–437 (1979)

[5] Amann, H.: Feedback stabilization of linear and semilinear parabolic systems. In: P. Clément et al. (eds.), Semigroup Theory and Applications, pp. 21–57. Marcel Dekker, New York (1989)

[6] Balogh, A., Krstic, M.: Infinite dimensional backstepping-style feedback transformations for a heat equation with an arbitrary level of instability. European Journal of Control **8**, 165–175 (2002)

[7] Balogh, A., Krstic, M.: Stability of partial differential equations governing control gains in infinite dimensional backstepping. Systems and Control Letters **51**, 151–164 (2004)

[8] Banks, H.T., Kunisch, K.: Estimation Techniques for Distributed Parameter Systems. Birkhäuser, Boston (1989)

[9] Baumeister, J., Scondo, W., Demetriou, M.A., Rosen, I.G.: On-line parameter estimation for infinite-dimensional dynamical systems. SIAM Journal of Control and Optimization **35**, 678–713 (1997)

[10] Bensoussan, A., Prato, G.D., Delfour, M.C., Mitter, S.K.: Representation and control of infinite-dimensional systems. Birkhäuser, Boston (2006)

[11] Bentsman, J., Orlov, Y.: Reduced spatial order model reference adaptive control of spatially varying distributed parameter systems of parabolic and hyperbolic types. International Journal of Adaptive Control and Signal Processing **15**, 679–696 (2001)

[12] Bohm, M., Demetriou, M.A., Reich, S., Rosen, I.G.: Model reference adaptive control of distributed parameter systems. SIAM Journal of Control and Optimization **36**, 33–81 (1998)

[13] Boskovic, D.M., Balogh, A., Krstic, M.: Backstepping in infinite dimension for a class of parabolic distributed parameter systems. Mathematics of Control, Signals, and Systems **16**, 44–75 (2003)

[14] Boskovic, D.M., Krstic, M.: Nonlinear stabilization of a thermal convection loop by state feedback. Automatica **37**, 2033–2040 (2001)

[15] Boskovic, D.M., Krstic, M.: Backstepping control of chemical tubular reactor. Computers and Chemical Engineering **26**, 1077–1085 (2002)

[16] Boskovic, D.M., Krstic, M.: Stabilization of a solid propellant rocket instability by state feedback. International Journal of Robust and Nonlinear Control **13**, 483–495 (2003)

[17] Boskovic, D.M., Krstic, M., Liu, W.J.: Boundary control of an unstable heat equation via measurement of domain-averaged temperature. IEEE Transactions on Automatic Control **46**, 2022–2028 (2001)

[18] Bresch-Pietri, D., Krstic, M.: Adaptive trajectory tracking despite unknown actuator delay and plant parameters. Automatica **45**, 2075–2081 (2009)

[19] Bresch-Pietri, D., Krstic, M.: Delay-adaptive full-state predictor feedback for systems with unknown long actuator delay. Submitted to IEEE Transactions on Automatic Control

[20] Burns, J.A., Hulsing, K.P.: Numerical methods for approximating functional gains in LQR boundary control problems. Mathematical and Computer Modeling **33**, 89–100 (2001)

[21] Carslaw, H.S., Jaeger, J.C.: Conduction of Heat in Solids. Clarendon Press, Oxford (1959)

[22] Christofides, P.: Nonlinear and Robust Control of Partial Differential Equation Systems: Methods and Applications to Transport-Reaction Processes. Birkhäuser, Boston (2001)

[23] Colton, D.: The solution of initial-boundary value problems for parabolic equations by the method of integral operators. Journal of Differential Equations **26**, 181–190 (1977)

[24] Courdesses, M., Polis, M.P., Amouroux, M.: On identifiability of parameters in a class of parabolic distributed systems. IEEE Transactions on Automatic Control **26**(2), 474–477 (1981)

[25] Curtain, R., Demetriou, M.A., Ito, K.: Adaptive compensators for perturbed positive real infinite-dimensional systems. International Journal of Applied Mathematics and Computer Science **13**(4), 441–452 (2003)

[26] Curtain, R.F.: Robust stabilizability of normalized coprime factors: The infinite-dimensional case. International Journal of Control **51**, 1173–1190 (1990)

[27] Curtain, R.F., Demetriou, M.A., Ito, K.: Adaptive observers for structurally perturbed infinite dimensional systems. In: Proceedings of the 36th Conference on Decision and Control, vol. 1, pp. 509–514. San Diego, Calif. (1997)

[28] Curtain, R.F., Demetriou, M.A., Ito, K.: Adaptive observers for slowly time varying infinite dimensional systems. In: Proceedings of the 37th Conference on Decision and Control, vol. 4, pp. 4022–4027. Tampa, Fla. (1998)

[29] Curtain, R.F., Zwart, H.J.: An Introduction to Infinite Dimensional Linear Systems Theory. Springer-Verlag, New York (1995)

[30] Demetriou, M.A., Ito, K.: Online fault detection and diagnosis for a class of positive real infinite dimensional systems. In: Proceedings of the 41st IEEE Conference on Decision and Control, vol. 4, pp. 4359–4364 (2002)

[31] Demetriou, M.A., Ito, K., Smith, R.C.: Adaptive techniques for the MRAC, adaptive parameter identification, and on-line fault monitoring and accommodation for a class of positive real infinite dimensional systems. International Journal of Adaptive Control and Signal Processing **23**(2), 193–215 (2009)

[32] Demetriou, M.A., Rosen, I.G.: On the persistence of excitation in the adaptive estimation of distributed parameter systems. IEEE Transactions on Automatic Control **39**, 1117–1123 (1994)

[33] Demetriou, M.A., Rosen, I.G.: Adaptive parameter estimation for degenerate parabolic systems. Journal of Mathematical Analysis and Applications **189**(3), 815–847 (1995)

[34] Demetriou, M.A., Rosen, I.G.: On-line robust parameter identification for parabolic systems. International Journal of Adaptive Control and Signal Processing **15**, 615–631 (2001)

[35] Demetriou, M.A., Rosen, I.G.: Variable structure model reference adaptive control of parabolic distributed parameter systems. In: Proceedings of the 2002 American Control Conference, vol. 6, pp. 4371–4376 (2002)

[36] Dochain, D.: State observation and adaptive linearizing control for distributed parameter (bio)chemical reactors. International Journal of Adaptive Control and Signal Processing **15**(6), 633–653 (2001)

[37] Dubljevic, S.: Optimal boundary control of cardiac alternans. International Journal of Adaptive Control and Signal Processing **19**, 135–150 (2009)

[38] Duncan, T.E., Maslowski, B., Pasik-Duncan, B.: Adaptive boundary and point control of linear stochastic distributed parameter systems. SIAM Journal of Control and Optimization **32**, 648–672 (1994)

[39] Fattorini, H.O.: Boundary control systems. SIAM Journal of Control and Optimization **6**, 349–385 (1968)

[40] Freeman, R.A., Kokotovic, P.V.: Robust Nonlinear Control Design: State-Space and Lyapunov Techniques. Birkhäuser, Boston (2008)

[41] Friedman, A.: Partial Differential Equations of Parabolic Type. Robert E. Krieger Publishing Co., Huntington, N.Y. (1993)

[42] Fuji, N.: Feedback stabilization of distributed parameter systems by a functional observer. SIAM Journal of Control and Optimization **18**, 108–121 (1980)

[43] Gunzburger, M., Hou, L.S., Ravindrah, S.S.: Analysis and approximation of optimal control problems for a simplified Ginzburg-Landau model of supercondutivity. Numerische Mathematik **77**, 243–268 (1997)

[44] Gunzburger, M., Lee, H.: Feedback control of Karman vortex shedding. Transactions of the ASME **63**, 828–835 (1996)

[45] Guo, B.Z., Shao, Z.C.: Regularity of a Schrödinger equation with Dirichlet control and colocated observation. Systems and Control Letters **54**, 1135–1142 (2005)

[46] Hardy, G., Littlewood, J., Polya, G.: Inequalities, 2nd ed. Cambridge University Press, Cambridge (1959)

[47] He, J.W., Glowinski, R., Metcalfe, R., Nordlander, A., Periaux, J.: Active control and drag optimization for flow past a circular cylinder: I. Oscillatory cylinder rotation. Journal of Computational Physics **163**, 83–117 (2000)

[48] Hong, K.S., Bentsman, J.: Application of averaging method for integrodifferential equations to model reference adaptive control of parabolic systems. Automatica **30**(9), 1415–1419 (1994)

[49] Hong, K.S., Bentsman, J.: Direct adaptive control of parabolic systems: Algorithm synthesis and convergence and stability analysis. IEEE Transactions on Automatic Control **39**, 2018–2033 (1994)

[50] Huerre, P., Monkewitz, P.: Local and global instabilities in spatially developing flows. Annual Review of Fluid Mechanics **22**, 473–537 (1990)

[51] Ioannou, P., Sun, J.: Robust Adaptive Control. Prentice Hall Englewood Cliffs, N.J. (1996)

[52] Jovanovic, M., Bamieh, B.: Lyapunov-based distributed control of systems on lattices. IEEE Transactions on Automatic Control **50**, 422–433 (2005)

[53] Kalman, R.E.: When is a linear control system optimal? Transactions of the ASME, Series D, Journal of Basic Engineering **86**, 51–60 (1964)

[54] Kanellakopoulos, I., Kokotovic, P.V., Morse, A.S.: A toolkit for nonlinear feedback design. Systems and Control Letters **18**, 83–92 (1992)

[55] Khalil, H.: Nonlinear Systems, 2nd ed. Prentice-Hall, Englewood Cliffs, N.J. (1996)

[56] Kim, J.Y., Bentsman, J.: Disturbance rejection in a class of adaptive control laws for distributed parameter systems. International Journal of Adaptive Control and Signal Processing **23**(2), 166–192 (2009)

[57] Kitamura, S., Nakagiri, S.: Identifiability of spatially-varying and constant parameters in distributed systems of parabolic type. SIAM Journal on Control and Optimization **15**(5), 785–802 (1977)

[58] Kobayashi, T.: Global adaptive stabilization of infinite-dimensional systems. Systems and Control Letters **9**(3), 215–223 (1987)

[59] Kobayashi, T.: Adaptive regulator design of a viscous Burgers' system by boundary control. IMA Journal of Mathematical Control and Information **18**, 427–437 (2001)

[60] Kobayashi, T.: Adaptive stabilization of the Kuramoto-Sivashinsky equation. International Journal of Systems Science **33**, 175–180 (2002)

[61] Kravaris, C., Seinfeld, J.H.: Identification of parameters in distributed parameter-systems by regularization. SIAM Journal on Control and Optimization **23**(2), 217–241 (1985)

[62] Krener, A.J., Kang, W.: Locally convergent nonlinear observers. SIAM Journal on Control and Optimization **42**, 155–177 (2003)

[63] Krstic, M.: Systematization of approaches to adaptive boundary stabilization of PDEs. International Journal of Robust and Nonlinear Control **16**, 812–818 (2006)

[64] Krstic, M.: Optimal adaptive control: Contradiction in terms or a matter of choosing the right cost functional. IEEE Transactions on Automatic Control **53**, 1942–1947 (2008)

[65] Krstic, M.: Adaptive control of an anti-stable wave PDE. In: Proceedings of the 2009 American Control Conference, pp. 1505–1510 (2009)

[66] Krstic, M., Deng, H.: Stabilization of Nonlinear Uncertain Systems. Springer-Verlag, New York (1998)

[67] Krstic, M., Guo, B.Z., Smyshlyaev, A.: Boundary controllers and observers for Schrödinger equation. In: Proceedings of the 46th IEEE Conference on Decision and Control, pp. 4149–4154 (2007)

[68] Krstic, M., Kanellakopoulos, I., Kokotovic, P.: Nonlinear and Adaptive Control Design. Wiley, New York (1995)

[69] Krstic, M., Smyshlyaev, A.: Adaptive boundary control for unstable parabolic PDEs. Part I. Lyapunov design. IEEE Transactions on Automatic Control **53**, 1575–1591 (2008)

[70] Krstic, M., Smyshlyaev, A.: Boundary Control of PDEs: A Course on Backstepping Designs. SIAM (2008)

[71] Ladyzhenskaya, O.A., Solonnikov, V.A., Uralceva, N.N.: Linear and Quasilinear Equations of Parabolic Type. AMS, Providence, R.I. (1968)

[72] Laroche, B., Martin, P., Rouchon, P.: Motion planning for the heat equation. International Journal of Robust and Nonlinear Control **10**, 629–643 (2000)

[73] Lasiecka, I., Triggiani, R.: Stabilization and structural assignment of Dirichlet boundary feedback parabolic equations. SIAM Journal on Control and Optimization **21**, 766–803 (1983)

[74] Lasiecka, I., Triggiani, R.: Optimal regularity exact controllability and uniform stabilisation of Schrödinger equations with Dirichlet control. Differential Integral Equations **5**, 521–535 (1992)

[75] Lasiecka, I., Triggiani, R.: Control Theory for Partial Differential Equations: Continuous and Approximation Theories. Cambridge University Press, Cambridge (2000)

[76] Lauga, E., Bewley, T.: h_∞ control of linear global instability in models of non-parallel wakes. In: Proceedings of the Second International Symposium on Turbulence and Shear Flow Phenomena. Stockholm, Sweden (2001)

[77] Lauga, E., Bewley, T.: The decay of stabilizability with Reynolds number in a linear model of spatially developing flows. Royal Society of London, Proceedings, Series A: Mathematical, Physical and Engineering Sciences **459**, 2077–2095 (2003)

[78] Lauga, E., Bewley, T.: Performance of a linear robust control strategy on a nonlinear model of spatially-developing flows. Journal of Fluid Mechanics **512**, 343–374 (2004)

[79] Li, Z.H., Krstic, M.: Optimal design of adaptive tracking controllers for nonlinear systems. Automatica **33**, 1459–1473 (1997)

[80] Liu, W.J.: Boundary feedback stabilization of an unstable heat equation. SIAM Journal on Control and Optimization **42**, 1033–1043 (2003)

[81] Liu, W.J., Krstic, M.: Adaptive control of Burgers' equation with unknown viscosity. International Journal of Adaptive Control and Signal Processing **15**, 745–766 (2001)

[82] Logemann, H.: Stabilization and regulation of infinite-dimensional systems using coprime factorizations. In: R.F. Curtain, A. Bensoussan, J.L. Lions (eds.), Analysis and Optimization of Systems: State and Frequency Domain Approaches for Infinite-Dimensional Systems, pp. 102–139. Springer-Verlag, Berlin (1993)

[83] Logemann, H., Martensson, B.: Adaptive stabilization of infinite-dimensional systems. IEEE Transactions on Automatic Control **37**(12), 1869–1883 (1992)

[84] Logemann, H., Townley, S.: Adaptive stabilization without identification for distributed parameter systems: An overview. IMA Journal of Mathematical Control and Information **14**, 175–206 (1997)

[85] Logemann, H., Zwart, H.: Some remarks on adaptive stabilization of infinite-dimensional systems. Systems and Control Letters **16**(3), 199–207 (1991)

[86] Machtyngier, E.: Exact controllability for the Schrödinger equation. SIAM Journal on Control and Optimization **32**, 24–34 (1994)

[87] Machtyngier, E., Zuazua, E.: Stabilization of the Schrödinger equation. Portugaliae Mathematica **51**, 243–256 (1994)

[88] Monkewitz, P.: Private communication (June 2002)

[89] Moylan, P.J., Anderson, B.D.O.: Nonlinear regulator theory and an inverse optimal control problem. IEEE Transactions on Automatic Control **AC-18**, 460–465 (1973)

[90] Nakagiri, S.I.: Identifiability of linear systems in Hilbert spaces. SIAM Journal on Control and Optimization **21**(4), 501–530 (1983)

[91] Nambu, T.: On the stabilization of diffusion equations: boundary observation and feedback. Journal of Differential Equations **52**, 204–233 (1984)

[92] Naylor, A.W., Sell, G.R.: Linear Operator Theory in Engineering and Science. Springer-Verlag, New York (1982)

[93] Orlov, Y.: Sliding mode observer-based synthesis of state derivative-free model reference adaptive control of distributed parameter systems. Journal of Dynamic Systems, Measurement, and Control **122**(4), 725–731 (2000)

[94] Orlov, Y., Bentsman, J.: Adaptive distributed parameter systems identification with enforceable identifiability conditions and reduced-order spatial differentiation. IEEE Transactions on Automatic Control **45**(2), 203–216 (2000)

[95] Park, D., Ladd, D., Hendricks, E.: Feedback control of a global mode in spatially developing flows. Physical Letters A **182**, 244–248 (1993)

[96] Park, D., Ladd, D., Hendricks, E.: Feedback control of von Kármán vortex shedding behind a circular cylinder at low Reynolds numbers. Physics of Fluids **6**, 2390–2405 (1994)

[97] Phung, K.D.: Observability and control of Schrödinger equations. SIAM Journal on Control and Optimization **40**, 211–230 (2001)

[98] Pierce, A.: Unique identification of eigenvalues and coefficients in a parabolic problem. SIAM Journal on Control and Optimization **17**(4), 494–499 (1979)

[99] Polyanin, A.D.: Handbook of Linear Partial Differential Equations for Engineers and Scientists. Chapman & Hall/CRC, Boca Raton, Fla. (2002)

[100] Praly, L.: Adaptive regulation: Lyapunov design with a growth condition. International Journal of Adaptive Control and Signal Processing **6**, 329–351 (1992)

[101] Prudnikov, A.P., Brychkov, Y.A., Marichev, O.I.: Integrals and Series, vol. 2: Special Functions. Gordon and Breach (1986)

[102] de Queiroz, M.S., Dawson, D.M., Agarwal, M., Zhang, F.: Adaptive nonlinear boundary control of a flexible link robot arm. IEEE Transactions on Robotics and Automation **15**, 779–787 (1999)

[103] Roussopoulos, K., Monkewitz, P.: Nonlinear modelling of vortex shedding control in cylinder wakes. Physica D **97**, 264–273 (1996)

[104] Russell, D.L.: Differential-delay equations as canonical forms for controlled hyperbolic systems with applications to spectral assignment. In: A.K. Aziz, J.W. Wingate, M.J. Balas (eds.), Control Theory of Systems Governed by Partial Differential Equations. Academic Press, New York (1977)

[105] Russell, D.L.: Controllability and stabilizability theory for linear partial differential equations: recent progress and open questions. SIAM Review **20**, 639–739 (1978)

[106] Seidman, T.I.: Two results on exact boundary control of parabolic equations. Applied Mathematics and Optimization **11**, 145–152 (1984)

[107] Sepulchre, R., Jankovic, M., Kokotovic, P.: Constructive Nonlinear Control. Springer (1997)

[108] Smyshlyaev, A., Krstic, M.: Closed form boundary state feedbacks for a class of 1D partial integro-differential equations. IEEE Transactions on Automatic Control **49**, 2185–2202 (2004)

[109] Smyshlyaev, A., Krstic, M.: Backstepping observers for a class of parabolic PDEs. Systems and Control Letters **54**, 613–625 (2005)

[110] Smyshlyaev, A., Krstic, M.: On control design for PDEs with space-dependent diffusivity or time-dependent reactivity. Automatica **41**, 1601–1608 (2005)

[111] Smyshlyaev, A., Krstic, M.: Adaptive boundary control for unstable parabolic PDEs. Part II. Estimation-based designs. Automatica **43**, 1543–1556 (2007)

[112] Smyshlyaev, A., Krstic, M.: Adaptive boundary control for unstable parabolic PDEs. Part III. Output feedback examples with swapping identifiers. Automatica **43**, 1557–1564 (2007)

[113] Smyshlyaev, A., Orlov, Y., Krstic, M.: Adaptive identification of two unstable PDEs with boundary sensing and actuation. International Journal of Adaptive Control and Signal Processing **23**, 131–149 (2009)

[114] Solo, V., Bamieh, B.: Adaptive distributed control of a parabolic system with spatially varying parameters. In: Proceedings of the 38th IEEE Conference on Decision and Control, pp. 2892–2895 (1999)

[115] Temam, R.: Infinite-Dimensional Dynamical Systems in Mechanics and Physics. Springer (1988)

[116] Triggiani, R.: Boundary feedback stabilization of parabolic equations. Applied Mathematics and Optimization **6**, 201–220 (1980)

[117] Vazquez, R., Krstic, M.: Explicit integral operator feedback for local stabilization of nonlinear thermal convection loop PDEs. Systems and Control Letters **55**, 624–632 (2006)

[118] Vazquez, R., Krstic, M.: A closed-form feedback controller for stabilization of the linearized 2D Navier-Stokes Poisseuille flow. IEEE Transactions on Automatic Control **52**, 2298–2312 (2007)

[119] Vazquez, R., Krstic, M.: Control of Turbulent and Magnetohydrodynamic Channel Flows. Birkhäuser, Boston (2007)

[120] Wen, J.T.Y., Balas, M.J.: Robust adaptive control in Hilbert space. Journal of Mathematical Analysis and Applications **143**, 1–26 (1989)

[121] Zheng, S.: Nonlinear evolution equations. Chapman & Hall/CRC, Boca Raton, Fla. (2004)

Index